The Virtual Laboratory

Springer
*Berlin
Heidelberg
New York
Barcelona
Hong Kong
London
Milan
Paris
Tokyo*

Jaap A. Kaandorp
Janet E. Kübler

The Algorithmic Beauty of Seaweeds, Sponges, and Corals

With 139 Illustrations, 51 in Color

 Springer

Jaap A. Kaandorp
Section Computational Science
University of Amsterdam
Kruislaan 403
1098 SJ Amsterdam
The Netherlands
jaapk@science.uva.nl

Janet E. Kübler
Biology Department
California State University
Northridge, CA 91330-8303
USA
janet.kubler@csun.edu

Series Editor
Przemyslaw Prusinkiewicz
Department of Computer Science
The University of Calgary
2500 University Drive N.W.
Calgary, Alberta T2N IN4
Canada
pwp@cpsc.ucalgary.ca

Photograph included in cover design:
Pacifigorgia agassizii, by Juan A. Sanchez,
courtesy of United States National Museum collection
(see Sect. 2.2.3, Fig. 2.29b)

ISBN 3-540-67700-3   Springer-Verlag Berlin Heidelberg New York

Library of Congress Cataloging-in-Publication Data applied for

Die Deutsche Bibliothek – CIP Einheitsaufnahme
Kaandorp, Jaap A.:
The algorithmic beauty of seaweeds, sponges, and corals / Jaap. A. Kaandorp ; Janet E. Kübler. With contributions by Edward Abraham .... – Berlin ; Heidelberg ; New York ; Barcelona ; Hongkong ; Mailand ; Paris ; Tokyo : Springer, 2001
(The virtual laboratory)
ISBN 3-540-67700-3

This work is subject to copyright. All rights are reserved, whether the whole or part of the material is concerned, specifically the rights of translation, reprinting, reuse of illustrations, recitation, broadcasting, reproduction on microfilm or in any other way, and storage in data banks. Duplication of this publication or parts thereof is permitted only under the provisions of the German Copyright Law of September 9, 1965, in its current version, and permission for use must always be obtained from Springer-Verlag. Violations are liable for prosecution under the German Copyright Law.

Springer-Verlag Berlin Heidelberg New York,
a member of BertelsmannSpringer Science+Business Media GmbH
http://www.springer.de

© Springer-Verlag Berlin Heidelberg 2001
Printed in Germany

The use of general descriptive names, registered names, trademarks, etc. in this publication does not imply, even in the absence of a specific statement, that such names are exempt from the relevant protective laws and regulations and therefore free for general use.

Data Conversion: LE-TeX, Jelonek, Schmidt & Vöckler GbR, Leipzig
Cover Design: KünkelLopka, Heidelberg

Printed on acid-free paper
SPIN: 10762183   45/3142/ud   5 4 3 2 1 0

This book is dedicated to our children
Iris, Lila, Sara, and Mikael.

We sincerely hope that in the future
they too can enjoy coral reefs
and all other marine organisms
in their full beauty.

Acknowledgements

The work described in this book could be completed thanks to the support and contributions of many persons. The whole project, writing a book on the state of the art of modeling and simulating the growth and form of seaweeds, sponges, and stony corals, was initiated in a meeting at the National Center for Ecological Analysis and Synthesis in Santa Barbara (California) that we organized in August 1999 (see below). This meeting was very inspiring and successful; afterwards we requested several participants to contribute sections in their field of expertise to this book. We and the contributors would like to thank many persons.

Bastien Chopard and Peter Sloot wish to thank Alfons Hoekstra and Benno Overeinder from the University of Amsterdam as well as Alexandre Masselot and Alexandre Dupuis from the University of Geneva for their contribution and assistance in preparing their section.

Ligia Collado Vides wants to thank Gerardo Rivas for the generation of the figures in her section.

Steve Dudgeon would like to thank Peter Edmunds for permitting use of the photographs shown in Fig. 2.23a and b.

Susanne Gatti wants to thank Julian Gutt (Alfred-Wegener-Institute for Polar and Marine Research, Bremerhaven, Germany) for permitting use of the photographs of Antarctic sponges (Figs. 6.1 and 6.2).

Jaap Kaandorp's work was partly carried out within the MPR (Massive Parallel Computing) project "The discreteness-continuity dichotomy in individual-based population dynamics using massively parallel machines" funded by the Dutch Foundation for Scientific Research.

Janet Kübler wishes to thank Bob Jackson (Surface Physics Laboratory, University of Maine, USA) for many inspiring conversations about mathematics and biology which, in part, made this whole collaboration possible.

Soyoka Muko wishes to thank the members of the meeting at the National Center for Ecological Analysis and Synthesis for their stimulating discussions; she would also like to thank Y. Iwasa, K. Sakai, and N. Shigesada for helpful suggestions during the preparation of her contribution.

Buki Rinkevich's work was conducted at the Minerva Center for Marine Invertebrate Immunology and Development Biology and was also supported by the AID-CDR.

We would like to thank Rob van Soest (Institute for Systematics and Population Biology, University of Amsterdam) for allowing us the use of collection material and for introducing us to the natural products from sponges, Louis van der Laan (Institute for Systematics and Population Biology, University of Amsterdam) for skillfully preparing the black and white photographs in this book, Christopher Lowe (Delft University of Technology) for his collaboration in the lattice Boltzmann simulations, and Mario de Kluijver (Institute for Systematics and Population Biology, University of Amsterdam) for his collaboration in the transplantation experiments with the sponges. J.E.N. (Charlie) Veron (Australian Institute of Marine Sciences) kindly permitted use of the photographs of *Pocillopora damicornis* shown in Fig. 1.1.

The authors would like to thank Robert Belleman, Kamil Iskra, Rafael Garcia Leiva, Frank Lakeman, Nico Rozemeijer, Hans Ragas (University of Amsterdam, Section Computational Science) for their assistance in creating many of the color pictures shown in Chapters 3 and 4, and Benno Overeinder (University of Amsterdam, Section Computational Science) for all his help during the preparation of the manuscript. Sean Spicer (Stanford University) kindly permitted us to use his software for producing the volume renderings shown in Fig. 3.17. Figure 2.4 was reprinted with permission of the Journal of Phycology. Eshel Ben-Jacob (School of Physics and Astronomy, Tel Aviv University) and Inon Cohen (School of Mathematical Sciences, Tel Aviv University) kindly permitted us to use a picture of a *Paenibacillus dendritiformis* colony (see Fig. 1.5a). We want to thank Sarita Ingalsuo-Kaandorp for her help in preparing several of the pictures. Without her support and her patience, at least the first author could never have written this book.

We would like to thank all contributors and we hope these collaborations may result in many new research projects on modeling the growth and form of marine sessile organisms. We want to thank Hans Wössner and Ingeborg Mayer from Springer-Verlag for their pleasant cooperation. We wish to thank the National Center for Ecological Analysis and Synthesis in Santa Barbara (California) for supporting the workshop on "Modeling Growth and Form of Marine Sessile Organisms" in August 1999 in Santa Barbara and their support during the writing of this book. Please note that quicktime movies, related to the topics discussed in this book, are made available at the homepage of one of the authors (http://www.science.uva.nl/~jaapk/).

Amsterdam and Northridge, July 2001   Jaap A. Kaandorp
                                        Janet E. Kübler

List of Contributing Authors

Edward Abraham
National Institute of Water and Atmospheric Research, Wellington, New Zealand

Rolf P. M. Bak
Netherlands Institute of Sea Research, Den Burg (Texel), The Netherlands

Dave J. Barnes
Australian Institute of Marine Science, Townsville, Australia

Robert Carpenter
Biology Department, California State University, Northridge, Calif., USA

Bastien Chopard
Département d'Informatique, University of Geneva, Switzerland

Ligia Collado Vides
Lab. Ficologma, Fac. Ciencias, Universidad Nacional Autónoma de México, Mexico City, Mexico

Peter Dodds
Earth, Atmos. & Plan Department, Massachusetts Institute of Technology, Cambridge, Mass., USA

Steven Dudgeon
Biology Department, California State University, Northridge, Calif., USA

Susanne Gatti
Alfred-Wegener-Institute for Polar and Marine Research, Bremerhaven, Germany

Brian Helmuth
Department of Biological Sciences, University of South Carolina, Columbia, S.C., USA

Mimi A. R. Koehl
Department of Integrative Biology, University of California, Berkeley, Calif., USA

Leo E. H. Lampmann
St Elisabeth Hospital, Tilburg, The Netherlands

Howard R. Lasker
Department of Biological Sciences, State University of New York, Buffalo, N.Y., USA

Janice Lough
Australian Institute of Marine Science, Townsville, Australia

Werner E. G. Müller
Institut für Physiologische Chemie, Abt. Angewandte Molekularbiologie, University of Mainz, Germany

Soyoka Muko
Department of Biology, Faculty of Science, Kyushu University, Fukuoka, Japan

Przemyslaw Prusinkiewicz
Department of Computer Science, University of Calgary, Canada

Buki Rinkevich
Minerva Center for Marine Invertebrate Immunology and Developmental Biology, Haifa, Israel

Juan A. Sanchez
Department of Biological Sciences, State University of New York, Buffalo, N.Y., USA

Peter M. A. Sloot
Section Computational Science, Faculty of Science, University of Amsterdam, The Netherlands

Ray B. Taylor
Australian Institute of Marine Science, Townsville, Australia

Mark J. A. Vermeij
Institute for Biodiversity and Ecosystem Dynamics, University of Amsterdam, The Netherlands

List of Contributions

Section 2.1.1 by Mimi A. R. Koehl, Robert Carpenter, and Brian Helmuth

The example of the analogy from Pettigrew (1908) between electric discharge patterns and the growth of the red coral, Subsection of Section 2.2.3 on the sponge *Raspailia inaequalis*, Sections 3.2 and 4.4 by Edward Abraham

Subsection of Section 2.2.3 on hydrozoans and Sections 4.7 and 5.3 by Steven Dudgeon

Subsection of Section 2.2.3 on octocorals by Howard R. Lasker and Juan A. Sanchez

Subsection of Section 2.2.3 on scleractinians by Mark J. A. Vermeij, Dave J. Barnes, and Soyoka Muko

Subsection of Section 2.2.4 on sponges by Werner E. G. Müller

Subsection of Section 2.2.4 on stony corals by Buki Rinkevich

Section 3.1 by Peter Dodds

Section 3.4 by Mark J. A. Vermeij, Jaap Kaandorp, Rolf P. M. Bak, and Leo E. H. Lampmann

Section 4.1.1 by Przemyslaw Prusinkiewicz

Section 4.1.2 by Ligia Collado Vides

Section 4.3 by Bastien Chopard and Peter M. A. Sloot

Section 5.1 on stony corals by Soyoka Muko and Brian Helmuth

Section 6.1 on sponges by Susanne Gatti

Section 6.2 by Dave J. Barnes, Ray B. Taylor, and Janice Lough

Table of Contents

Chapter 1	**Introduction**	1
1.1	About the Book	13
Chapter 2	**Environmentally Driven Plasticity**	15
2.1	The Physical Environment	15
2.1.1	Growing and Flowing	17
2.2	The Case Studies	30
2.2.1	Case Studies of Environmentally Driven Plasticity: Seaweeds	30
2.2.2	Morphological Plasticity in Sponges	34
2.2.3	Morphological Plasticity in Colonial Cnidarians	43
2.2.4	Biologically Inherent Regulation of Morphogenesis	56
Chapter 3	**Measuring Growth and Form**	67
3.1	Metrics for Branching Networks	67
3.1.1	Branch Ordering	68
3.1.2	Horton Statistics	69
3.1.3	Tokunaga Statistics	70
3.2	Morphological Analysis of a Branching Sponge	72
3.2.1	Image Processing	72
3.2.2	Horton Analysis	73
3.2.3	Fractal Analysis	75
3.3	Two-dimensional Morphological Analysis of Ranges of Growth Forms	75
3.3.1	Sampling Growth Forms Along a Gradient of Increasing Water Movement	77
3.3.2	Morphological Measurements in a Range of Growth Forms	78
3.3.3	A Comparison of the Morphological Measurements in a Range of Growth Forms of the Three Species	84
3.4	Three-Dimensional Morphological Analysis of Growth Forms of *Madracis Mirabilis* (Preliminary Results)	87

| Chapter 4 | **Simulating Growth and Form** | 91 |

4.1	L-systems	91
4.1.1	Introduction to Modeling Using L-systems	91
4.1.2	Examples of L-systems for Modeling Seaweed	94

| 4.2 | Example of a Simple Model of Plasticity in Algal Morphology | 97 |

4.3	Modeling Fluid Flow Using Lattice Gases and the Lattice Boltzmann Model	99
4.3.1	Cellular Automata as Models for Fluid Flow	100
4.3.2	Transport, Erosion, Deposition, and Hydrodynamic Forces	105
4.3.3	Transport and Sedimentation	108

4.4	A Laplacian Model of Branching Growth	109
4.4.1	Laplacian Growth	109
4.4.2	The Numerical Model	110
4.4.3	Model Results	112

4.5	Growth by Aggregation	114
4.5.1	Morphological Plasticity and the Influence of Hydrodynamics	114
4.5.2	Modeling the Nutrient Distribution	115
4.5.3	Growth by Aggregation in a Monodirectional Flow	116
4.5.4	Growth by Aggregation in a Bidirectional (Alternating) Flow	119
4.5.5	Comparison Between the Range of Aggregates and the Growth Forms	123

4.6	Accretive Growth	125
4.6.1	Surface Normal Deposition in Marine Sessile Organisms	125
4.6.2	A Model of Surface Normal Accretive Growth	127
4.6.3	Accretive Growth Using an Approximation of Actual Deposition Velocities and the Amount of Contact with the Environment	128
4.6.4	A Model of Accretive Growth Driven by the Local Amount of Available Nutrient and the Influence of Hydrodynamics	130
4.6.5	A Model of Accretive Growth Driven by Local Light Intensities	137
4.6.6	A Model of Accretive Growth Driven by Local Nutrient Availability and Regulated by a Growth-Suppressing Isomone	139
4.6.7	A Comparison Between the Accretive Model and the Growth Forms	139

| 4.7 | Gastrovascular Dynamics of Hydractiniid Hydrozoans | 144 |

Chapter 5	**Verifying Models**	147
5.1	Transplantation Experiments with Stony Corals	147
5.2	Transplantation and Other Perturbation Experiments with the Sponge *Haliclona oculata* and a Comparison to the Simulation Models	151
5.3	Colonial Hydrozoans: Perturbation Experiments of Gastrovascular Physiology and Effects on Colony Development	155
Chapter 6	**Applications**	159
6.1	Antarctic Sponges	160
6.2	Coral Records	163
Chapter 7	**Epilogue**	169
7.1	Conclusion	172
References		173
Subject Index		191

1. Introduction

Growth and form of marine organisms inhabiting hard substrata, the "marine sessile organisms", is characterized by a number of remarkable properties. One remarkable feature of these organisms is that many of them can be characterized as modular organisms. Modular organisms are typically built of repeated units, the modules, which might be a polyp in a coral colony or a frond in seaweeds. In most cases, the module has a distinctive form, while the growth form of the entire colony is frequently an indeterminate form. Indeterminate growth indicates that the same growth process may result in an infinite number of different realizations of the growth form. This is in contrast to unitary organisms such as vertebrates and insects, in which a single-celled stage develops into a well-defined, determinate structure. In many cases the growth process in modular organisms leads to complex shapes, which are often quite difficult to describe in words. In most of the biological literature these forms are only described in qualitative and rather vague terms, such as "thinly branching", "tree-shaped" and "irregularly branching".

Another major characteristic of marine sessile organisms is that there is frequently a strong impact of the physical environment on the growth process, leading to a variety of growth forms. Growth by accumulation of modules allows the organism to fit its shape to its environment i.e., have plasticity. In many seaweeds, sponges, and corals, differences in exposure to water movement cause significant changes in morphology. A good example of this plasticity is the Indo-Pacific stony coral *Pocillopora damicornis* (Veron and Pichon 1976) shown in Fig. 1.1. In very sheltered environments, this species has a thin-branching growth form. The growth form gradually transforms to a more compact shape when the exposure to water movement increases. Without knowledge of this relationship between the physical environment and the growth process, the enormous variety in growth forms found in this species does not make any sense. Early taxonomists were often confused by this diversity; very often different growth forms of the same species were classified as different species. Ortmann (1888) wrote about this genus:

"haben wir hier ein Chaos von Formen"[1]

since he was not able to decide to which of the described species his specimens could be attributed.

People have been studying the growth and form of marine sessile organisms for many decades. For example, the classic work by D'Arcy Thompson (1917) describes the packing of the cups containing the polyps, the corallites on the surface of a coral colony. In the work of Jackson (1979) six basic shapes

---
[1] We have a chaos of forms here.

Fig. 1.1a–f. Range of growth forms of *Pocillopora damicornis* from sites with different intensity of water movement. Form (*a*) is from the most exposed site, form (*f*) from the most sheltered site. In the range (*a–f*) the exposure to water movement gradually decreases.

are distinguished that can be found in marine sessile animals from very different taxonomical groups, such as sponges and stony corals. He presented a model in which each of these recurrent basic shapes is characterized by size and shape, and analyzed the ecological significance of the six morphological strategies. The property that many of these organisms are built of repeating units was used in the study by Lindenmayer (1968). In this paper the branching pattern of the red algae *Callithamnion roseum* (see Fig. 1.2) was captured in an algorithmic form and described in a formal language, using the iterative architecture of this seaweed.

In this book we give an overview of how simulation models can provide insights into the growth and form of seaweeds, sponges, corals, and some other marine sessile organisms. One fundamental question in biology is how the interplay between the genome and the physical environment controls morphogenesis. Even now, knowledge about how the genetic information is translated into a physical form is scarce, although major advances have been made in the developmental biology of model organisms such as the fruit fly *Drosophila melanogaster* and the plant *Arabidopsis thaliana*. Even less is currently known about how the enormous diversity of life forms has evolved. One of the remarkable properties of a coral reef and many other marine ecosystems is the high diversity of growth forms that coexist in virtually the same (at least at a first superficial view) physical environment.

Fig. 1.2. The red seaweed *Callithamnion roseum* (after Rosenvinge 1923)

Indeterminate growth patterns are fundamental to the diversity of life forms on earth. The sessile, colonial way of life is represented in all kingdoms. A glimpse of the overwhelming variety and beauty of marine sessile organisms is presented in Fig. 1.3 (after Margulis 1993) which shows only a few representatives of the kinds of organisms which we discuss in this book.

The most primitive marine sessile growth forms are stromatolites, organosedimentary structures principally formed by cyanobacteria. Red, brown, and green algae all include species with flexible and with rigid, calcified sessile growth forms, the coralline algae. Sessile, modular growth occurs in several animal phyla and in different classes within phyla. Sessile growth forms are also found in plants and fungi but we focus our discussion on algae and animals, the dominant macroscopic organisms in the marine environment. We will briefly discuss stromatolites since these structures are, from a modeling point of view, very interesting. In the animal kingdom we will restrict ourselves to sessile growth forms in sponges and cnidarians, with a relatively simple developmental biology. Sessile growth forms have also developed in many other phyla in the animal kingdom. Beautiful sessile growth forms, for example, developed in the bryozoans, an animal phylum with an enormous diversity in growth forms (see McKinney and Jackson 1991) and a more complex developmental biology. It is also important to realize that most of the growth forms depicted in Fig. 1.3 cannot be simply considered the product of a single organism. Most of the growth forms are the result of a firm symbiosis between several organisms, either colonies of a single species or symbiosis of multiple species or both. Mixtures of microorganisms, mainly cyanobacteria, form stromatolites by trapping sediment and chemical deposition. Many sponges live in symbiosis with bacteria or algae. Most stony corals live in symbiosis with algae, the zooxanthellae. The symbionts often play a crucial role in the energy supply of the organism and consequently in the emergence of the growth form. A good introduction to the full diversity of marine sessile organisms would require a full course in invertebrate zoology (Barnes 1974), phycology (Graham and Wilcox 2000) and stromatolites (Walter 1976).

A remarkable property of the forms of the organisms displayed in Fig. 1.3 is that although they originate in tremendously distantly related organisms, there are recurring themes, or resemblances, in overall shape. This phenomenon of recurrent forms in different taxonomic groups was also observed by Jackson (1979). Even an experienced marine biologist might find it difficult to decide which taxonomic group a certain specimen should be attributed to, based only on their overall shape. This figure illustrates two of the basic themes in this book. First, similar growth forms found in very different taxonomic groups seem to suggest that there is a deeper physical reason why different organisms sometimes converge on just a few "styles". Second, there

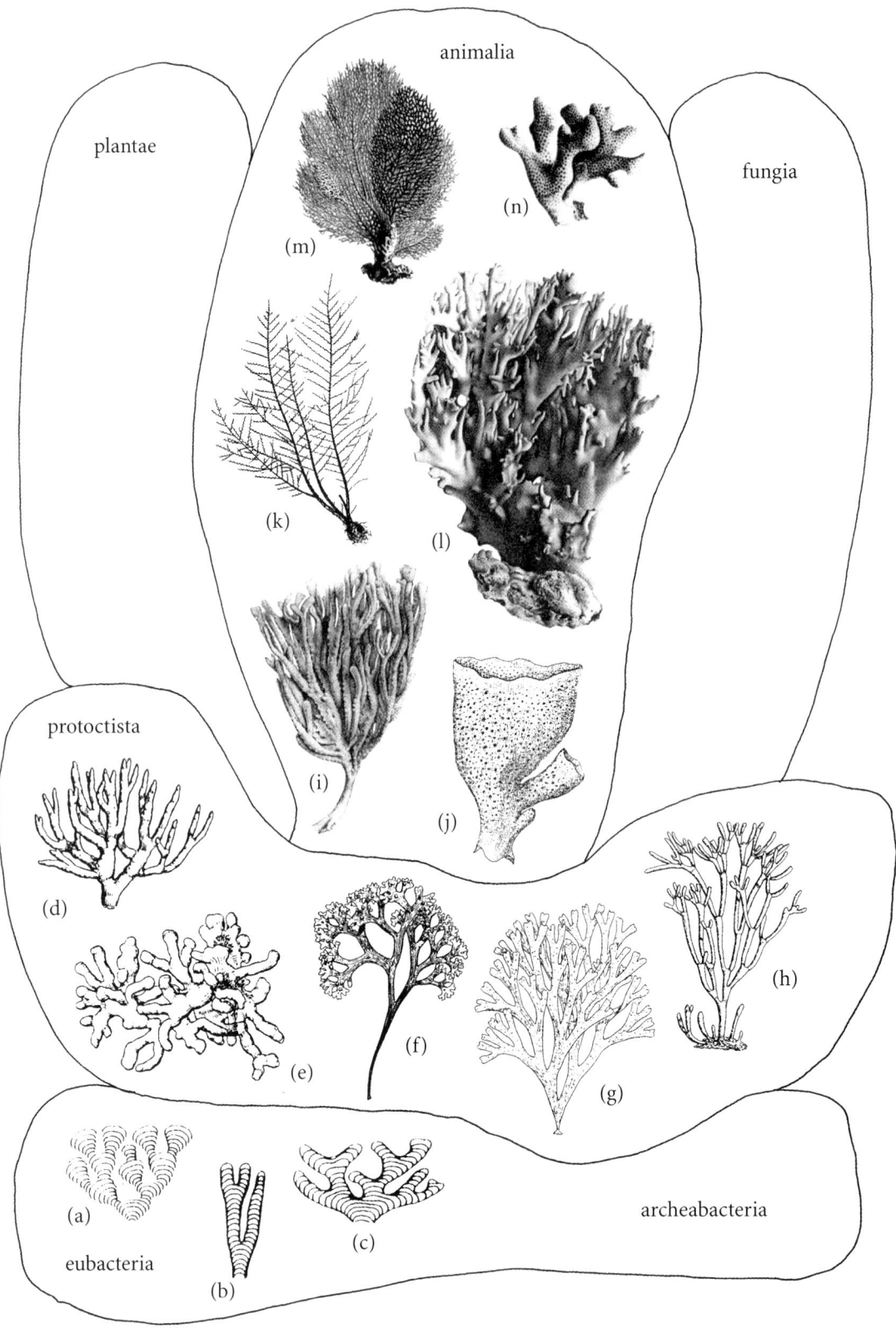

must be some way to characterize and compare these forms quantitatively. Comparing the shapes of organisms without looking at the finer biological details might seem a rather academical exercise, however in palaeontology in many cases only a fossilized macroscopic form is available without any finer structures for further study.

In this book we want to show that in addition to field (in vivo) and laboratory (in vitro) approaches, there is another option for the study of growth and form in marine sessile organisms. That is by using simulation models (the "in silico" option). Development of practical simulation models of the growth and development of marine sessile organisms is now both possible and important. It is possible because their growth process is relatively simple and with the enormous progress that has been made toward understanding the basic developmental mechanisms of model organisms, we might expect to find some basic mechanisms for the development of all organisms. In addition, it seems feasible to capture some vital features of the marine environment with mathematical models.

Using the in silico option for studying growth of marine sessile organisms is important because, while marine ecosystems are valuable sources of known and undiscovered resources, they are also increasingly degraded and endangered. We have only begun to appreciate the aesthetic and ecological value of these systems. New chemical and biological products, some of significant medical or commercial value, are being isolated from marine organisms daily. Some sessile marine organisms produce long-term records of their interactions with their environment, recorded in their accreting layers of growth. Very-long lived organisms with stable skeletons, such as stony corals and coralline algae, can provide bioarchives of conditions over the preceding hundreds or thousands of years. Most importantly, simulation models can help us better understand the beautiful and intricate ecology of marine ecosystems by allowing us to focus on the most promising hypotheses in much less time than growth experiments would require and without disturbing the living system.

The best example of simplicity of a sessile growth form is probably from the stromatolites. Stromatolites are relatively well studied, since these structures belong to the oldest known fossils. A stromatolite (see Fig. 1.4a) grows by the deposition of material on top of the previous growth stages, which remain unchanged. This growth process closely resembles a physical deposition process. Some authors (Grotzinger and Rothman 1996) even argue that it is not possible to distinguish whether the form emerged from a biotic or an abiotic growth process. Stromatolites represent a transition between these two. In a number of other cases, such as the Tungussiform, Bacaliform, and Gymnosoleniform stromatolites shown in Fig. 1.3, branching forms develop which at least superficially resemble other marine sessile organisms. Although branching patterns may emerge in layered physical deposition processes, for example in the growth of ammonium chloride crystals (Brener et al. 1992), there are clearly biological processes at work in the growth of umbrella-shaped stromatolites which orient themselves toward sunlight.

A similar transition case between abiotic and biotic growth is found in the growth patterns of many other bacterial colonies (Matsuyama and Matsushita 1993, Ben-Jacob 1993 and 1997). For example, a colony of *Paenibacillus dendritiformis* can closely resemble growth patterns found in electro deposition, electric discharge patterns, air bubbles pressed between glass plates

◀ Fig. 1.3a–n. A phylogenetic overview of seaweeds, sponges, and corals: A. diagrammatic drawings of stromatolites: (*a*) Baicaliform, (*b*) Gymnosoleniform, (*c*) Tungussiform (after Hoffman, 1976); B. red seaweeds (Rhodophyta), coralline algae: (*d*) and (*e*) (growth forms of *Lithothamnion calcareum*), red seaweeds (Rhodophyta): (*f*) *Chondrus crispus*; C. brown seaweeds (Phaeophyta): (*g*) *Dictyota dichotoma*; D. green seaweeds (Chlorophyta): (*h*) *Codium tomentosum* (all seaweeds after Newton 1931); E. sponges: (*i*) demosponge *Haliclona oculata* (after Bowerbank 1876), (*j*) hexactinellids *Rhabdocalyptus mollis* (after Schulze 1887); F. cnidarians: (*k*) hydrozoan *Halecium halecinum* (after Hincks 1868), (*l*) calcified hydrozoan *Millepora alcicornis* (after Agassiz 1880), (*m*) octocoral *Rhipidigordia flabellum* (after Agassiz 1880), (*n*) scleractinian (stony coral) *Porites furcata* (after Agassiz 1880)

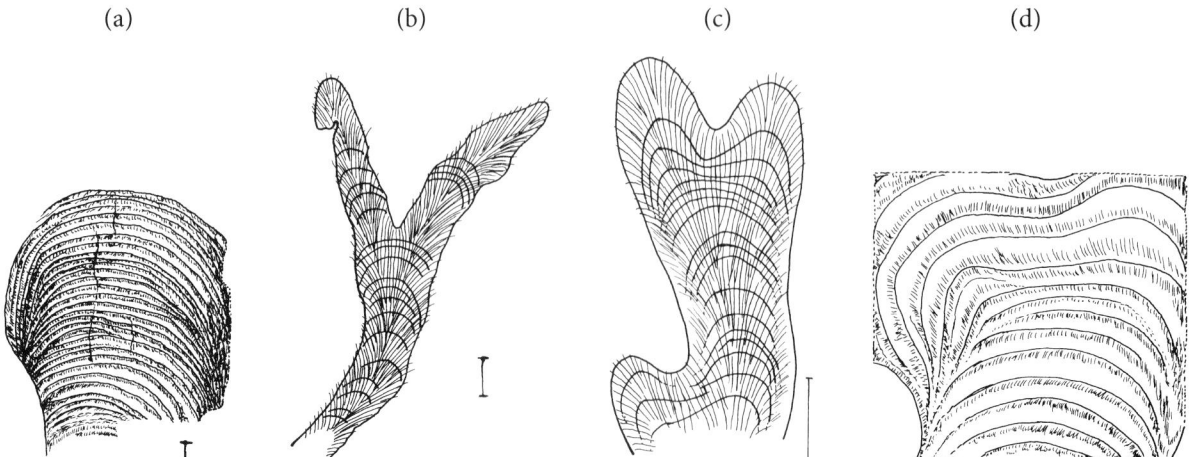

Fig. 1.4. (*a*) Section through an unbranched stromatolite (redrawn from a photograph of an Altyn stromatolite in Horodyski 1976); (*b*) Section through a branch of the sponge *Haliclona oculata*; (*c*) Section through a branch of the stony coral *Porites porites* (redrawn from a photograph in Le Tissier et al. 1994); (*d*) Section through a branch of the coralline algae *Lithothamnion coralliodes* (redrawn from a photograph in Bosence 1976). The scale bars in (*a*), (*b*) and (*c*) indicate 1 cm, while the scale bar in (*d*) represents 1 mm.

and so on. In Fig. 1.5a the growth pattern of a *Paenibacillus dendritiformis* colony is shown. Fig. 1.5b shows a growth pattern formed in an experiment where air displaces a high-viscosity fluid between two glass plates. In this experiment irregular branching patterns known as "viscous-fingering" are formed (Feder 1988). In a number of other cases it has been demonstrated (Ben-Jacob, 1997) that some forms of bacterial colonies develop through the interaction between the individual bacteria. Stromatolites and bacterial colonies are crossing the border between abiotic and biotic growth processes. The analogy between electric discharge patterns and the growth of marine sessile organisms was also made by the Scottish anatomist J. Bell Pettigrew in the illustrated volume 'Design in Nature' (1908). Pettigrew was an anatomist who was interested in the relationships between physical and biological growth. He presented examples of branched structures, including discharge patterns and red coral (*Corallium rubrum*) as shown in Fig. 1.6, in order to illustrate the similarities in the underlying design of physical and biological objects. Quite remarkably, Pettigrew's search for underlying physical mechanisms for explaining growth patterns in nature was driven by religious and anti-darwinistic motivations.

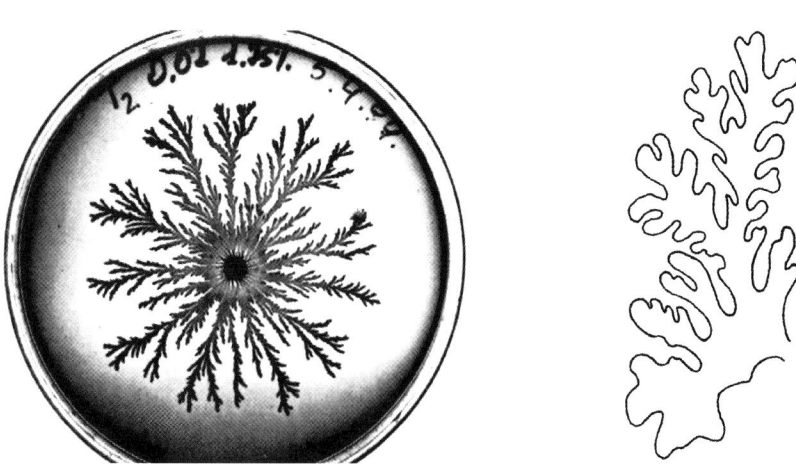

Fig. 1.5. (*a*) Growth pattern of a colony of the bacteria *Paenibacillus dendritiformis*. There are about 10^9–10^{10} bacteria in this 88 mm petri dish (picture by E. Ben-Jacob and I. Cohen); (*b*) An example of a physical deposition pattern: air displaces a high-viscosity fluid between two glass plates and a viscous-fingering pattern is formed.

In some algae, as well as many sponges, hydro corals, and stony corals, the growth process can be described as a relatively simple deposition process, if we study the growth at the level of deposition of individual skeleton elements. In Fig. 1.4 drawings are shown of sections made through respectively an unbranched stromatolite, branches of a sponge (*Haliclona oculata*), a stony coral *Porites porites*, and a coralline algae *Lithothamnion coralliodes*. Growth of the branches of these organisms proceeds by the deposition of new layers on top of the previous growth stages, which remain unchanged.

The drawings in Fig. 1.4 provide only macroscopic and phenomenological information about the growth process. Regarding the genetical regulation at the molecular level of the growth process, enormous progress has been made within the developmental biology of animals, the metazoans. In, for example, the fruitfly *Drosophila* detailed insights have become available into regulation by developmental control genes, including a cluster called the *hox* genes (see St Johnston and Nüsslein-Volhard 1992, Nüsslein-Volhard 1996, Wolpert et al. 1998). In *Drosophila* it is demonstrated that gradients of morphogens, which shape the embryo, are controlled by the *Hox* genes. This

Fig. 1.6a,b. Illustrations from 'Design in Nature' showing the similarity between the branched form of an electrical spark and a red coral (*Corallium rubrum*). The original caption to (*b*) was that the coral 'shows the tree-like branching form observable in lightning flashes'.

(a)

(b)

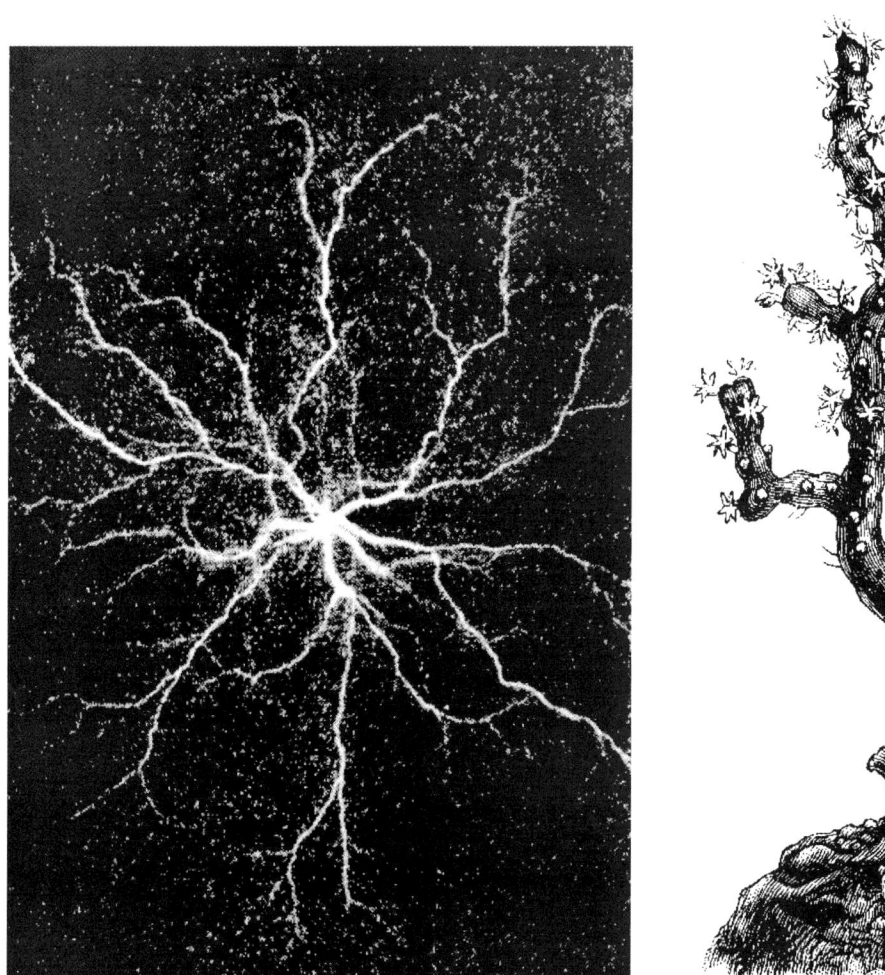

Fig. 1.7. (*a*) Diffusive patterning in *Drosophila*. Diagrammatic drawing of the embryo sac in which a gradient of the morphogen Bicoid is formed. (*b*) Concentration of morphogen along the posterior–anterior axis of the embryo (drawings after Nüsslein-Volhard 1996)

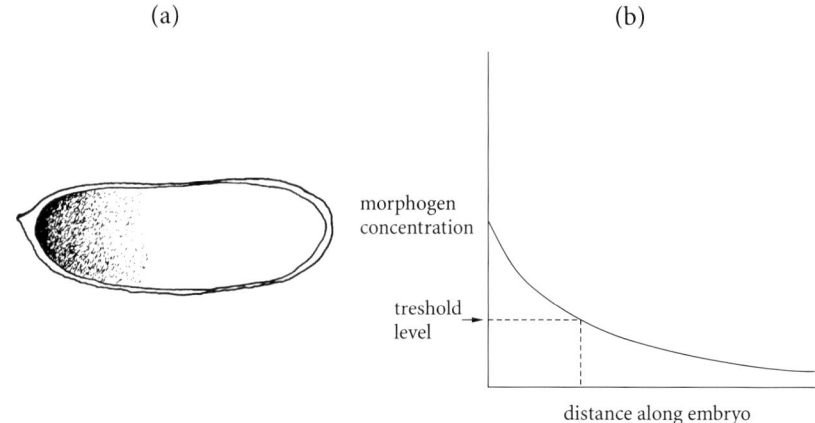

mechanism has been demonstrated in the formation of body axes in the embryo sac of *Drosophila*. At a certain point in time an unequal distribution of maternal RNA in the yolk of the fertilized *Drosophila* egg switches on the *bicoid* gene, which produces the protein Bicoid. This Bicoid diffuses through the embryo sac (see Fig. 1.7a); so the concentration will become the highest in a region which is destined to become the head of the embryo and declines gradually. Bicoid is an unstable protein, the concentration at remote points will become lower and a persisting gradient emerges. At the point in the embryo where the Bicoid is below a certain threshold (Fig. 1.7b) another gene is switched on, the so-called *hunchback* gene. This gene also produces a protein, which again produces a gradient. This gradient divides the embryo into a region which will become the head and a region that will become the thorax and abdomen. The *bicoid* gene expression results in the specification of an anterior–posterior axis. Similar mechanisms based on morphogen gradients have been demonstrated to determine the dorso–ventral axis, the segmentation of the embryo, the specification of the position of the appendages, etc. The body plan of the developing embryo is being controlled by a cascade of successive expressions of *Hox* genes and other regulatory genes.

One of the most remarkable discoveries of the past few years is that developmental regulatory genes, of which the *Hox* genes are one of the best studied classes, control development throughout the animal kingdom (see Raff 1996, Erwin et al. 1997). Furthermore it has been found that these regulatory genes are highly conserved in animals and that similar regulatory genes are used over and over within different phyla within the animal kingdom to specify the different body axes and the body plan. The body plan in the developing embryo emerges under control of these regulatory genes; furthermore it can be demonstrated that modification in the body plan in the different phyla can be linked with modification in these regulatory genes. Consequently the evolution of the different phyla can be related to modifications in the body plan and to the developmental biology. There is a close link between the number of developmental regulatory genes and the overall complexity of the body plan. In the evolutionary tree depicted in Fig. 1.8 some of the basic steps in the evolution of metazoans are shown. Sponges represent the group with the most simple body plan and were the first multicellular metazoans to evolve; there are no body axes specified and there is (at least) one *Hox* gene. In the cnidarians there are three *Hox* genes; these are the first metazoans in which

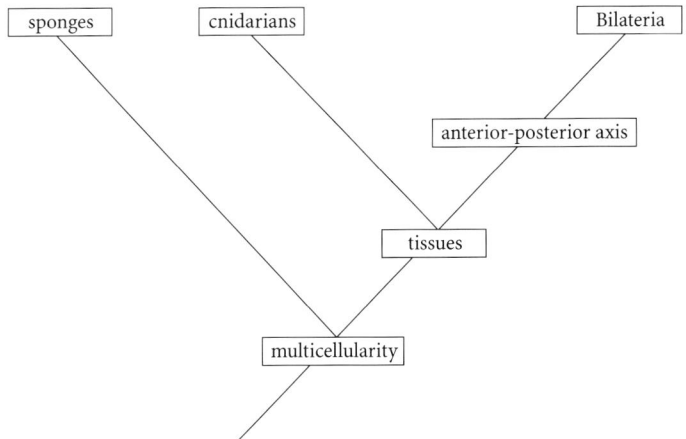

Fig. 1.8. Simplified evolutionary tree of the metazoans with the major regulatory events (after Raff 1996). 1. multicellularity; 2. tissues; 3. anterior–posterior axis

cells are organized in tissues and exhibit a fundamental radial symmetry. In the next step in the evolutionary tree an anterior–posterior axis is specified and the Bilateria develop, consisting of all other metazoans including *Drosophila* and mammals which are characterized by a bilateral symmetry. This all seems to indicate that the developmental biology of sponges is the most simple, followed by the development of cnidarians. Unfortunately detailed knowledge of morphogen gradients regulating the morphogenesis, comparable with those for *Drosophila*, is scarce in sponges and cnidarians. Knowledge about these mechanisms is crucial for including models of genetical regulation in the simulation models and could potentially bridge the gap between the genetical information and the physical shape of the organism.

The second reason why marine sessile organisms are an excellent case study for developing simulation models of morphogenesis is that the marine environment is in many ways more predictable, more constant, and more uniform than the terrestrial environment. In many of the marine sessile organisms, as was demonstrated in Fig. 1.1, the physical environment has a major impact on the growth process. Under a number of assumptions, it is feasible to make approximations of the marine environment in simulation models and to capture the influence of water movement, transport of food particles through diffusion and hydrodynamics, and local available light intensities. We can here benefit enormously from methods recently developed in computational science. In, for example, computational physics methods such as the lattice gas and the lattice Boltzmann method (see Chopard and Droz 1998, Rothman and Zaleski 1997) have been developed, capable of modeling transport, diffusion, flow, and mechanical stress in complex three-dimensional geometries. In computational physics these methods have for example successfully been applied to model flow and diffusion processes in three-dimensional porous media. A major problem in studying growth and form of marine sessile organisms and the influence of hydrodynamics, is that usually forms develop in the simulations with a high degree of geometrical complexity which can, in most cases, be represented adequately only in three dimensions. These simulations of growth and form in three dimensions typically require large-scale computing and computing techniques suitable for objects with complex shaped boundaries. With the recent advances in computational science and the availability of large-scale computing facilities, simulation of the physical environment, including hydrodynamics and local light intensities, has become feasible. The advances in developmental biology may even make it possible that diffusive patterning mechanisms and gradients of morphogens to be included and represented by similar computational models.

There exists a fairly large amount of information on experiments done in vivo and in vitro with marine sessile organisms and the impact of the physical environment. Water flow has a strong influence on the distribution of food particles in the immediate environment of suspension feeders. In several studies (for example Frechette et al. 1989; Buss and Jackson 1981; Pile et al. 1997) it was demonstrated that locally around a sessile suspension feeder areas may occur which are depleted in food particles. There is also a (species-specific) relation between the ability of species to capture food particles and the flow rate. A review of the effects of flow on suspension feeders is given by Shimeta and Jumars (1991). Water movement also has an important mechanical impact on marine sessile organisms (see Koehl 1998, Vogel 1994), which might have very different effects on flexible organisms, for example many seaweeds, and rigid organisms, for example the stony corals. Mechanical forces may lead to breakage and abrasion in the organisms, but also strongly influence the capture of food particles. In the photosynthetic marine sessile organisms, for example the seaweeds and many of the stony corals, the available local light intensity is a dominant environmental parameter (Goreau et al. 1971). In many of these organisms the physiology is adapted to a certain range of light intensities and wave lengths, allowing the species to inhabit a certain range of depths.

A third reason why simulation models of growth and form are required is that this type of model might be useful to study the regeneration capabilities and the impact of changes in the global climate on the growth of, for instance, reef organisms. Recently coral reefs have been affected severely, on a worldwide scale, by so-called "coral bleaching". Coral bleaching occurs when the coral colony expels its photosynthetic symbionts, the zooxantellae, leading to a white-colored "bleached" colony. The process is reversible, but if this bleaching event continues for a longer period, the colony will ultimately die. There is currently much evidence that elevated sea temperatures, in combination with increased ultraviolet radiation, is the main cause of mass bleaching events (see Lesser 1996). In 1998 the most severe coral bleaching event until now was observed worldwide. The bleaching event coincided with elevated sea temperatures during a strong El Niño period in 1998. The consequences of this event were summarized by the International Tropical Marine Ecosystems Management Symposium:

> "A summit meeting on coral bleaching by world experts on coral bleaching held in Townsville on 24 November 1998 released the following statement on the status of reefs following the 1998 global coral bleaching event."
>
> "Tropical sea surface temperatures in 1997/98 have been higher than at any other time in the modern record. Record sea surface temperature increases over the tropics in the past 15 years are not explained by existing climate models. The coral bleaching associated with the high sea surface temperatures has affected almost all species of corals. Loss of some corals more than 1000 years old indicates the severity of this event. Associated reef invertebrates have been severely affected by unusually high sea temperatures."

Another reason why marine sessile organisms are an important case study is that growth and form is closely linked with the state of the physical environment. There exist potentially many applications in biomonitoring research, where the growth form in combination with simulation models is analyzed to assess the state of the physical environment. In the series of growth forms of *Pocillopora damicornis* in Fig. 1.1 it is potentially possible to assess the state of the environment, in this case the amount of water movement from the growth forms. In, for example, a study by Bosence (1976) the morphology of coralline algae was correlated to the amount of water movement. Coralline algae represent an important part of the fossil record and Bosence in his paper argues that these different morphologies enable palaeontologists to make detailed interpretations of the palaeoenvironment. Because of their enormous age (up to hundreds or thousands of years) and because of their morphological plasticity due to the environmental influence, some of the marine sessile organisms, for example stromatolites, coralline algae, sponges, and stony corals can be used as bioarchives, in which information on the state of the environment is stored during the growth process. A very good example of this was given in a paper by Adkins et al. (1998) in which they analyzed carbon isotopes in the deep sea coral *Desmophyllum cristagalli* (see Fig. 1.9), which lives at a depth from 500–2000 m. Similar to reef building corals, this solitary coral, consisting of one cup with a polyp, exhibits a layered growth process, as shown in Fig. 1.4, and paired light and dark density bands are formed in the skeleton. The deep sea coral has a growth rate of about 0.2–1.0 mm per year. The specimen used in their study had an estimated age of approximately 15500 years and lived for about 160 years. By tracing isotopes along the growth layers of the coral, the authors observed a decline in ^{14}C going from the oldest to the youngest part of the object. This decline indicates, according to the authors, a change in ocean circulation within the period of 160 years when the coral was alive, and when the coral was "suddenly" exposed to older water (containing relatively less ^{14}C) which had not seen the surface for a long time. By using such bioarchives, provided that enough specimens from all over the world of this species are available, it is now possible to study physical oceanography in the past. This could not be done before and may have major consequences for global climate studies. In the case of *Desmophyllum cristagalli* there is a relatively simple growth form, in which a transect of samples in the growth layers can be determined relatively easily; in more complex shaped objects, for example the branching forms depicted in Fig. 1.3, a simulation model that connects successive growth layers and growth form could be very useful.

Fig. 1.9. The solitary deep sea coral *Desmophyllum cristagalli*

Finally one more area in which simulation models of growth can be applied is the setting up and designing of aquacultures of marine sessile organisms. Numerous important chemical agents are produced by these organisms, which have many applications in medicine. One of the first drugs for successfully treating cancer, cytosine arabinoside, was isolated from a sponge (see Cohen 1963). Setting up aquacultures of marine sponges is a notoriously difficult problem (see Osinga et al. 1999). Major pitfalls could be the type of food particles absorbed by the sponges, and hydrodynamical effects. Absorption of food particles in combination with hydrodynamics could very well be studied through modeling and simulation.

Until relatively recently the study of growth and form was mainly a descriptive and experimental one, but with the advent of new computers and

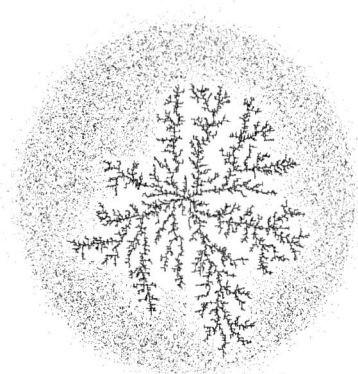

Fig. 1.10. The Diffusion Limited Aggregation model

computing techniques, the in silico option has become important. In many cases it will not be possible to construct a mathematical model for the growth and form problem and solve the problem analytically. Unfortunately this is true even for very simple growth processes, for example the simulation of the growth of a Diffusion Limited Aggregation model (Witten and Sander 1981) shown in Fig. 1.10. The growth form, the branching cluster, is represented in this simulation by lattice sites in a square lattice. Growth of this object proceeds by releasing particles from a circle surrounding the cluster. The particles make a random walk through the square lattice, and the random walk stops as soon as a particle sticks to the growth form. The simulated growth process results in an irregular branching object. This type of simulation can be used to model growth of the bacteria colony and the deposition pattern shown in Fig. 1.5. Even in this relatively simple growth process it can be demonstrated theoretically (see Machta 1993) that the problem of predicting the growth form at a certain point in time is intractable, thus the only way to find the growth at a certain point in time is by explicit simulation, simulating every growth step. There is, unfortunately, no way to predict the growth form analytically.

In most growth processes found in marine sessile organisms, the only way to predict growth forms in models is to simulate every growth step in the mathematical model. To make things even worse in many cases, with the possible exception of encrusting growth forms, to approximate the actual process as closely as possible requires three-dimensional simulations, since many aspects of growth in the marine environment can be captured adequately only in three-dimensional models. However two-dimensional models may also provide insight into morphogenesis of marine sessile organisms and may be used as "toy-models" for a first exploration of the parameter space. Consequently in many cases large-scale computing techniques will be needed in these simulations.

Simulation models supplement experimental observations for the study of growth and form. Experimental observations are limited by logistics and expense, especially in remote and submerged environments. Furthermore the spatial and temporal scales may be too small or too large to study in field or laboratory experiments. Some of the organisms, especially the organisms interesting from a point of view of bioarchives, may be thousands of years old. For example, in sponges from the Antarctic (see Dayton 1978), the growth velocity is so low that even after observing growth forms for a period of three years no significant change in size was observed. There are also several phenomena, for instance micro flow patterns between branching organisms and micro absorption patterns of food particles, which due to the small temporal and spatial scales are difficult to study in vivo or in vitro experiments. In simulations this type of measurement can be done with a high accuracy. A similar observation is made in studies on porous media (see Heijs and Lowe 1995): microscopic phenomena such as diffusion and flow in porous media are very difficult to access experimentally, but can be studied in great detail in simulations.

For the simulation models it is often necessary to do additional experiments to verify assumptions made in the model or to detect crucial information missing from the growth process. In creating simulation models very specific information on the growth process is required, which is in general not available in the biological literature. For example, in many marine sessile organisms detailed information on growth velocities is often not present, there are often no time series available in which the growth of a certain organism is followed for a long period, for instance a photographic registration. The creation of simulation models more or less forces researchers to look systematically through all information available on the growth process.

Simulation experiments can also be used to indicate interesting field or laboratory experiments. In simulation models it may be possible to predict certain changes in growth forms due to changes in external parameters and this may lead to a series of focused field or laboratory experiments. Simulation experiments may serve to study the underlying emergent behavior of the system. Growth processes are typically examples of complex systems, and such a system may be defined as a population of basic elements (for example the corallites in stony corals, spicules in sponges, cells or meristems in algae) which exhibit a certain interaction with each other. The population of these basic elements shows a certain emergent behavior, the growth form, which cannot be predicted from studying isolated individual elements. To obtain insight into this emergent behavior, simulation models are needed.

Another option, to refrain from expensive and complicated laboratory and field experiments and to restrict oneself to simulation experiments, is we believe a very bad option. Losing the coupling between simulation models and actual observations leads to a very empty non-biological world. At this point we believe it is relevant to quote the physicist Alexander Polyakov (van Klaveren and Verhoeven 1995), who worked on turbulence:

> "History shows that science without experiments often tends to lose its way."

1.1 About the Book

In August 1999 we had the opportunity to organize a workshop entitled "Modeling Growth and Form of Marine Sessile Organisms" at the National Center for Ecological Analysis and Synthesis in Santa Barbara, California. For this workshop, to our knowledge the first on this topic ever organized, we invited scientists from a wide range of disciplines including developmental biology, paleontology, ecology, computer science, physics, and mathematics, who have research interests in modeling the development of stromatolites, algae, sponges, octocorals, hydrozoans, and stony corals. The aim of this workshop was to provide a state-of-the-art overview of modeling growth and form of marine sessile organisms and also to get an overview of what pieces of knowledge are still missing. One of the results of this workshop is this book, where we have asked the participants to contribute sections which correspond to their specific expertise.

In Chap. 2 we will zoom in on the marine physical environment (light and hydrodynamics) and its impact on the growth of marine sessile organisms. In the second part of Chap. 2 more details about the biology of the different organisms will be given, where we will focus on the organisms that will be used as case studies in the simulations. In this part an overview will be given of what is known about genetic regulation, the growth processes, and morphological plasticity due to the influence of the physical environment.

In Chap. 3 we will discuss how to measure, quantify, and compare growth forms. In Chap. 4 we will focus on how to construct simulation models of growth and form. In this Chapter we will discuss how Lindenmayer systems can be applied to model algae, how hydrodynamics can be modeled with the lattice Boltzmann method, and how Laplacian growth models, aggregation models, and three dimensional accretive growth models can be used to simulate growth. In Chap. 5 we will discuss how the models can be verified and whether we can do predictions with the model. Chapter 6 will be about the applications, how we can use simulation models to interpret information stored in bioarchives and why knowledge about growth processes is crucial in regeneration of marine ecosystems and setting up aquacultures for producing various types of chemical agents. In Chapter 7 we will end with some main conclusions and an overview of all the missing pieces of knowledge.

2. Environmentally Driven Plasticity

2.1 The Physical Environment

The two major environmental parameters which have the greatest impact on the growth forms of marine sessile organisms are light, required for photosynthesis, and hydrodynamics. A full discussion of the physics of underwater light distributions and hydrodynamics could easily cover a few textbooks. In Box 2.1 the basic hydrodynamic laws are summarized, together with two dimensionless parameters, the Reynolds number Re (2.3) and the Péclet number Pe (2.4), which can be used to characterize the impact of the flow on the organism. In Sect. 2.1.1 "Growing and flowing" we will focus on the biomechanical impact of hydrodynamics on the growth process and try to construct a number of laws for the biomechanical impact of hydrodynamics using an engineering approach. In Sect. 4.3 we will return to the topic of hydrodynamics, from a modeling point of view and try to construct a computational method capable of capturing the influence of hydrodynamics in models of growth and form of marine sessile organisms. In Box 2.2 the basic equations, in a highly simplified form, of underwater light distributions are shown. To a certain extent, in contrast with the hydrodynamic equations, these simplified equations can more or less straightforwardly be included in computational models; this will be discussed in Chap. 4.

Box 2.1 Hydrodynamics

Flow of water in space can be captured through two fundamental equations:

$$\frac{\partial \rho}{\partial t} + \nabla \cdot \rho U = 0 \tag{2.1}$$

$$\frac{\partial U}{\partial t} = -(U \cdot \nabla)U - \frac{1}{\rho}\nabla P + v\nabla^2 U \tag{2.2}$$

In these equations ρ represents the mass density, t the time, U the flow velocity, P the pressure, and v the kinematic viscosity. The symbols ∇ and ∇^2 are respectively the del and Laplacian operator (in three dimensions respectively: $\partial/\partial x + \partial/\partial y + \partial/\partial z$ and $\partial^2/\partial x^2 + \partial^2/\partial y^2 + \partial^2/\partial z^2$). The first equation (the continuity equation) expresses the conservation of mass, and states that the density can change at a point in space only due to a net in- or outflow of matter. The second equation, the Navier–Stokes equation, expresses the conservation of momentum, and states that the flow velocity U changes in time in response to convection $(U \cdot \nabla)U$, spatial variations in pressure ∇P, and viscous forces $v\nabla^2 U$

Two important dimensionless parameters, characterizing the flow, are:

$$Re = \frac{\bar{u}l}{v}, \qquad (2.3)$$

$$Pe = \frac{\bar{u}l}{D} \qquad (2.4)$$

The Re parameter is the Reynolds number and is the ratio between the inertial forces $\bar{u}l$ and the viscous forces, where $\bar{u}$ is the average flow velocity and l a characteristic length in the system (for example the height of the organism). High values of Re indicate that the flow becomes turbulent and the flow will not attain a steady state ($\partial U/\partial t = 0$), while the flow will reach a steady state for low Re numbers (laminar flow).

The Pe parameter is the Péclet number and is the ratio between the inertial forces $\bar{u}l$ and the Brownian forces, where D is the diffusion coefficient. The value of D represents, for example, the diffusion coefficient of a suspended food particle. Low Pe numbers indicate that particles are mainly distributed by diffusion, while high Pe numbers indicate that food particles are mainly dispersed through hydrodynamics.

Box 2.2 Underwater light intensities

The change of light intensity, due to the attenuation in the water column, is described for monochromatic light by the Lambert–Beer Law:

$$I_d = I_0 e^{-\gamma d} \qquad (2.5)$$

where I_d and I_0 are respectively the light intensities at depth d and just below the water surface. The attenuation coefficient γ is the sum of the absorption coefficient and the scattering coefficient. The light absorption depends upon the wavelength of the monochromatic light, minimal absorption takes place in clear water at 465 nm, absorption of blue light increases by the presence of soluble yellow humus-like substances in the water. The scattering coefficient depends upon the amount of suspended particles in the water.

The Lambert-Beer Law gives the light intensity at a certain depth. The amount of light received by a unit area can be calculated by considering the angle of incidence of light, which gives a more detailed description of the local light intensities at the surface of an organism. A (highly simplified) light model (Foley et al. 1990) is described by the equation:

$$I = I_S \cdot \cos \theta \qquad (2.6)$$

The light intensity I (W/m^2) on a surface is determined by $\cos \theta$, where θ is the angle of incidence between the direction of the light beam and the surface normal and I_S is the intensity of the light source (at a certain depth). In this equation it is assumed that the light source is constant, the light direction corresponds to the vertical, and there is no diffuse reflection from the environment.

2.1.1 Growing and Flowing

Marine sessile organisms, such as macroalgae and colonial animals, risk being dislodged or broken by ambient water currents and waves, yet they also depend on that moving water for transport (reviewed in Koehl 1982, 1986, 1999, Denny 1988, 1999, Vogel 1994). Ambient water motion is responsible for dispersing the spores or larvae, and in many cases the gametes, of sessile organisms. Waste products and sediments are also carried away by moving water. Attached algae and animals depend on moving water for the transport of dissolved materials such as nutrients and gases, while suspension-feeding animals depend on ambient currents to bring particulate food to them, and in many cases to ventilate their filters. Body designs that enhance an organism's interaction with the water flowing around it enhance not only transport, but also hydrodynamic forces.

The first step in studying how sessile organisms interact with the water flowing around them is to determine their hydrodynamic microhabitats (e.g. Koehl 1977a, Denny 1988). Many benthic organisms in deep water and in protected bays and estuaries encounter unidirectional currents or tidal currents that flow in one direction for several hours, and then in the opposite direction. Attached organisms in shallow coastal habitats are also exposed to waves. When a wave passes over an organism on the substratum where water depth is less than one half of the crest-to-crest distance between waves, the water flow along the bottom is back-and-forth with a period of seconds. When fluid flows along a solid surface, such as the substratum or the surface of an organism's body, the layer of fluid in contact with the surface does not slip with respect to it. Therefore, a velocity gradient (the boundary layer) develops in the fluid between the surface and the freestream flow. The greater the distance a fluid flows across a surface, the thicker the boundary layer becomes. In marine habitats, the benthic boundary layer can be a meter or more thick, although the steepest velocity gradient occurs within a few centimeters of the substratum (reviewed in Jumars and Nowell 1984). Although a thin sublayer (mm's thick) of laminar flow occurs along the substratum, water flow in the benthic boundary layer is turbulent, so mass and momentum are mixed between the freestream flow and the bottom by swirling eddies. Since it takes time for a boundary layer to build up when water begins to flow over a surface, the benthic boundary layer in the back-and-forth sloshing water of waves is much thinner than in unidirectional flow (e.g. Denny 1988). Local topography and neighboring organisms can have a profound effect on the water flow encountered by a benthic organism, hence the hydrodynamic microhabitat of an attached animal or plant can be very different from the freestream flow over the site where it occurs (Koehl 1977a, Koehl and Alberte 1988). As a sessile organism grows, it can encounter more rapid water movement as it sticks up higher in the benthic boundary layer and becomes larger relative to its neighbors.

Biomechanical studies have shown that general physical rules that apply across taxa can permit us to understand and predict how organisms interact with their physical environments. Such an approach provides a useful framework for considering the consequences of shape and size on the hydrodynamic forces and on the transport experienced by sessile animals and macrophytes as they grow.

Hydrodynamic forces

UNIDIRECTIONAL WATER CURRENTS. Drag is the hydrodynamic force tending to push an organism downstream. The drag on macroscopic organisms is due to the pressure difference across the body that occurs when a wake forms behind the organism (form drag), and to the viscous resistance of the fluid in the boundary layer along the surface of the body to being sheared (skin friction drag) (e.g. Vogel 1994). Drag D on a macroscopic body is given by:

$$D = 0.5\, \rho C_D U^2 S \qquad (2.7)$$

where D is drag, ρ is the density of the fluid, C_D is the drag coefficient of the body (which depends on its shape and surface texture), U is the water velocity relative to the body, and S is a relevant plan area of the body. The convention for relatively undeformable organisms is usually to use the projected area of the organism at right angles to the flow for S (e.g. Koehl 1977a, Vogel 1994, Denny 1988), whereas the convention for very flexible organisms such as macroalgae is to use the maximum plan area of the thallus (Koehl 1986, Carrington 1990, Gaylord et al. 1994) This simple equation points out important features of any benthic organism that affect the drag it experiences. Since drag is proportional to the square of velocity, as organisms grow and encounter more rapid water motion up away from the substratum, they experience disproportionately larger drag. Any morphological characteristics that decrease the size of the wake that forms on the downstream side of a macroscopic organism reduce drag. Such features include orientation parallel to the flow direction, streamlined shape (i.e. a shape that is long and tapered on the downstream side), and porosity (i.e. gaps between branches or lobes that permit water to flow through the structure) (e.g. Koehl 1977a, Vogel 1994).

Most macroalgae and some colonial animals (e.g. arborescent hydroids and bryozoans; gorgonian sea whips and sea fans, Fig. 2.1) are flexible and are passively reconfigured by ambient water currents into more streamlined shapes as flow velocity increases. Such passive reorientation or reconfiguration of flexible organisms by flowing water reduces the size of the wake

Fig. 2.1. Gorgonian sea fan reconfigured by ambient water current

downstream of a body, thereby reducing form drag (e.g. Koehl 1977a, 1986, Vogel 1984, Koehl and Alberte 1988, Carrington 1990). Vogel (1984) has proposed an index, the "figure of merit" B, to describe the relative reduction in drag experienced by flexible structures as they reconfigure as flow velocity increases, where B is the slope of a log-log plot of speed-specific drag D/U^2 as a function of velocity; the greater the absolute value of the negative slope, the greater the relative drag reduction experienced with an increase in velocity. In addition, if the flexibility of a blade-like sessile organism permits it to be pushed down close to the substratum, the underside of the blade encounters slower flow than the upper surface, thereby reducing the shear and thus the skin friction drag on that surface (Koehl 1986). Although shape can affect the hydrodynamic forces which flexible organisms experience (e.g. Koehl 1977a, Koehl and Alberte 1988, Johnson and Koehl 1994), Carrington (1990) found that a variety of very flexible blade-like, branching, and bushy seaweeds converged onto similar drag coefficients when subjected to high water velocities at which they all were compacted into similar streamlined bundles.

Benthic organisms in a water current can also experience lift, the hydrodynamic force acting at right angles to drag. When water speeds up locally as it moves over and around an obstacle, such as an organism, the local pressure on the organism is lower where the flow along its surface is faster; a pressure difference across the body of an organism can thus develop. Organisms protruding above the substratum are pulled up by lift, and organisms that present an asymmetric shape to the oncoming current are pulled laterally if the water speeds up more to move around one side of the organism relative to the other (Denny 1988, Vogel 1994). Lift (L) is given by

$$L = 0.5\, \rho C_L U^2 S \qquad (2.8)$$

where C_L is the coefficient of lift (which depends on shape) and S is a relevant area (usually planiform area normal to the direction in which the lift acts). Thus, as with drag, lift increases as an organism grows, not only because its S increases, but more importantly because it encounters higher water velocities. Even a symmetric structure, such as the cylindrical branch of an animal colony, can experience transient lateral lift, alternating from side to side as vortices are shed in the wake behind it (explained in Denny 1988). If such structures are flexible, they wobble side-to-side as water flows past. Furthermore, if flexible organisms are pushed over by drag, and pulled back up by lift, they can flutter like a flag; the wake behind a fluttering organism can be bigger and the drag force higher than on a body of similar shape and orientation that does not flutter (Koehl and Alberte 1988). When ambient currents encounter a branching structure, such as a coral colony, some of the water flows between the branches, but most of it is diverted above the colony. The lower pressure that occurs above a coral colony as water speeds up to flow over it not only subjects the colony to lift, but also can draw the slowly-moving water between the branches up and out of the colony, thereby reducing the stagnation of flow that can occur in the middle of colonies as they grow larger (Chamberlain and Graus 1977).

WAVES. Sessile organisms exposed to waves experience back-and-forth water motion. Since the velocity changes with time, the instantaneous lift and drag that the organisms experience (which are proportional to the square of the instantaneous velocity) also vary with time. In addition to lift and drag, organisms in the accelerating flow in waves also are subjected to the acceleration reaction force (*A*),

$$A = \rho C_M \frac{dU}{dt} V \tag{2.9}$$

where C_M is the inertia coefficient (which depends on shape), $\frac{dU}{dt}$ is the instantaneous water acceleration, and *V* is the volume of the organism (Koehl 1977a, Denny 1988). Bodies with shapes that deflect the path of the water moving around them a lot (such as stiff, planar colonies normal to the flow) have higher C_M's than do bodies that do not deflect the flow as much (such as streamlined bodies). Since *A* depends on the volume of an organism (*V* is proportional to length^3), a small increase in body length can lead to a very large increase in *A*. If water is trapped between the branches or blades of an algal thallus or animal colony, the functional volume of the organism that affects *A* is the volume of that water in addition to the volume of the organism's body (Gaylord et al. 1994). Since *A* is proportional to the instantaneous water acceleration, it varies with time as the water flows back and forth past an organism; when water is speeding up, *A* acts in the same direction as drag, but when water is slowing down, it acts in the opposite direction from drag. The instantaneous net force on an organism in waves is the sum of the acceleration reaction, drag, and lift at that instant.

Since the hydrodynamic forces on an organism depend on the magnitude of the velocity and acceleration of the water relative to the organism, a flexible organism that can move along with the water in waves can avoid being pulled by hydrodynamic forces until it is fully extended in the direction of flow and the water moves past it. The longer the organism relative to the distance the water moves before it flows back the other way, the more likely the organism is to avoid flow forces at times when accelerations and velocities are high (Koehl 1999). However, a flexible wave-swept organism moving with the flow can be jerked to a halt if it reaches the end of its rope before the water in a wave begins to flow back in the opposite direction; when this occurs, a brief inertial force (proportional to the mass, and hence to the *V*, of the organism) pulls on the organism (Denny et al. 1998). The length of a flexible macrophyte or colony relative to the distance *X* that the water in a wave flows in one direction before stopping and accelerating in the opposite direction can have a profound effect on the forces the organism experiences in waves because it determines when in the wave cycle the organism is jerked to a halt and begins to experience flow relative to its body. As flexible organisms that are short relative to *X* grow, the total force they experience in waves increases (e.g. Gaylord et al. 1994). However, once organisms grow long enough relative to *X* that they reach the end of their rope only after the water in a wave has begun to decelerate, further growth does not lead to an increase in force on the holdfast, as demonstrated by experiments with model organisms in an oscillating-flow tank and by measurements of forces on real kelp on waveswept shores (Koehl 1999). Stretchy tethers such as the stipes of kelp can act as shock absorbers whose stretching absorbs mechanical work, thereby permitting the kelp to withstand the transient high loads they encounter in turbulent or wave-swept habitats (Koehl and Wainwright 1977). Mathematical models suggest that the tuning of the time-dependent material properties of stretchy tethers relative to the frequencies at which these structures must resist high inertial loads in wave-swept environments can have a significant effect on their likelihood of experiencing large forces (Denny et al. 1998).

While some sessile organisms (such as stony corals) are very stiff, and some (such as thin blade-like algae) are very flexible, others (such as sea fans, Fig. 2.1, or stipitate kelp) are of intermediate stiffness. Measurements of hydrodynamic forces experienced in a wave tank by models of organisms of the same blade-like shape, but different flexural stiffnesses, showed that as the stiff models "grew" the peak force increased, whereas as the very flexible models "grew" the force remained low (Koehl 2000). A third type of size-dependent behavior was shown by the models of intermediate stiffness: lengthening increased hydrodynamic forces on short models, had no effect on models of intermediate length, and decreased forces on long models. Since the deflection of the free end of an organism attached to the substratum (like a cantilever) depends on length^4, the longer the models of intermediate stiffness become, the more they bend over and go with the flow (Koehl 2000).

CONSEQUENCES OF HYDRODYNAMIC FORCES AS ORGANISMS GROW. Hydrodynamic forces can deform sessile organisms (which can in turn affect performance of functions such as light or food interception), and can break or rip them off the substratum. How much the tissues in a body deform, and whether or not they break, depends not only on the stiffness and strength of the tissues, but also on the stresses to which they are subjected (stress is the force imposed on a material divided by the cross-sectional area of material bearing that force). The stresses within an organism's body when being stretched, bent, or compressed by a hydrodynamic force can be calculated using the same techniques engineers use to calculate the stresses in man-made structures (for details, see e.g. Roark and Young 1975, Wainwright et al. 1976). Such analyses reveal that, not only do the shape and size of attached organisms determine the magnitude of the hydrodynamic forces which they experience when exposed to ambient currents, but shape and size also determine the distribution and magnitude of stresses within their bodies when bearing those forces (e.g. Koehl 1977b, 1986).

Whether organisms grow geometrically (i.e. maintain the same shape as they get bigger) or allometrically (i.e. change shape as they grow) determines whether or not the local stresses in their tissues increase, decrease, or stay the same during their ontogeny. Although the scaling of the proportions of terrestrial and locomoting organisms of different sizes has received much attention, the scaling of attached sessile organisms is less well studied (Denny 1988, 1999, Johnson and Koehl 1994, Koehl 2000). The "safety factor" of a structure is the ratio of the strength of the material composing it to the maximum stress it experiences during its lifetime. Since ambient water flows on sessile organisms often vary with season and since the size, shape, and mechanical properties of their tissues can change with age, the "environmental stress factor" is a biological version of safety factor that relates the ability of organisms at their particular stages in ontogeny to resist breakage relative to the maximum loads that they experience in nature at those stages. For example, the giant kelp *Nereocystis luetkeana* adjusted their shapes and material properties as they grew in different hydrodynamic habitats in such a way that the "environmental stress factor" was the same in all the habitats and was maintained as the kelp grew during the summer months, but decreased during the winter, after the kelp had reproduced and when they were subjected to storms (Johnson and Koehl 1994).

Whether or not waves impose a mechanical upper limit to the sizes of attached organisms is still being explored (reviewed by Denny 1999). However, breakability is not necessarily a "bad" feature preventing organisms from succeeding in wave-swept environments if those broken organisms can regrow. For example, when bits of a sessile organism or colony break off, the hydrodynamic forces on the part of the structure that remains can be reduced, hence partial breakage can prevent total destruction (e.g. Black 1976). Furthermore, if the broken-off pieces of an organism or colony can reattach and grow, breakage can be a mechanism of asexual reproduction and dispersal, as has been shown for a number of species of coral (Highsmith 1982).

To be able to include the impact of hydrodynamic forces in simulation models of growth processes, as will be discussed in Sections 4.5 and 4.6, it is required to be able to compute local forces exerted by the fluid on the growing object. In most cases these objects will have a typical complex-shaped and branching geometry and are usually characterized by a rough and fractal-like surface. In (2.8) and (2.9) all morphological details are "hidden" in the coefficients C_l and C_M. In morphological simulations of growth processes, where local hydrodynamic forces are included, in many cases a more microscopic description of forces will be needed to be able (for example) to simulate partial breakage. In Sect. 4.3.1 we will discuss how microscopic estimates can be derived from simulated hydrodynamics about complex-shaped obstacles.

Mass transfer

BOUNDARY LAYERS AND MASS FLUX. Organisms such as corals and seaweeds rely on uptake of nutrients and gases across the surfaces of their tissues. Such an exchange of mass is subject to the physical laws of diffusion and convection which are mediated by both properties of the organism surface and hydrodynamic characteristics of the fluid environment. Mass and heat transfer at surfaces have been addressed rigorously in the engineering literature (White 1988, Kays and Crawford 1993). Engineering correlations have been used successfully to describe mass transfer processes at the seafloor (Dade 1993) and for various organisms and communities (Patterson et al. 1991, Bilger and Atkinson 1992, Baird and Atkinson 1997). As fluid moves over surfaces, momentum is extracted from the fluid through friction and a gradient in flow speed is established over the surface that is called the momentum boundary layer (Fig. 2.2a). Analogously, if mass is transferred at a surface by the uptake of a compound, gas, or ion from the bulk fluid, a gradient in concentration is established over the surface; this is the diffusive boundary layer (Fig. 2.2b). Delivery of mass to the surface by diffusion is described by Fick's 1st Law of Diffusion

$$flux = -D_m \, dC/dX \qquad (2.10)$$

where D_m is the coefficient of diffusion for the compound, gas, or ion in the fluid and dC/dX is the concentration gradient over the diffusive boundary layer. As a result, at a constant bulk concentration, flux is inversely proportional to the thickness of the diffusive boundary layer. The relationship

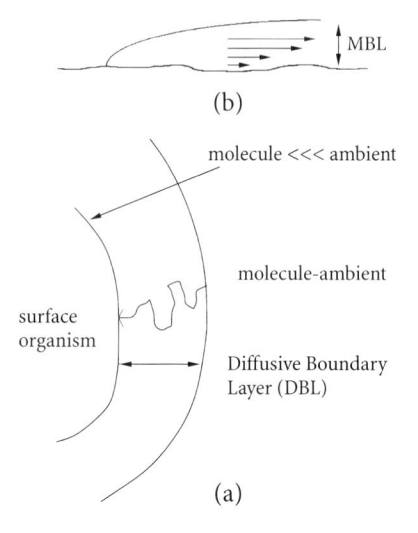

Fig. 2.2. (*a*) Diagram of the formation of a momentum boundary layer (*MBL*) as a fluid flows over a surface. The *MBL* is a gradient in velocity from zero at the substratum to 99% of the freestream velocity. (*b*) The diffusive boundary layer (*DBL*) is a gradient in concentration of some molecule of interest, from an ambient concentration far from the surface of an organism to a much lower concentration adjacent to the organism surface where it is taken inside and used in metabolism. Molecule transport across the *DBL* is by diffusion

between the momentum (*MBL*) and diffusive boundary (*DBL*) layers is given by

$$DBL/MBL = Sc^{-0.33} \qquad (2.11)$$

where *Sc* is the Schmidt number, which is the ratio of the diffusivity of momentum to the diffusivity of the compound, gas, or ion in question. For molecules that are relevant to photosynthesis and respiration, values of *Sc* in seawater at 25 °C are 797 for HCO_3^- and 410 for O_2, resulting in diffusive boundary layers that are approximately 1/7 th to 1/10 th, respectively, as thick as the momentum boundary layer.

Momentum and diffusive boundary layer characteristics are related to the type and speed of fluid motion, the distance the fluid moves over the surface, the surface roughness, and the steadiness of the flow (White 1994). Both types of boundary layers decrease as flow speed increases but how they decrease depends on whether the flow is laminar or turbulent. Characteristics of the flow surrounding an object are related to the Reynolds number *Re* (see (2.3)). Fluids with $Re < 10^5$ (over a smooth flat plate) usually are laminar and become turbulent at $Re > 5 \cdot 10^5$. As a fluid flows steadily over a surface, the boundary layer grows and its thickness is a function of the local *Re* (Re_x), defined as $U_\omega x/\nu$, where *x* is the distance downstream from the leading edge of the surface and where $U\omega$ is the freestream flow speed. The thickness of a boundary layer over a smooth flat surface is $\approx 5(Re_x)^{-0.5}$, and in turbulent flow is $\approx 0.37(Re_x)^{-0.2}$. Laminar boundary layers become turbulent when $Re_x > 10^5$–10^6. Over smooth surfaces turbulent boundary layers consist of a thin viscous sublayer adjacent to the surface, a transition zone, and an outer region that is fully turbulent. In the presence of surface roughness the viscous sublayer disappears. Unsteady flows (e.g. oscillatory) introduce a temporal component to boundary layer formation and growth and can result in the periodic disruption of established boundary layers (White 1994). All other things being equal, boundary layers (both momentum and diffusive) will be thicker over organisms with smooth, bladelike shapes compared with organisms whose surfaces have projections and/or arehack highly branched.

Diffusive boundary layers may represent a significant resistance to the flux of mass to and from the surfaces of benthic organisms. If the compound, gas, or ion is taken up and used immediately in a metabolic process, then diffusion across the boundary becomes the rate-limiting step and the process is mass transfer limited (Bilger and Atkinson 1992). In this case, the metabolic rate should be a function of flow speed to either the 0.5 (laminar) or 0.8 (turbulent) power (Fig. 2.3).

MODELS OF MASS FLUX. To make analyses dimensionless, previous approaches to relate rates of mass transfer to fluid motion have used the Sherwood number (*Sh*, Patterson et al. 1991) or the Stanton number (*St*, Bilger and Atkinson 1992). *Sh* is defined as $h_m W/D_m$, where h_m is the mass transfer coefficient, *W* is the characteristic dimension of the organism, and D_m is the coefficient of diffusion for the compound, gas, or ion. The mass transfer coefficient is calculated from the metabolic rate per unit area divided by the concentration gradient $C_b - C_o$ between the bulk fluid C_b and the wall (C_o, site of exchange). *Sh* represents the metabolic rate in a dimensionless form and is the ratio of convection-assisted mass transfer to exchange by

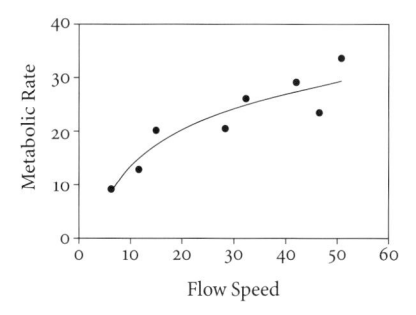

Fig. 2.3. Theoretical relationship between flow speed and the rate of metabolism of mass transfer limited processes. Axis units are arbitrary. The curve fit represents a power function of the form: $MR = aFL^b$

diffusion alone. *Sh* is related to *St* as

$$Sh = St\, Sc\, Re \qquad (2.12)$$

where:

$$St = m/U_\omega (C_b - C_o) \qquad (2.13)$$

where *m* is the metabolic rate or uptake rate per unit area. As a result, either *Sh* or *St* can be used to examine relationships between metabolism and water motion.

These engineering correlations are for mass transfer to hydrodynamically smooth surfaces. However, seaweeds and corals often have projections (e.g. hairs, bullae, calices) on their surfaces.

For rough surfaces, a more appropriate formulation of *St* is given by Kays and Crawford (1993). Alternatively, in the cases where a detailed estimation is required of the local mass transfer at the surface of complex-shaped growth forms and to be able to include mass transfer in simulation models of growth processes, estimates of the local mass transfer in a simulation of hydrodynamics can potentially be made, by estimating local flow velocities and concentration gradients from a simulation. In Sections 4.3.1, 4.5, and 4.6, methods will be discussed for approximating flow fields and concentration gradients through simulation.

EFFECTS OF THE FLOW ENVIRONMENT ON ORGANISMAL METABOLISM: EXAMPLES. Several previous studies have quantified the effect of increasing water flow on rates of nutrient uptake and rates of organismal and community metabolism (Parker 1981, Carpenter et al. 1991, Patterson et al. 1991). A few studies have also examined how organismal morphology interacts with flow to alter boundary layer dynamics and mass flux (Koehl and Alberte 1988, Hurd et al. 1996, Hurd and Stevens 1997).

Patterson and Sebens (1989) used an engineering approach to examine the effects of water flow on rates of respiration of a temperate species of octocoral (*Alcyonium*) and a species of sea anemone (*Metridium*). For both species they found a positive relationship between water flow and *Sh* (based on respiration rate), suggesting that mass transfer of gas exchange limits the metabolic rate. They concluded that organism shape, the local flow environment, and the resulting boundary layer dynamics were important determinants of organismal function.

Seaweeds vary in morphology, both within and between species. Koehl and Alberte (1988) investigated the effects of morphological variation in the bull kelp, *Nereocystis luetkeana*, on boundary layer thickness and rates of photosynthesis of low and high flow morphs under different flow environments. *Nereocystis* has strap-like blades that might be expected to develop thick boundary layers under low flow conditions. Their results indicate that variation in blade morphology allows the low flow morph to flap at a lower flow speed, thereby increasing the flow relative to the blade, resulting in higher rates of photosynthesis. The narrow, flat blades of the high flow morph collapse into a bundle more readily, reducing the drag force experienced, but likely also reducing rates of photosythesis due to self-shading and perhaps increased boundary layer thickness between the blades. This study provides a good example of how seaweed morphology is often a trade-off between the costs and benefits of interaction of the thallus with the physical environment.

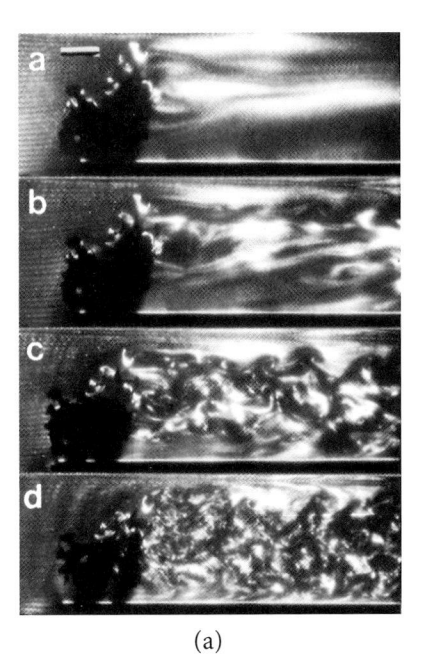

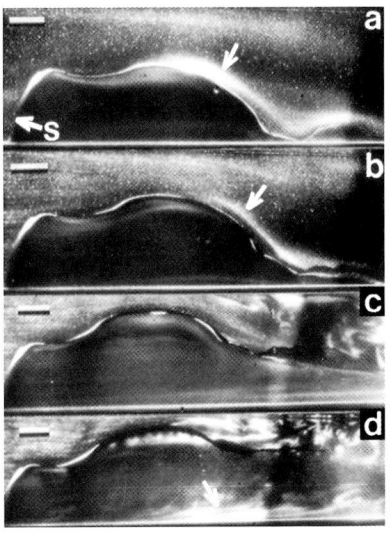

Fig. 2.4. (a) Visualized flow around *Fucus gardneri*. (b) Visualized flow over *Laminaria setchellii*. In both (a) and (b) the flow velocity, and consequently turbulence, increases from the top picture to the bottom (pictures after Hurd and Stevens 1997).

(a) (b)

Patterns of variation in seaweed morphology across flow environments and the resulting effects on metabolic processes have been investigated by Hurd et al. (1996) and Hurd and Stevens (1997). In their initial work, they found that kelp (*Macrocystis integrifolia*) morphology varied between low flow and high flow locations and that nutrient uptake for both forms increased as a function of flow speed. However, estimates of the diffusive boundary layer over the low flow morph blades were no different from those estimated over high flow morph blades, and in this case, changes in blade morphology did not result in higher nutrient uptake at a given flow speed. Hurd and Stevens (1997) used reflective particles and photography to visualize the flow fields around eight taxa of seaweeds that varied in gross morphology, from flat blades (*Laminaria setchellii*) to highly branched thalli (*Gelidium coulteri*), over a range of flow speeds from 0.5 to 5 cm/s. From the photographs (Figs. 2.4a and b) they were able to determine whether the flow over the blades or thalli was laminar or turbulent and concluded that under field flow conditions, it is likely that flow is turbulent over most of the morphologies examined. There were two cases where this conclusion did not apply. Flow over a frond of *Macrocystis* that included multiple blades (the normal condition in the field) was less turbulent than flow over an isolated blade, again suggesting that flow-induced changes in morphology resulting from blades compressing may have important effects on the local flow environment. The second case was for highly-branched thalli where turbulence was reduced and flow always exited the thalli as laminar. For branched thalli, it is the flow environment between branches that determines boundary layer dynamics and these results suggest that boundary layers may be much thicker within the tangle of branches of highly dissected thalli (Carpenter et al. 1991). Given that the demand for nutrients and gases should be related positively to the surface area:volume ratio SA/V of the thallus (Littler and Littler 1980), rates of metabolism of seaweed taxa with high SA/V should be the most flow-dependent. Data collected to date support this prediction (Carpenter et al. 1991, Stewart 1999).

A final example of the effects of flow on organismal metabolism is how uptake of phosphorus by coral reef communities is mediated by hydrodynamic processes (Atkinson and Bilger 1992). Using the St number approach outlined above, phosphorus uptake by coral reef benthos (corals, algal turfs, and macroalgae) arranged in a flume over a range of flow speeds indicated that uptake was mass transfer limited and occurred through turbulent boundary layers. Furthermore, Atkinson and Bilger concluded that rates of phosphorus uptake are 6–7 times higher than predicted from theory and this might result, in part, from the fractal nature of coral reef surfaces and their high surface area (of organisms) per planar area.

FLOW, FORM, AND FUNCTION. The examples given above illustrate the complex relationships between organism morphology, the physical environment, and physiological function (and presumably growth). For seaweeds, some morphologies perform better in particular flow environments with shape and form resulting from a trade-off between the positive and negative effects of water motion. However, the morphology is not constrained entirely by genetics and phenotypic changes in morphology are common.

Size may also have important effects on mass transfer. As the size of an organism increases, the flow environment that it experiences changes. Larger benthic organisms generally extend further from the substratum and away from the benthic boundary layer and experience higher flow that may influence rates of mass transfer. Conversely, larger organisms present more surface area over which local boundary layers can develop, perhaps negating any beneficial effects of growing out of the benthic boundary layer. Combined with the lower SA/V of larger organisms (assuming isometry), the overall ability of larger organisms to take up materials from the surrounding fluid relative to the demand for these materials, may be inversely related to size. Clearly, the interplay between water motion, seaweed morphology and size, and the resulting effects on mass transfer are complex. Attempts to model the growth and form of seaweeds must incorporate, as far as possible, the effects of the flow environment on organismal function, short-term, flow-induced changes in morphology, and the numerous and intimate interactions between morphology and the physical environment.

Particle Capture

Suspension-feeding invertebrates rely, to varying degrees, on the movement of water for the delivery of plankton and other particulate matter to their feeding surfaces. As is the case for the processes of mass flux and momentum transfer, the delivery of food is often strongly affected by the characteristics of water moving past the organism's surface, which in turn can be significantly modified by the organism's shape and position within the benthic boundary layer. Furthermore, particulate capture is also dependent on the size, density, and, in the case of zooplankton, behavior of the food item being captured.

The general theories underlying the mechanisms of particle capture in the marine environment have been thoroughly examined by previous reviews (Rubenstein and Koehl 1977, LaBarbera 1984, Shimeta and Jumars 1991, Wildish and Kristmanson 1997). These mechanisms are often divided into general categories, each reflecting the relative importance of factors such as the relative size of the particle and the filtering apparatus, the density (and thus momentum) of the particle in the fluid, the difference in charge between the particle and the filter, and the tendency of the particle to sink due to gravity. The size and even species of prey item captured from the water column by a sessile suspension feeder is thus to a large extent determined both by the characteristics of the particle and by the interaction of the filter feeder with the characteristics of the ambient flow environment. Both theoretical and empirical approaches have been undertaken to address the role of organism morphology in driving prey capture by sessile invertebrates. While theoretical approaches can offer considerable insight into the factors most likely to affect particle capture, extrapolation to organisms living in the field can be difficult. In contrast, empirical measurements account for more of this natural variability, but in doing so can reduce our ability to generalize. Both levels of approach are therefore necessary and complementary in order to gain a better understanding of the interactions of coral and sponge architecture with the hydrodynamic environment.

SPONGES. As active filter feeders, sponges are able to generate currents through the action of flagella, which line the walls of the interior of the sponge. Bacteria and other microscopic food particles are absorbed from the moving water and incorporated into food vacuoles, and then transported into the main body of the cell. As a result of this active filtration, sponges are often able to thrive in areas of low to moderate flow, where more passive feeders are excluded (Reiswig 1974, Leichter and Witman 1997). Nonetheless, the overall morphology and architecture of sponges can also have a significant impact on particle transport. The intake of water, and thus particulate matter, into the sponge occurs through myriad pores called ostia. The fluid then travels through chambers of varying length and complexity to a central chamber, where the water is expelled through a large opening called an osculum. Generally, the total combined surface area of the incurrent pores exceeds that of the exhalant osculum (Bidder 1923, Vogel 1974). The effect of the reduced surface area is thus akin to a jet, and water exiting the sponge is accelerated, reducing the chances that the sponge will refilter the water that it just processed.

Sponges and other sponge-like organisms also benefit from water movement not induced by flagellar action. As water flows around and over the top of a sponge, the fluid is accelerated. This faster-moving water induces a region of lower pressure, which in concert with viscous entrainment within the fluid serves to induce flows out of the top of the sponge, further enhancing the transport of water throughout the organism (Vogel 1974). The movement of water outside the sponge also serves to replenish nutrient- and particle-depleted water, as sponges at high densities can compete for food resources with one another (Buss and Jackson 1981).

CORALS. While corals display some ciliary activity within the coelenteron, and in some cases have been shown to generate weak currents (Helmuth and Sebens 1993), particle capture is to a large degree dependent on the delivery of zooplankton and particulate matter directly to the coral's tentacles, where the particle is ensnared by a series of harpoon-like nematocysts. The interaction of ambient flow with a coral's morphology thus can have a significant impact on rates of particle capture, as can the presence of neighboring organisms.

Quantifying the interaction of coral morphology with flow is, however, a very complex undertaking, especially given the wide array of flow

28 2. ENVIRONMENTALLY DRIVEN PLASTICITY

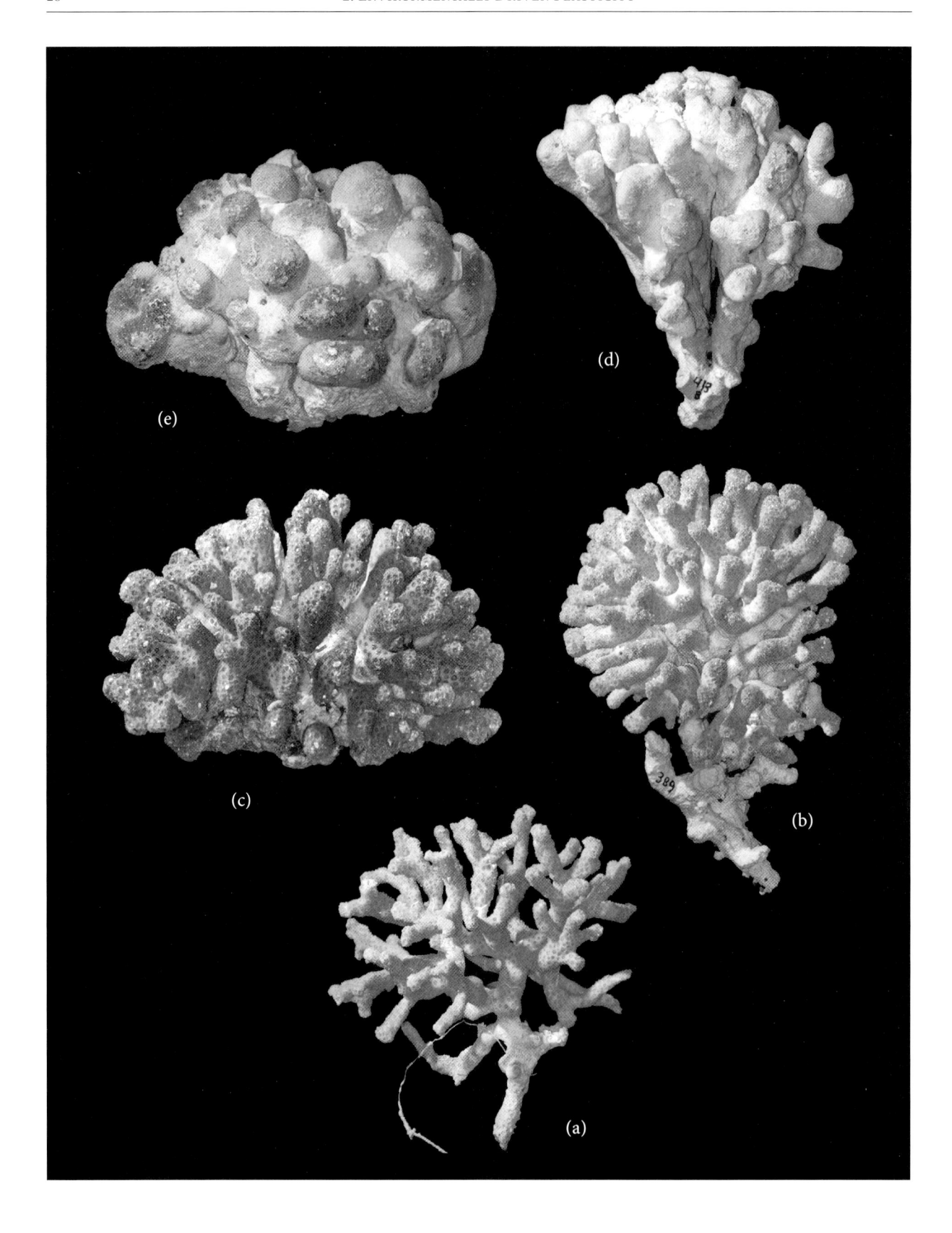

regimes normally found on most coral reefs. Both theoretical and empirical approaches have been undertaken to address this issue. Models of coral feeding under simplified, laminar flow conditions (Abelson et al. 1993) have suggested that the size and type of particle captured is dependent on the height of the organism above the substrate relative to the width of the organism in the direction of flow. Corals which extend above the substrate are expected to feed on finer particles which are resuspended from the bottom, whereas corals which lie close to the bottom were predicted to feed primarily on heavier, bed-load transported particles (Abelson et al. 1993). While this model was supported by measurements conducted using physical models, it has yet to be tested using corals living under more realistic flow conditions.

Empirical measurements of particle capture by corals and other suspension feeding invertebrates have also been employed under both artificial (laboratory flume) and semi-natural (field flume) conditions, and have shown that patterns vary significantly with coral morphology (Heidelberg et al. 1997). Sebens and Johnson (1991) measured rates of feeding by two species of scleractinian corals using brine shrimp cysts as food particles. They found that the branching, cylindrical species of coral, *Madracis decactis* (see Fig. 2.5), showed an increase in particle capture with increasing flow speed. In contrast, a flat species of coral, *Meandrina meandrites*, showed no effect of flow speed on particle capture, due to the tendency of the coral's tentacles to flatten under ever higher flow speeds (Johnson and Sebens 1993). Helmuth and Sebens (1993) examined particle capture by several morphotypes of *Agaricia agaricites*, and found that particle capture in unidirectional flow increased with flow speed up to a velocity of approximately 30 cm/s, but then decreased at flows above this level. Similarly, Sebens et al. (1997) found that feeding by solitary branches of the coral *Madracis mirabilis* (see Fig. 2.5, a cylindrical coral that exists almost exclusively in aggregations of clonemates) experienced maximum rates of food capture at a flow speed of 10–15 cm/s. These studies show that while increasing flow speeds do increase the rates of particle delivery to the coral surface, particle capture efficiency often decreases with increasing flow due to the tendency for a coral's tentacles to flatten under high flows, rendering them unable to capture the particles moving across their surfaces (Patterson 1984, Lasker 1981, Johnson and Sebens 1993, Sebens et al. 1997). Thus, particle capture rates are often highest at intermediate flow speeds, where particle delivery rate is high, but tentacles are still capable of retaining particles.

Some species have apparently been able to at least partially circumvent this limitation through the formation of aggregations in which the spacing between ramets varies as a function of ambient flow (McFadden 1986, Sebens et al. 1997). For example, McFadden (1986) found that under low flow conditions, the presence of neighboring colonies reduced the rate of particle capture by the soft coral *Alcyonium*. However, at higher flow speeds, particle capture rates were enhanced by the presence of neighbors. Sebens et al. (1997) found that branches within aggregations of the cylindrical coral *Madracis mirabilis* were more widely spaced in slower moving water than in areas with higher average water velocities. Feeding trials in a laboratory flume confirmed that particle capture increased with branch spacing in low flows, but decreased with branch spacing in higher flows, suggesting that plasticity in branch spacing represents a means of acclimatizing to the characteristics of the local flow environment. Thus, living in aggregations may

◀ Fig. 2.5a–e. The stony coral *Madracis decactis* collected at different depths, sample (*e*) originates from a depth of 6 m, (*d*) was collected at a depth of 15 m. The stony coral *Madracis mirabilis*, samples (*c*), (*b*), and (*a*), were collected at depths of respectively 6, 8, and 20 m.

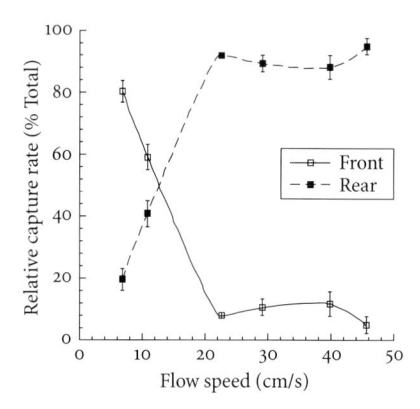

Fig. 2.6. Capture rates of a single branch of a *Madracis mirabilis* colony at the front (upstream) and the rear (downstream) region (after Sebens et al. 1997)

serve as a means of dampening fluctuations in ambient flows (Chamberlain and Graus 1977, Sebens et al. 1997, Helmuth et al. 1997).

Measurements of particle capture by corals have also shown that the location of capture on a colony, and within aggregations, can vary consistently depending on local flow conditions. For example, several studies have shown that particle capture at low flow speeds tends to be highest on upstream regions of colonies. As flow speed increases, areas of maximum capture rate are shifted to the downstream region of the colony or aggregation (Patterson 1984, Helmuth and Sebens 1993, Sebens et al. 1997). In Fig. 2.6 the capture rates of a single branch of a *Madracis mirabilis* colony at the front and rear sites are compared; for high flow speeds the maximum capture rate is shifted to the downstream region of the branch. However, most of these studies were conducted under conditions of unidirectional, low-turbulence flow, and it is unclear whether or not these patterns remain under conditions of highly turbulent flow. Patterson (1984) found highly asymmetrical patterns of particle capture by a soft coral under conditions of laminar flow; however, these differences disappeared under higher levels of turbulence. Hunter (1989) studied feeding by a hydroid under oscillatory flow, and found that patterns in intracolony particle capture disappeared when compared with those observed during feeding in unidirectional flow, but also that capture rates under conditions of alternating flow could not be predicted given measurements in unidirectional flow. While both oscillatory and unidirectional flows occur on coral reefs (Sebens and Johnson 1991, Helmuth and Sebens 1993), the relationship between colony morphology and particle capture under highly complex flow regimes requires much more detailed study before we can explicitly relate patterns in flow over large scales to patterns in food uptake by corals.

2.2 The Case Studies

In this section the different organisms, used as case studies, will be discussed in more detail. Information will be provided regarding the growth processes, which will be later used in the simulation models. This section will focus on the level of organization varying from molecular genetics to the level of the modules and give an overview of the morphological plasticity of each organism.

2.2.1 Case Studies of Environmentally Driven Plasticity: Seaweeds

Seaweeds include a vast variety of evolutionarily distant organisms with the single unifying feature of being macroscopic algae which live in the sea. Their myriad variety of growth forms is produced by a number of very different growth processes which are, in general, associated with different evolutionary lines.

Seaweeds occur in three different divisions, or phyla: reds (Rhodophyta), browns (Phaeophyta), and greens (Chlorophyta). The three groups are not closely related to each other and one, the green algae, are more closely related to land plants than they are to other seaweeds. Red and green seaweeds are evolutionarily ancient, first appearing in the midproterozoic, 1600 million years ago, before the earliest traces of invertebrates. Brown seaweeds are part of a very large and diverse group which includes unicellular algae such as diatoms which date to 900 million years ago, but the earliest putative brown seaweed appeared only 490 million years ago in the ordovician epoch (Graham and Wilcox 1999).

All three groups include unicellular and multicellular forms, filaments, sheets, balloons and branching structures while each group has, with a few exceptions, a different characteristic mode of producing new growth (Fig. 2.7).

Fig. 2.7. (*a–c*): red seaweeds: (*a*) *Bangia spp.* (*b*) *Mazaella splendens* (*c*) Heteromorphic generations of *Mastocarpus stellatus*. The upright fronds are the haploid phase and the dark colored crust on the rock, in the foreground, is the diploid phase of the same species. (*d–f*): brown seaweeds: (*d*) *Ectocarpus siliculosus* (*e*) *Postelsia palmaeformis* (*f*) *Fucus evanescens* (*g–i*) green seaweeds: (*g*) *Cladophora*, (*h*) *Ulva lactuca* (*i*) *Caulerpa racemosa* (photos (*a*), (*d*), (*g*), and (*i*) courtesy of R. Sheath, C. O'Kelly, J. Graham, and R. Carpenter, respectively)

Fig. 2.8a–d. Branching red seaweeds. (*a*) *Nitophyllum punctatum* (*b*) *Polysiphonia sp.* (*c*) *Bonnemaisonia asparagoides* (*d*) *Stenogramme interrupta* (after Harvey 1869)

(a)

(b)

(c)

(d)

Green seaweeds have diffuse growth, meaning that cell divisions occur all over the thallus as the whole organism grows. Red and some brown seaweeds have apical growth. Like land plants, they grow outward from the edges, or apices. In many of the red seaweeds branching growth forms are found. An example is *Callithamnion roseum* shown in Fig. 1.2, and selected examples of branching red seaweeds are depicted in Fig. 2.8. Most brown seaweeds grow outward from a zone of tissue within the thallus. In the case of kelps, the growth zone is at the base of the blades. Stipe tissue is produced from one edge of the growth zone while frond tissue is produced on the other.

Individual species, of a given general morphology, may express subtle or extreme morphological variations either within or between genotypes. There are four different kinds of variation within species; two of which involve drastic switches between morphologies so different that they look like they should be classified in different genera or even orders (1 and 2, below) and two of which represent continua of plasticity within a bodyplan (3 and 4, below). These are: (1) heteromorphic alternation of generations, (2) environmentally driven morphological switches, (3) ecotypic or geographic variation, and (4) phenotypic plasticity.

Most seaweeds have separate, free-living, macroscopic haploid and diploid phases of their life cycles. The structures of the two phases may be similar or strikingly different. The latter case is called the heteromorphic alternation of generations. One example is *Mastocarpus stellatus* (Fig. 2.7c) which has a flat, encrusting phase formerly classified as the genus *Petrocelis* and a leafy, branched upright phase. The giant kelps, *Macrocystis sp.* which have complex internal and external morphology in the diploid stage with individuals up to 100 m long, alternate with a microscopic, branched, haploid filament. Unless the spores produced by each phase are grown through the complete life cycle, there is nothing in the gross morphology to indicate that these incredibly different shapes are produced by a single genotype. Although heteromorphic alternation of generations occurs in many species of red and brown seaweeds, the biological basis of the biphasic morphogenesis is unknown.

Switches in morphology within a genotype can also be elicited by environmental conditions in some species (Lüning 1990). For example, the brown seaweed, *Scytosiphon lomentaria*, grows as a clump of elongated tubes or as a felty encrusting sheet, formerly known as *Ralfsia sp.*, depending on the temperature and photoperiod during its development (Dring and Lüning 1975). Similarly, there are seaweeds which do or do not grow branches, whorls, or hairs depending on the color of light they are exposed to. The complex red seaweed species *Bonnemaisonia hamifera* develops into an alternate, filamentous form when the water temperature is below 13 °C. The two morphotypes of this species were originally thought to be different species with overlapping geographic distributions. The green alga *Urospora wormskioldii* exhibits three different morphologies – a filament, a hollow ball, or a globular multicellular shape – depending only on the temperature during its development from zoospores (Bachman et al. 1976; Lüning 1990).

As is the case for most species of both modular and unitary organisms, many seaweeds vary over their geographic range with "stress tolerant" morphologies occurring at the limits of its distribution. Geographic variation is probably due to differential selection of morphologies in local habitats. For example, species of the brown, fucoid algae have distinct miniature morphs in the Baltic Sea, at the limit of their salinity tolerance. *Ascophyllum nodosum* has long, straight fronds attached firmly to the substrate over most of its range in the Northern Atlantic but exhibits a tightly curled, free living form in the lochs and fjords of Northwest Europe. Some examples of geographic variation in seaweed morphology can be found in the work of Hanisak and Samuel 1987 and Chopin et al. 1996.

The most intensively studied type of morphological variation in seaweeds is phenotypic plasticity in response to environmental conditions including light color and intensity, temperature, water flow and nutrient availability. These responses are akin to the changes in coral or sponge morphology in response to light supply and water flow which are described in the following sections. In all these cases, sessile organisms are adjusting their morphology according to the prevailing conditions in their habitat. There are numerous examples of seaweeds, especially brown seaweeds, developing spiky, hairy, and ruffled structures when water motion is low and more streamlined shapes when water motion is intense (see Fig. 2.4 and Sect. 2.1.1). Similarly, changes in surface complexity or surface-area-to-volume ratio can be driven by temperature or by light intensity (Kübler and Dudgeon 1996). All of these morphological plasticities appear to be reversible over the lifetime of the organism. Therefore, these kinds of changes in morphology can form a record of environmental changes occurring over the lifetime of an accreting, modular organism (see Sect. 6.2) and these are the types of morphological variation which are currently best understood and most amenable to simulation (see Chap. 4).

2.2.2 Morphological Plasticity in Sponges

SPONGES: THE BRANCHING SPONGE *RASPAILIA INAEQUALIS*. The sponge *Raspailia inaequalis* is endemic to northern New Zealand, where it grows on subtidal reef flats (Fig. 2.9). The species has a fan-like branching form, with an indeterminate branching pattern (the name *inaequalis* refers to the unequal lengths of the branches). The skeleton is axially symmetric, with each branch having a dense spicule and spongin core. This well-defined skeletal structure makes the species ideal for morphological studies, as the branching pattern can be accurately determined by digitizing photographs. In the study described here, a number of individuals growing in the Sponge Garden (within the Goat Island Marine Reserve) were photographed several times over the course of a year, allowing their growth to be followed in detail. Photographs were taken using an underwater camera with a fixed focal-length macro lens. A clear perspex sheet was mounted in the focal plane and this was pushed against the specimen, holding it against a white background.

Fig. 2.9. The Sponge Garden at the Goat Island Marine Reserve in northern New Zealand. The two fan-like sponges are *Raspailia inaequalis*, which is a common species on this reef. The sponges are growing so that the fans are at right-angles to the prevailing currents, which are predominantly swell- and tide-driven

Fig. 2.10a,b. Two photographs of the same sponge taken six months apart, showing how the branch tips have grown: (*a*) The background is marked in 2 cm squares. This sponge is a small specimen, only 5 cm high. (*b*) The growth of the branch tips over the six month period is shown by the numbers (mm). The growth rate of these sponges is only 1 cm per year.

(a)   (b)

The images of the sponges were traced, and the tracings were digitized and skeletonized using the procedure described in Sect. 3.2. The branch lengths could then be calculated and used to determine the growth rates. An example showing the growth of a small specimen over a three month period is given in Fig. 2.10. By looking at these sequential photographs in detail many general features of the growth process can be derived.

- *Growth is occurring only at the branch tips.* Because of the spicule and spongin skeleton it was expected that the growth of *Raspailia inaequalis* would occur only at the tips of the branches. This was found to be the case, suggesting that the growth response of only the tips needs to be incorporated into a model.
- *Branching is usually symmetric.* When they branch, the tips flatten and divide symmetrically. Some tips appeared to have formed from the side of a pre-existing branch, but this was much less common.
- *Branch tips turn away from the substrate.* The long branch to the left of the specimen in Fig. 2.10 is curving upwards. This was seen on a number of specimens, and suggests that there is a vertical environmental gradient which the sponge is able to sense and respond to.
- *The sponge does not rapidly turn away from itself.* The small interior branch in Fig. 2.10 turns upwards only after it has grown into another branch. Whatever the mechanism which controls growth, it does not allow branches to take an advance action to avoid growing into one another.

Fig. 2.11a,b. The branches to the left of this sponge all grow by the same amount over the six months between these photographs. The tip third from the left does not appear to be affected by being crowded.

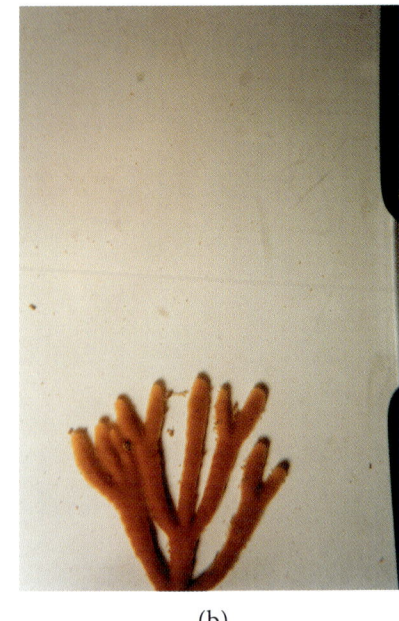

(a) (b)

- *Growth rate is higher away from the substrate* (Figs. 2.10 and 2.11). The observation that the tip-growth is higher away from the substrate is consistent with the suggestion that the sponges are responding to the flow, as water movement increases away from the bottom. It is also possible that tip growth-rate is regulated by a hormonal gradient between the tips and the base of the sponge.
- *Even very crowded branch tips continue growing* (Fig. 2.11). Growth rate does not appear to be affected by branch density. This suggests that the

(a) (b)

Fig. 2.12a,b. The small branch-tip initially has a symmetrically pointed shape, but by the time of the second photograph it has become flattened. These sponges are flexible and the flattening has been caused by the tip touching the neighboring branches.

flow around a branch does not directly influence the growth speed of the corresponding tip.
- *There appears to be genetic variation in the branching pattern.* The majority of the specimens show a dichotomous pattern with branches forming symmetric pairs. The specimen in Fig. 2.11 is unusual because it has a triple branching structure, the branch-tips dividing into three symmetric branches. Triple branchings were otherwise very rare among the eighty sponges which were photographed and it seems likely that this is an example of genetic variation between the individuals.
- *Mechanical interaction is important* (Fig. 2.12). In some cases tips showed evidence of being knocked against neighboring branches. The sponge is flexible so mechanical interaction between branches will be common, particularly when the sponges become larger. It is possible that this is an important factor in structuring the branching pattern. The increased flexibility of the longer fingers may then explain why the spacing between branches tends to increase with distance from the base of the fan.
- *Growth may be prevented by other objects* (Fig. 2.13). This photograph shows a *Raspailia inaequalis* sponge which has been prevented from growing into a normal fan-like shape by another sponge of a related species, *Raspailia topsenti*, that happened to be directly above it. The tips on the left have not grown around the obstacle but have simply stopped growing. The branches to the right of the sponge are developing normally.
- *Damage is important in modifying the branching pattern* (Figs. 2.14 and 2.15). The models of sponge growth which have been developed to date do not incorporate changes to the branching pattern caused by

Fig. 2.13. A photograph showing an interaction between two different sponge species. The growth of the left-hand side of the fan-like *Raspailia inaequalis* sponge has been blocked by a horizontal branch of *R. topsenti*. The fan sponge does not grow out of its plane in order to get around the obstacle.

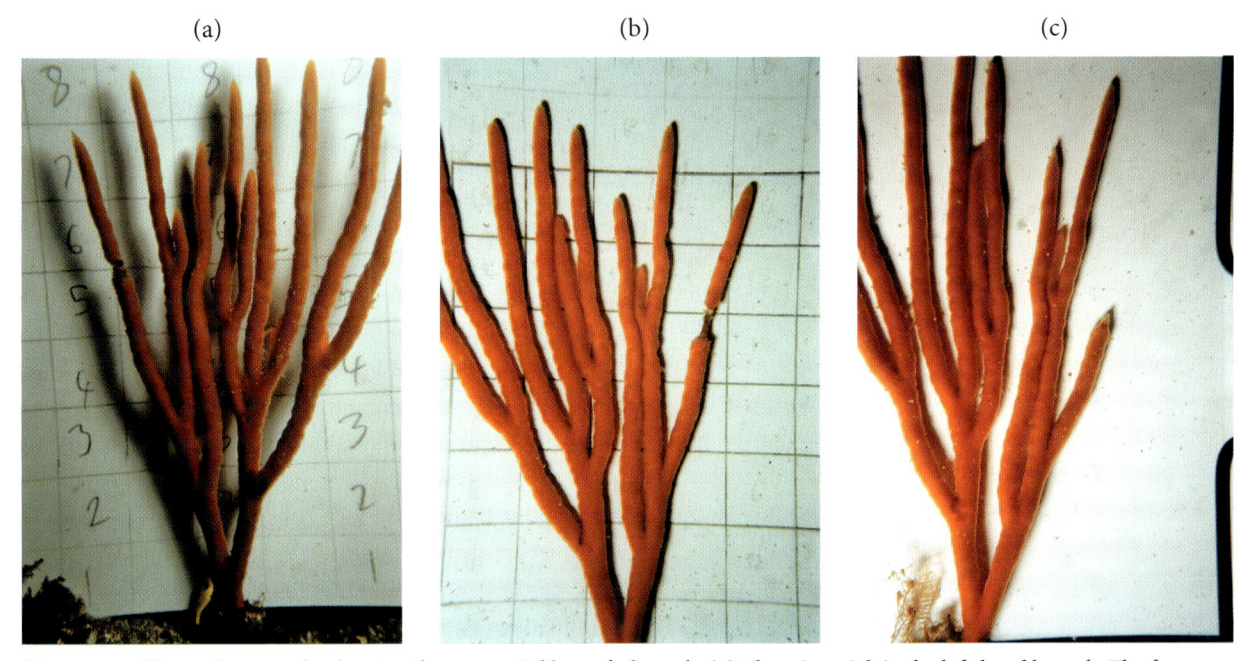

Fig. 2.14a–c. Three photographs showing the sequential loss of a branch: (*a*) There is a nick in the left-hand branch. The fuzz near the damaged point is detrital material which has settled onto the sponge's surface. (*b*) The tissue has become infected and has fallen away, exposing the spicule core. (*c*) The top of the branch has fallen off. The tip appears to be healing and it is likely that this branch will have continued to grow.

Fig. 2.15a,b. This sponge has become diseased, losing many of its branches: (*a*) The tissue on the left-hand side has wasted away and many of the other branches are discolored, indicating that the disease is widespread. (*b*) The infected branches have fallen away. The remaining sponge appears to be healthy.

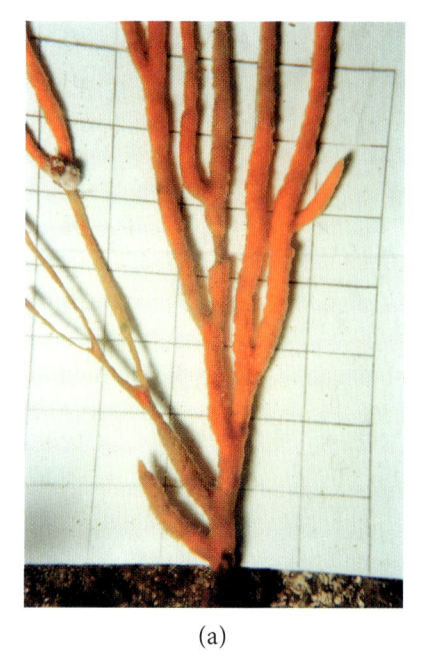

(a) (b)

damage and disease. The observed branching pattern of an older sponge will reflect both the growth and the loss processes. Modeled forms and actual sponges may compare poorly if the damage is not included, even if the growth process is fully understood. The tip to the right-hand side of Fig. 2.15 grew by less than 1 mm during the six months between the photographs. Even though it does not appear to have been infected, the stress to the sponge has stopped its growth suggesting that resources for growth are shared around the organism.

Sequential observations of the same individuals reveal many features of the growth process, which may be used to help inform models of branching growth. The photographs demonstrate the complexity of the processes which must be included in a realistic model. Unfortunately, the photographs do not give any information on how the external environment is sensed within the sponge, or on how the growth is regulated so as to produce the observed growth patterns. A more detailed experimental approach would be needed to gather data of this kind.

Sponges: the Branching Sponge *Haliclona oculata*. An example of a sponge where the growth form develops in a surface normal deposition process (see also Fig. 1.4) is the sponge *Haliclona oculata*. This sponge shows a strong morphological plasticity due to the influence of the physical environment. In Fig. 2.16 examples from the range of growth forms are shown. The growth form of *Haliclona oculata* varies within a range from very thin-branching growth forms found in conditions sheltered from water movement (form A, collected in a tideless salt-water lake) to plate-like growth forms found in exposed conditions as shown in Fig. 2.16c. In general the thickness of the branches and the branching rate increase with the rate of exposure to water movement, while the overall growth form of the sponge tends to be more or less flattened, where the largest plane of the growth form develops

Fig. 2.16a–c. ▶ Growth forms of the sponge *Haliclona oculata*. Growth form (*a*) originates from a site sheltered from water movement, while (*c*) was collected at an exposed site, and sample (*b*) is from a site with an intermediate exposure to water movement.

(a)

(b)

(c)

Fig. 2.17. (*a*) Section through a branching tip of the sponge *Haliclona oculata* showing the skeleton architecture. (*b*) Drawing of the section shown in (*a*)

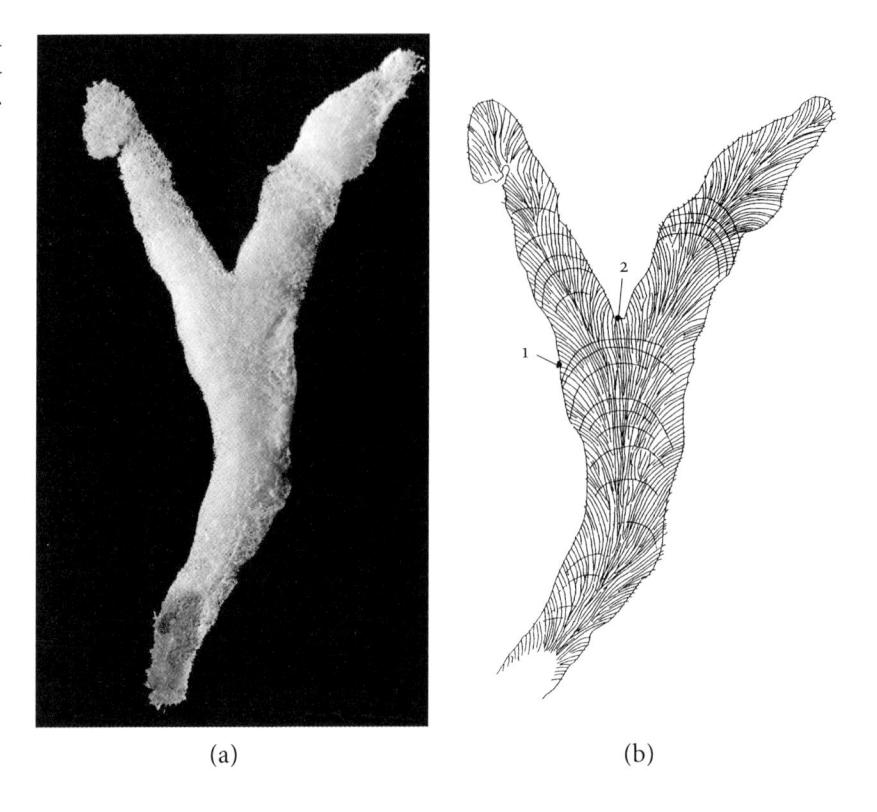

(a)   (b)

Fig. 2.18a–c. ▶ Microscopic views of the tangential skeleton architecture showing the arrangement of the individual spicules in the sponges: (*a*) *Haliclona oculata* and (*b*) *Haliclona simulans* (after De Weerdt 1986). (*c*) Growth form of the sponge *Haliclona simulans* (after Bowerbank 1876)

in a plane perpendicular to the governing flow direction. Forms (a) an (c) are two extremes between which all kind of intermediates can be found.

In sponges, only species with a certain kind of skeleton architecture can develop erect tree-like growth forms. The skeleton architecture of *Haliclona oculata* is shown in Fig. 2.17, where a branching tip of the sponge was sectioned, in this section all living tissue was removed, and only the skeleton, consisting of small silicium elements known as the spicules, is visible. The skeleton of *Haliclona oculata* consists of discrete identical skeleton elements (the spicules) which are connected by spongin and consolidated in a three dimensional mesh, where a distinction between ascending and interconnecting fibers of spicules can be made. In Fig. 2.17 especially the ascending fibers are visible as are some of the successive growth layers which were deposited on top of the previous growth stages in the growth process. It can be observed in this section that the ascending fibers or longitudinal elements are set perpendicular to the previous growth layer. This type of skeleton architecture is known as "radiate accretive" (terminology after Wiedenmayer 1977). The structure of the growth layer itself becomes clearly visible in a tangential view of the surface of the sponge as shown in Fig. 2.18a. In this microscopic view of the surface of the skeleton it can be seen that the spicules are arranged in 4- to 6- (infrequently 3-) sided polygons. The length of a side of a polygon is about the size of one spicule. The surface of the sponge can be considered to be tessellated with a pattern consisting mainly of pentagons and hexagons. The three-dimensional mesh of spicules in a tip of this sponge possesses a radial symmetry: a longitudinal section (parallel to the axis of the tip) will always show about the same structure; the tips however may be only somewhat flattened (Fig. 2.16). The radiate accretive architecture is the reflection

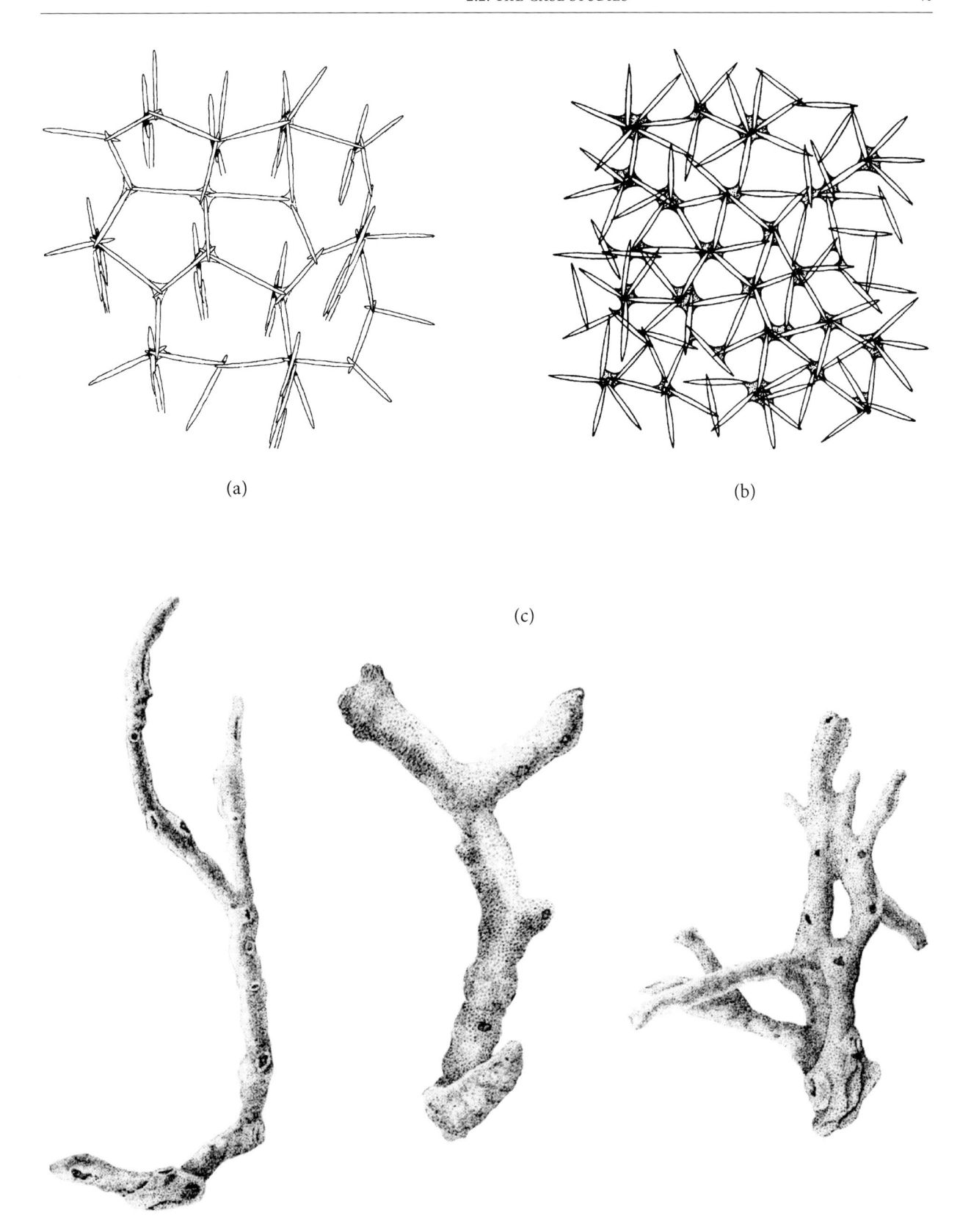

(a)

(b)

(c)

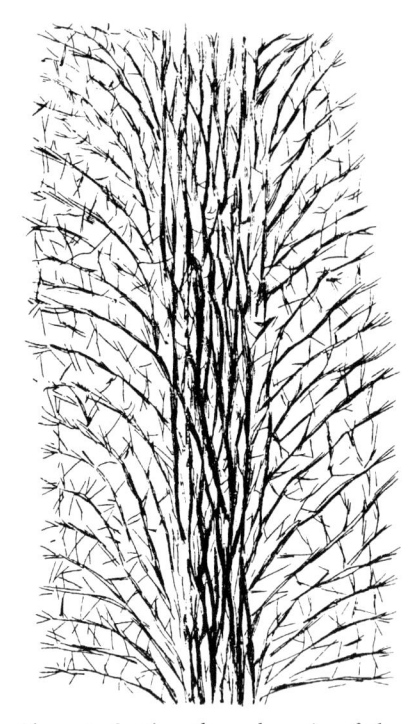

Fig. 2.19. Section through a tip of the sponge *Axinella polypoides*, related to the sponge *Raspailia inaequalis*, showing a skeleton architecture with an axial condensation of spicules (after Vosmaer 1912)

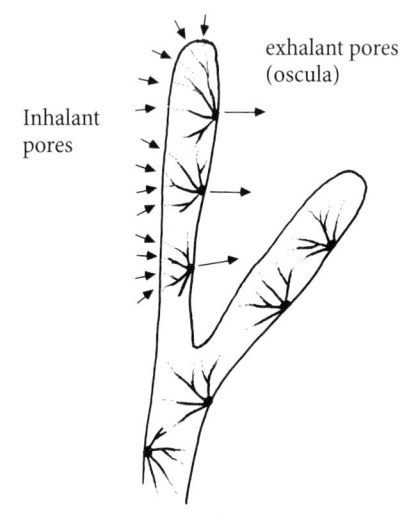

Fig. 2.20. Diagram of the aquiferous system in *Haliclona oculata*. The aquiferous system consists of inhalant pores where water together with suspended material enters the sponge.

of a growth process in which a new layer of material is added at the tip of the branch. The tangential fibers correspond to surfaces of earlier growth stages. In the closely related sponge *Haliclona simulans*, also with a radiate accretive architecture, branching growth forms are formed (see Fig. 2.18c). Quite remarkable in this species is the arrangement of the spicules as seen in a tangential view in Fig. 2.18b, showing an arrangement of the spicules in a pattern of triangles, which can be considered to be organized in a pattern consisting mainly of pentagons and hexagons.

Erect branching growth forms can also develop in sponges with a very different skeleton architecture. An example is the sponge *Raspailia inaequalis*, discussed in earlier in this section. In Fig. 2.19 a section through a branch of the sponge *Axinella polypoides*, related to *Raspailia inaequalis*, is depicted, showing a dense central axis of spicules with side branching fibers of spicules. Branches in *Raspailia inaequalis* seem to be formed in a process of splitting of the central axis. The overall differences in growth forms of *Raspaila inaequalis* (see Fig. 2.9) and *Haliclona oculata* are obvious for an expert, but difficult to describe in words. The branching forms of *Raspaila inaequalis* tend to be more "stiff-looking", while sponges with a radiate accretive structure tend to exhibit a more "viscous-fingering-like shape", resembling the shape of branching air bubbles pumped between glass plates (see Fig. 1.5b). The type of skeleton architecture has a predominating impact on the type of growth forms which can develop in the growth process. This observation is very well demonstrated in sponges with a halichondrid skeleton (Wiedenmayer 1977), where the spicules are oriented randomly, as found for example in *Halichondria panicea*. Such sponges usually develop quite irregular (often encrusting) growth forms and seldom exhibit tree-like forms.

The growth process of *Haliclona oculata* can be followed experimentally by marking experiments. The surface of the sponge can be marked with minute stainless steel needles. The needles are pushed into the living sponge, the ends of the needles corresponding with the original surface. The growth lines can be reconstructed by interpolating the ends of the needles (Kaandorp and de Kluijver, 1992). In longitudinal sections through skeletons of the marked tips the growth process can be traced. An example of such a section is shown in Fig. 2.21. From this type of experiment it can be derived that the growth velocity of *Haliclona oculata* is in the range of 1.0–1.5 cm in a period of about 10 weeks. Furthermore it can be observed in Fig. 2.21 that the left branch, after being marked, has overgrown a left branch and split into two new branches.

Another major component in the growth process of a sponge is the pump system with which suspended material is collected from the environment and transported through the sponge tissue, the aquiferous system. Fig. 2.20 shows a diagram of the aquiferous system in *Haliclona oculata*. The aquiferous system consists of inhalant pores where water together with suspended material enters the sponge. For sponges the typical size of the food particles is in the range of 10^{-4}–10^{-6} m (Brien et al. 1973). The filtered water leaves the sponge again through the oscula, the exhalant apertures of the sponge. The aquiferous system of *Haliclona oculata* is poorly developed in comparison to a related species such as *Haliclona simulans* (see Fig. 2.18c) where the oscula are very clearly visible as holes in the growth form. In *Haliclona oculata*, only close to the oscula macroscopic evidence of canals is found. In *Hali-*

clona simulans the aquiferous system is far more evolved and visible as an extensive system of canals. Probably the aquiferous system of *Haliclona oculata* is strongly supported by external water movements as well (compare Vogel 1974). In conditions with strong water movements plate-like growth forms are possible, whereas in sheltered conditions a decrease of food supply will occur in the tissue, unless it is in short-distance contact with the environment; in this case only thin-branching forms will occur. In the related species *H. simulans* with a more evolved aquiferous system, relatively wide branches and more globular forms are found. The development of the aquiferous system is a species-specific pattern which is another major component determining the resulting growth form.

2.2.3 Morphological Plasticity in Colonial Cnidarians

Cnidarians are, after the sponges, the organisms within the animal kingdom with the simplest organization; this is the first animal phylum in which cells are organized in tissues (see also Fig. 1.8). In the cnidarians, both sessile life forms, consisting of individual polyps or colonies of polyps, and free living life forms (jellyfish) have developed. The basic module in the sessile cnidarians, the polyp (Fig. 2.22), consists of three basic tissue layers: an outer layer of epidermis, an inner layer of cells lining the gastrovascular cavity, and between both a layer called the mesoglea. There is an internal space for digestion, the gastrovascular cavity, along the polar axis of the polyp which opens at one end to the exterior to form a mouth. The mouth is surrounded by a circle of tentacles which are used for capturing and ingestion of food particles. The presence of a mouth and digestion cavity permits the use of a much larger range of food particle sizes, than for example is possible in sponges. The body plan of the polyp is characterized by a radial symmetry.

In the three sections below we will focus on three groups of cnidarians in which sessile and colonial growth forms have developed. In the first group, the hydrozoans, in most species colonial growth forms are found. In most of the species the polyps are embedded in a supporting and flexible, nonliving, chitinous skeleton (see for example Fig. 1.3k). In two groups of hydrozoans, for example *Millepora alcicornis* shown in Fig. 1.3l, an external calcified skeleton is secreted. In the second group, the octocorals, the polyp always possesses eight tentacles, and the radial symmetry of the module is characterized by four axes of symmetry. The octocorals are colonial cnidarians with relatively small polyps. In third group, the stony corals or scleractinians (also sometimes called hexacorals), the mouth of the polyp is surrounded by six tentacles, and the radial symmetry of the module in stony corals is basically characterized by three axes of symmetry. Some of the stony corals are solitary (see for example Fig. 1.9) with polyps reaching 25 cm in diameter; the majority are colonial with very small polyps, with a diameter of 1–3 mm.

HYDROZOANS: COLONY INTEGRATION AND THE CONTROL OF MORPHOLOGICAL PLASTICITY. Many hydrozoan species are small, inconspicuous members of the habitat in which they live. Nevertheless, they are fascinating subjects for the study of growth and form. This is because they exhibit great diversity, in for example, polyp polymorphism, skeletal (or matrix) development and hence, colonial integration. There are also several advantages to using hydrozoans for studies of growth and form. First, their small

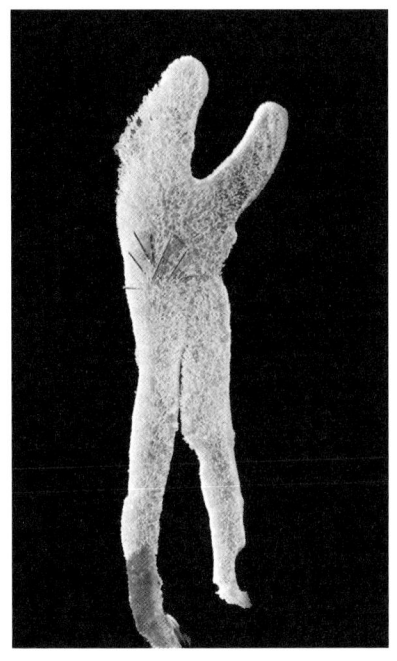

Fig. 2.21. A longitudinal section through a tip of the sponge *Haliclona oculata*, where the surface of a preceding growth stage is marked with needles. The growth lines can be reconstructed by interconnecting the ends of the needles (1.5 month experiment).

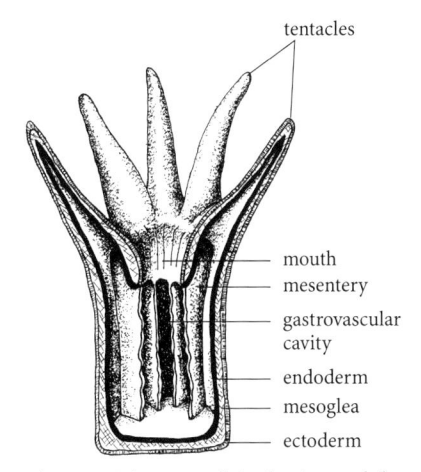

Fig. 2.22. Diagram of the basic module, the polyp, in sessile cnidarians

Fig. 2.23a–c. Variation in colony morphology due to differences in the spatial arrangement of polyps and stolons in hydrozoans: (*a*) *Allopora* (*b*) *Aglaophenia* (*c*) *Hydractinia* (pictures in (*a*) and (*b*) by P. Edmunds)

Fig. 2.24. *Hydractinia symbiolongicarpus* colony showing two (gastrozooid and male gonozooid) of the four polyp polymorphs (gastrozooids for feeding, gonozooids for reproduction, dactylozooids for capturing eggs of the hermit crab host, and tentaculozooids for defense). The stolonal network is encased between two continuous layers of ectoderm.

size and the ease of culturing them render them suitable for many types of experimental analysis. Because most colonies lack both a thick mesoglea and a calcium carbonate skeleton, there are no tissues obscuring the view of polyps and the interior of stolons; instead colonies are transparent, especially in the two-dimensional encrusting growth forms. This enables physiological behavior and its subsequent effect on the plasticity of growth and form to be observed relatively easily.

Benthic hydroid colonies exhibit relatively simple morphological construction. A colony is composed of polyps coupled to one another by a network of stolons, some of which are adherent to the substrate. Following metamorphosis to a primary polyp, colonies grow by linear extension of stolon tips, the branching of new stolon tips (that may fuse with other stolons), and the budding of polyps on stolons. Colony morphology is defined by the spatial arrangement of polyps and stolons. The different arrangements of stolon growth and branching patterns and the placement of polyps on stolons can generate arborescent (bush-like), pinnate (feather-like), or encrusting forms in which polyps arise singly from entirely adherent stolons (Fig. 2.23).

Within this morphological construction, the external anatomy of hydrozoans varies considerably with respect to polyp polymorphism and the development of a skeletal or connective tissue matrix. Polymorphic polyps and the presence of ectodermal tissue that encases the stolonal network are two key features associated with increasing colony integration (Cartwright et al. 1999, Cartwright and Buss 1999). This is because polyp specialization and either a stolonal mat (as in the encrusting species, *Hydractinia*, see Fig. 2.24) or a calcium carbonate skeleton (as in the arborescent hydrocoral *Millepora alcicornis* shown in Fig. 1.3f) results in an increased division of labor and physiological interdependence among the modules.

All hydrozoan colonies are physiologically integrated by their simply constructed internal anatomy, a common gastrovascular system. The gastrovascular system is the only physiological system of hydrozoans whose behavior is known to be manifested colony-wide. It is a system of fluid-filled endodermal canals (i.e. the lumens of stolons) that runs throughout the stolonal network and joins the gastric cavities of polyps (Fig. 2.25). One can idealize stolons as being oriented either radially, extending from the colony center to the periphery, or circumferentially, growing around the colony center and joining adjacent radial canals. Oscillations by polyps (expansion/contraction cycles) import and export fluid to and from the gastric cavity and transport it through the lumen of stolons to other polyps and growing stolon tips, resulting in the colony-wide exchange of nutrients and dissolved gases (Dudgeon et al. 1999). Fluid flows alternately in each direction within a stolon.

The variation in morphology and integration of hydrozoan colonies is best characterized in the encrusting hydractiniid genera, *Podocoryne* and *Hydractinia*. In *Podocoryne*, the stolons are free, thus, the layout of the gastrovascular system (i.e. stolons and polyp gastric cavities) in space is the morphology of the colony (Fig. 2.26). In both genera, colony morphology varies continuously between those with widely spaced polyps along long, sparsely branched stolons (called "runners") and those with closely packed polyps along short, highly branched stolons (called "sheets"). Colonies of *Podocoryne* generally are more runner-like than colonies of *Hydractinia* and in both taxa colony form is heritable (Blackstone and Buss 1991). In addition to free stolons in *Podocoryne*, polyps are unspecialized: feeding polyps (or gastrozooids, or hydranths) also bud medusae from the body column just below the tentacles. In contrast, in *Hydractinia* there are four polyp polymorphs (gastrozooids for feeding, gonozooids for reproduction, dactylozooids for capturing eggs of the hermit crab host, and tentaculozooids for defense) and the stolonal network (except for the longest, most peripheral stolons) is encased between two continuous layers of ectoderm (Fig. 2.24). Moreover, in older portions of a colony the ectodermal tissue contacting the substrate deposits a protective skeleton of chitinous spines.

As is expected for sessile organisms with an indeterminate lifespan, continuous growth and development, and living in spatially and temporally patchy environments, hydrozoans, like most cnidarians, display great morphological plasticity. In fact, runner genotypes can be transformed into sheet phenotypes and sheet genotypes can be transformed into runner phenotypes (Dudgeon and Buss 1996). Nevertheless, sheet colonies appear to be less plastic than runners in the relative investment of polyps and stolons by virtue of

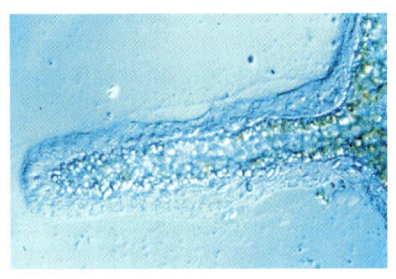

Fig. 2.25. A portion of the gastrovascular system in a hydrozoan colony viewed at 400× magnification. The diameter of the expanded stolon lumen is approximately 30 microns.

Fig. 2.26a,b. Colony morphology of the hydractiniid hydrozoan *Podocoryne carnea*: (*a*) shows detail, (*b*) gives an overview of the colony.

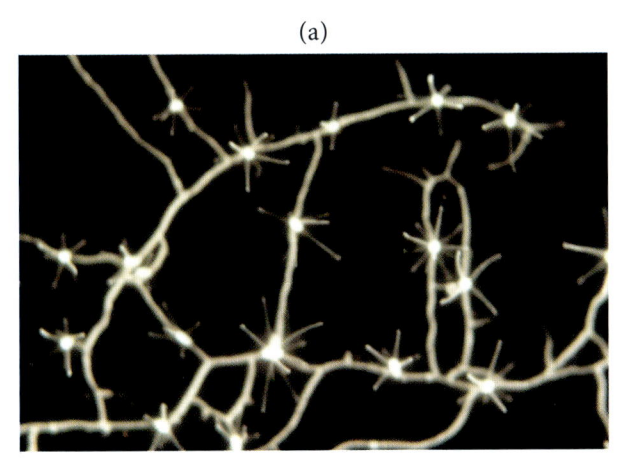

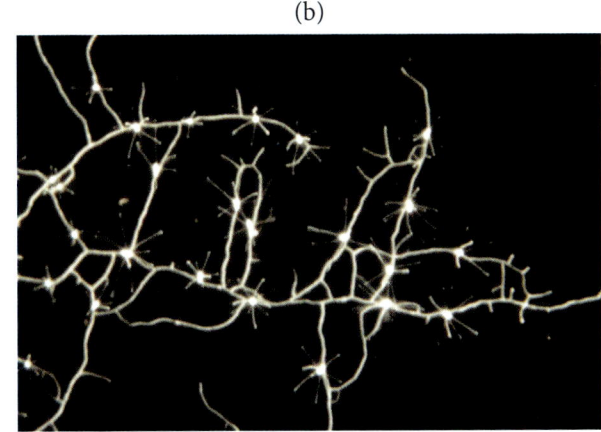

the greater extent of stolon branches that connect different parts of a colony and mitigate the effect of a spatially patchy food supply.

Changes in the timing and placement of polyps and stolon branches are manifest by experimentally manipulating either the flow rate or the distribution of flow throughout a colony (Blackstone and Buss 1993, Dudgeon and Buss 1996). Likewise, colony development can be affected by manipulation of redox state independent of gastrovascular flow patterns (Blackstone 1999). The relationship between gastrovascular transport and redox state in effecting a morphogenetic response is complex and it is likely that they act in concert.

Both mechanisms hypothesized to regulate morphological plasticity are products of polyp feeding behavior. The ingestion of food triggers oscillations by a polyp that pumps fluid and eventually the digested prey to the rest of the colony (Dudgeon et al. 1999). These oscillations are caused by contraction of epitheliomuscular cells of the polyp that shifts their redox state such that active polyps become oxidized relative to inactive polyps (Blackstone 1999). The behavior of polyps clearly varies over time in response to food availability. Therefore, the dynamics of polyp feeding behavior are an integral component to models of the regulation of morphological plasticity via the interplay of gastrovascular flow and redox chemistry.

The importance of polyp oscillations to colony development makes it necessary to characterize the range of polyp behaviors, both alone and when interacting with other polyps. If polyp behavior is sufficiently simple and predictable under specified conditions, then quantitative models of that behavior may enable prediction of gastrovascular flow and redox gradient patterns within a colony. Polyp behavior can be characterized from the dynamics of their oscillations with respect to the timing of ingestion of food. Dudgeon et al. (1999) characterized oscillations in terms of changes in polyp length and volume over a time course beginning prior to ingestion and terminating after regurgitation of undigested material. Polyps prior to feeding, or long after regurgitation, behave simply. Contractions are infrequent, they lack periodicity, and the amplitude of volume exchange with the stolon is small (Fig. 2.27). The behaviors of polyps between ingestion and regurgitation can be classified into three distinct phases. These phases of behavior reflect differing input-output relationships between the polyp and either the external (via the mouth) or internal (via the polyp–stolon junction) environment.

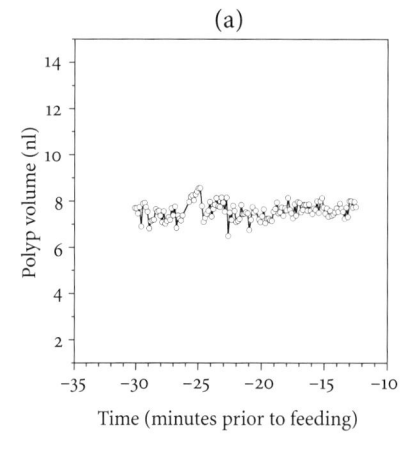

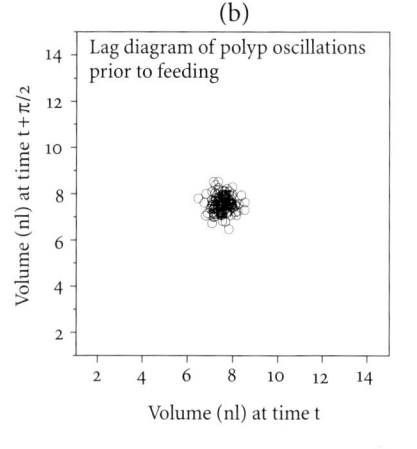

Fig. 2.27. (*a*) Volume of the polyps of *Podocoryne carnea* prior to feeding. (*b*) Lag plot of the oscillations in polyp length and volume of the polyps of *Podocoryne carnea* prior to feeding

Moreover, these phases carry the signatures of characteristic dynamical behavior that can be expressed in terms of frequencies and amplitudes of polyp and stolon oscillations (Dudgeon et al. 1999).

Food ingested by a polyp elicits oscillatory behavior within 5–15 minutes (phase 1). The elicitor of oscillatory behavior may be either mechanical strain on polyp tissue associated with ingestion, or the presence of nutrients released from the prey as digestion proceeds. The 5–15 minute delay prior to the onset of oscillations is consistent with the hypothesis that nutrients released from the food trigger polyp contractions. Nevertheless, prey in the gastric cavity stretches the body column of a polyp laterally and confines oscillatory behavior to only the hypostome. During this phase contractions of only the hypostome do not generate sufficient pressure to export either fluid or digested particulates into the stolon.

As the food is broken down into particulates, the constraint on shape is removed and oscillations in length increase in amplitude up to a limit set by the size of the polyp. At this point, length oscillations are of constant amplitude and presumably of sufficient pressure to export a small volume of fluid to the stolon as well as a limited exchange of small, digested particulates (phase 2). During this second phase, polyp behavior is semi-autonomous and most of the digestion process occurs in the gastric cavity of the polyp. Elicitors of polyp oscillations accumulate in the gastric cavity and begin to be exported into the gastrovascular system.

When food items are sufficiently solubilized, the third and final phase of feeding behavior is initiated by the export of dense streams of particulate material from the polyp into the stolon. In the case of multi-polyp colonies, a threshold concentration of elicitors in the gastrovascular system is reached upon export inducing oscillations by other (unfed) polyps. For either single or multi-polyp colonies, this phase represents the period of maximal volume flow rate through the gastrovascular system as evidenced by two behaviors: (1) the large amplitude oscillations by polyps, and (2) the greatest average lumen diameter and minimum amplitude of stolon oscillation (Fig. 2.28). Because gastrovascular, and presumably metabolic, activity are maximal in this phase, it has been the focus of data collection in all models hypothesizing a relationship between gastrovascular dynamics and the growth and form of hydrozoan colonies. Phase 3 is terminated by the regurgitation of undigested food and a return to pre-feeding behavior.

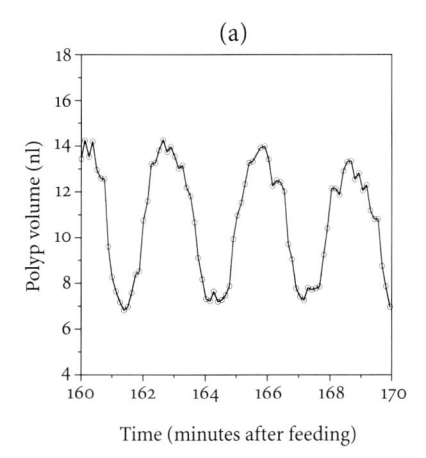

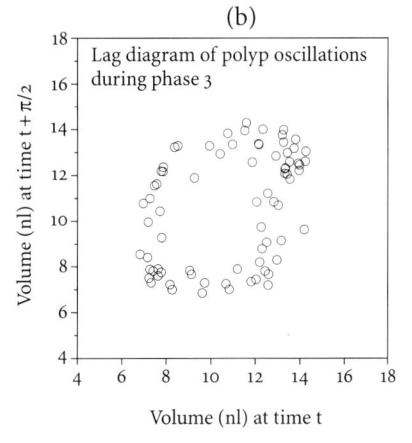

Fig. 2.28. (*a*) Volume of the polyps of *Podocoryne carnea* after feeding. (*b*) Lag plot of the oscillations in polyp length and volume of the polyps of *Podocoryne carnea* after feeding

Morphological plasticity enables rapid, adaptive responses to local environmental variation, such as supplies of space and food, and the adaptive significance of this plasticity has long been recognized. The physiological processes that regulate plasticity of colony form and function are now beginning to be understood largely due to the work by Dr. Neil Blackstone. In hydractiniids, the transport of gastrovascular fluid interacts with redox chemistry to generate an adaptive morphogenetic response (Blackstone and Buss 1992, Blackstone 1996, 1997, 1998, 1999, Dudgeon and Buss 1996) Low rates of flow in the gastrovascular system or a relative oxidation state of cells in a colony typically increase stolon branching and hydranth production, generating sheet growth (Blackstone 1998, 1999). In contrast, high rates of flow or a relative reduction state of cells reduce hydranth and stolon branch formation, thereby generating runner growth. Redox gradients arise within a colony associated with the pumping activity of polyps and the availability of metabolic substrate (Blackstone 1999). Factors regulating the production of other polyp types (gonozooids, dactylozooids, and tentaculozooids) are presently unknown.

Environments occupied by hydrozoans commonly vary in the availability of space, frequency of competitive encounters, and food supply. This environmental information is translated by polyp and stolon behaviors into emergent patterns of gastrovascular flow and redox gradients that specify local "physiological states" within the colony. These physiological states regulate adaptive plastic morphogenetic responses. Colony form itself influences characteristics of gastrovascular flow and may affect redox chemistry by influencing the pumping behavior of polyps. In this way, the architecture of a colony determines its future morphological trajectory. Colony form, gastrovascular transport and redox chemistry are clearly inextricably linked.

Octocorals: Modular Growth in Two and Three Dimensions. The two most abundant octocoral groups are the alcyonaceans, which reach their greatest abundance and diversity on Indo-Pacific coral reefs, and the gorgonians, which are most common on Caribbean coral reefs (Fig. 2.29 shows a montage of two species). Although they range in form from delicate branching structures to massive forms, octocorals all share a simple body plan of individual polyps embedded in a matrix of tissue, the coenenchyme. Colony form is maintained by some combination of support from calcium carbonate sclerites embedded in the coenenchyme, hydrostatic forces created by pumping of water into the polyps, and among the gorgonians a proteinaceous axial skeleton. In octocorals colony form is controlled by a feedback that occurs between growth and the environment. Among octocorals the end product of that interaction ranges from lacy forms that are restricted to protected waters to massive colonies that blanket tens of square meters and can withstand heavy surf.

The expansion of encrusting and nodular colonies typically occurs through the production of polyps along the colony margin. This gives colonies the potential of continuous expansion and monopolization of space. However, among most species this process is functionally inhibited by grazing and disturbance, and by allometric factors that may constrain resource supply. A solution to allometric constraints among some of these species is the formation of new colonies from fragments of colonies. In some cases the

Fig. 2.29a–c. Photographs of sea fans. (*a*) *Gorgonia ventalina* (Colombia Caribbean, photo J. Garzon-Ferreira); (*b*) *Pacifigorgia agassizii* (USNM collection, photo J.A. Sanchez); (*c*) *Pacifigorgia sp.* detail (USNM collection, photo J.A. Sanchez)

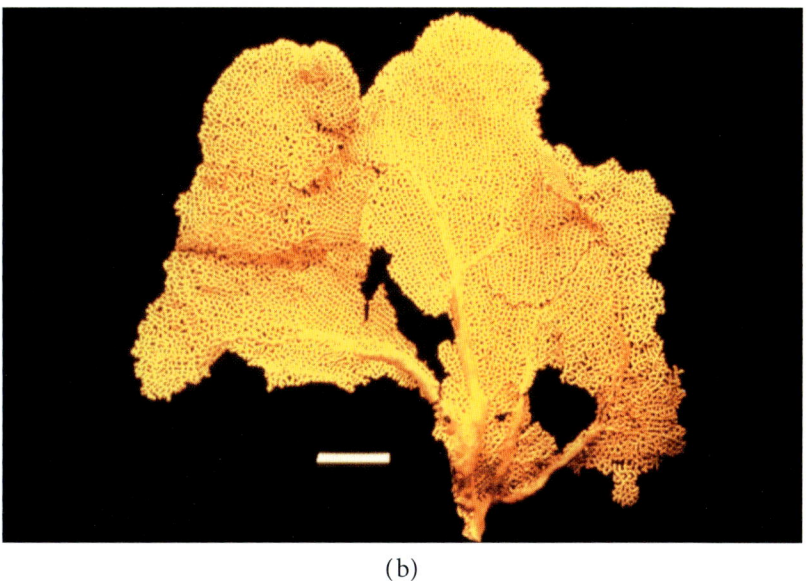

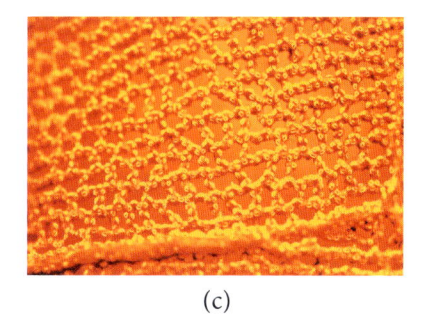

fragments are simply the portions of colonies that survive a disturbance, but in others the fragments appear to develop from fission of otherwise intact colonies. As among stony corals this level of modularity can produce large groups of genetically identical colonies each of which can then grow and generate new colonies.

In octocorals it is the pattern by which branches are added to the colony that controls colony form. The pattern of branching is controlled, in part, by environmental cues, the pattern of resource supply across the colony, and hydrodynamic forces. As among encrusting and nodular forms the mechanical disturbance of colonies as well as fission processes are also employed by some species to generate new colonies, allowing some species to monopolize large areas.

The feedback that occurs between modular growth and the environment creates high levels of morphological plasticity in virtually all octocorals. Although the mechanisms involved are generally unknown, responses of colony form to currents and wave action are well documented. Hydrodynamic forces are particularly important among upright structures, where excessive drag on the colony can lead to failure of either the colony (Lasker 1984) or of the colony's attachment to the substratum. Many species have characteristic orientations when they develop in unidirectional current/wave action (Wainwright and Dillon 1969). Sea fans, for instance, are almost always oriented perpendicular to the primary current/wave action (Fig. 2.29 shows sea fans growing parallel to each other). Some species such as plexaurid gorgonians may vary from bush-like colonies where flows are slow or do not have a preferred direction to candelabrum-like colonies in environments with unidirectional flow (Jordan and Nugent 1978). Other colony traits such as axial core composition and stiffness also vary within species due to the level of water movement (Lewis et al. 1992).

Form also affects the overall energetic/nutritive balance of a colony. Like many of the bottom-dwelling taxa of coral reefs, most of the reef-dwelling octocorals harbor symbiotic dinoflagellate algae, zooxanthellae. As in the stony corals, the zooxanthellae probably provide much of the colony's nutrition, but quantitative estimates of that contribution are lacking. Flow by controlling the exchange of essential inorganic nutrients such as bicarbonate probably affects primary production across the colony as has been observed among scleractinians (Helmuth et al. 1997). Furthermore, among densely branched colonies, form may affect productivity via self-shading, as light level within even loosely branched species is attenuated along the main axis when compared with the colony edge (Goulet and Coffroth 1997).

Gorgonians and alcyonaceans have not been considered particularly effective feeders on zooplankton because of their lack of true stinging cells (Mariscal and Bigger 1975). Some species can capture easily handled prey (Lasker 1981), but most heterotrophic feeding may be concentrated on small, passive particulates (Lasker 1981, Fabricius et al. 1995, Ribes et al. 1999). Mariscal and Bigger (1975) also suggest that extensive villi on polyp surfaces may enhance the uptake of dissolved organic molecules. The relative importance of the different modes of feeding are unknown. However, it is clear that body form by controlling water flow through the colony will affect colony-wide nutrition.

Encrusting forms by virtue of their veneer-like form escape many of the surface-to-volume ratio restrictions that inevitably limit the size of unitary (non-modular) organisms, but even among these forms size may be restricted by the depletion of nutrients as water flows over the surface of a colony. McFadden (1986) for instance suggested that the pattern of fission among *Alcyonium sp.* allowed colonies to capture more prey than if the colony grew as one continuous sheet. The presence of some form of environmentally induced inhibition of growth among branching forms is suggested by the experiments of Kim and Lasker (1997). They found that the growth of interior branches of colonies was adversely affected by the presence of neighboring branches, even when the adjacent branches were no longer attached to the colony.

Flow acting through the hydrodynamic forces on colonies and through the supply of essential nutrients to colonies generates variation in colony form among octocorals. The functional importance of those architectures is suggested by examples of both parallel (Caribbean and Indo-Pacific gorgoniidae) and convergent evolution (the pinnate structures in the Caribbean genera *Pseudopterogorgia* and *Muriceopsis*, Fig. 2.30).

SCLERACTINIAN STONY CORALS (HEXACORALS). Corals are most common in shallow, clear, oligotrophic, tropical waters where they help form coral reefs. The taxonomic classification of corals relies traditionally upon the morphological characteristics of coral skeletons and, especially, upon fine scale (1 mm) morphological characteristics which were believed to be nonvariable within species. This classification is still the backbone of today's coral identification where corals are seen as stable morphological entities that show a degree of plasticity due to environmental influences. For example, reef corals were shown to change their morphology from massive or branching forms in shallow water to plate-like colonies in deeper water. An example of this is the colony shapes of the Caribbean coral *Montastrea annularis* (see Fig. 2.31), where the colony gradually transforms from hemispherical form into a plate-like colony at deep locations. The branching stony coral *Porites sillimaniani*, a common species in a variety of reef environments of the tropical Pacific, displays a striking variation of the whole colony morphology with respect to light availability. Fig. 2.32a shows a typical branching morphotype from a shallow site, which gradually changes into a plate-like growth form at deeper sites. Another example of this morphological plasticity was shown in Fig. 1.1, where the growth form of the Indo-Pacific stony coral *Pocillopora damicornis* gradually transforms from

Fig. 2.30a–f. Photographs of black corals (*a* and *f*) and gorgonian corals (*b–e*). (*a*) *Stichopathes lutkeni*; (*b*) *Ctenocella (Viminella) barbadensis*; (*c*) *Pseudopterogorgia americana*; (*d*) *Muriceopsis flavida*; (*e*) *P. acerosa*; (*f*) *Antipathes pennacea* (all photos from Colombia Caribbean by J.A. Sanchez)

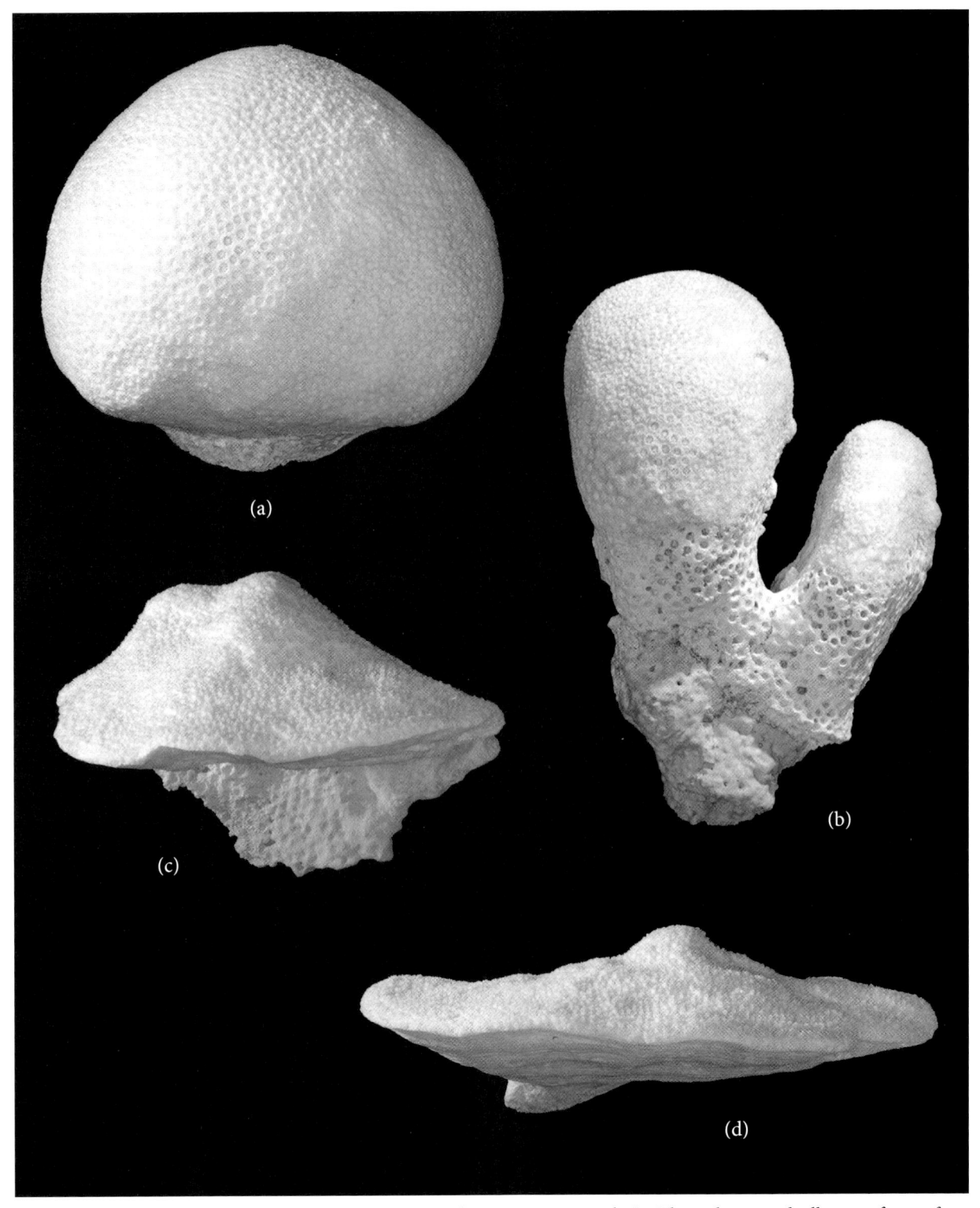

Fig. 2.31a–d. Range of colony shapes of the stony coral *Montastrea annularis*. The colony gradually transforms from hemispherical (*a*), column-shaped (*b*), and tapered forms (*c*) to a substrate covering plate (*d*) when the light intensity decreases.

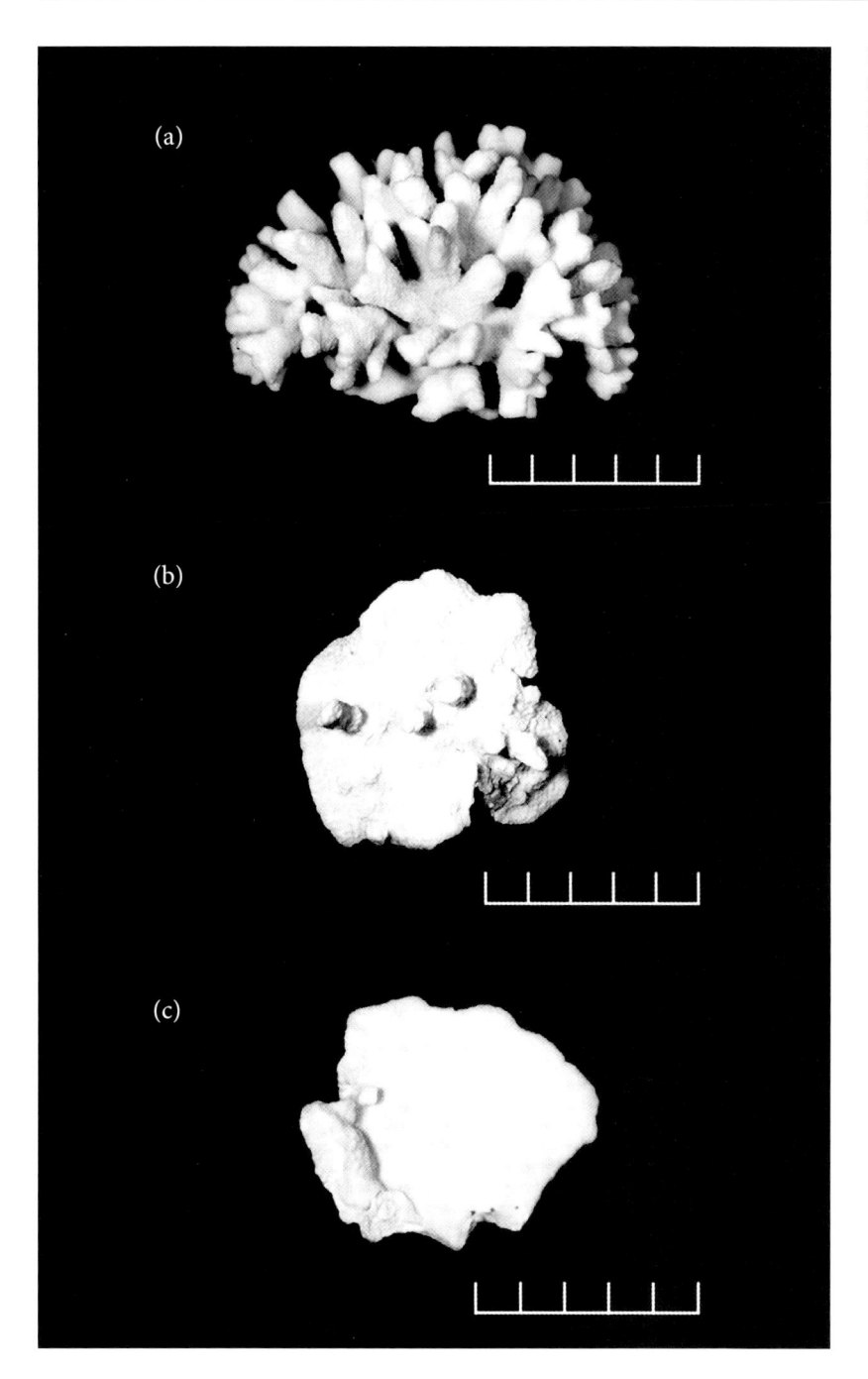

Fig. 2.32a–c. Range growth forms of the stony coral *Porites sillimaniani* (after Muko et al. 2000). (*a*) shows a typical branching morphotype from a shallow site, (*c*) is a plate-like growth form originating from deeper sites, while (*b*) is a colony form from an intermediate site.

a thin-branching growth form into a more compact shape when the exposure to water movement increases. Since it was well known that most reef corals contained symbiotic algae within their tissues, this shift in morphology was thought to optimize light capture as light diminished with depth. These observations were often projected to other species as well, causing simplifications and generalizations in coral behavior. Scuba diving broadened coral research and permitted studies of coral reef ecology and coral physiology and reproduction in addition to the traditional taxonomic studies. This led to the finding that morphological species boundaries did not necessarily correspond to the ecological or behavioral differences found in the field. In several Indo-Pacific coral species reproductive boundaries did not correspond to morphological classifications. Furthermore hybridization between presumed morphological species is found frequently (Veron 1995) which indicates a need to allow for much greater skeletal plasticity than is permitted by classical coral taxonomy within presumed species.

Corals are not restricted to shallow, well-lit tropical seas. They survive and grow without light. The branching coral *Lophelia pertusa* grows at depths around 250 m in the North Sea, most commonly in Norwegian waters (70° N). It forms frameworks up to 10 m high. Some species grow in abyssal Antarctic waters and have to extend their tissues around their entire skeleton to prevent skeletal dissolution at high pressure. Although these examples illustrate their adaptive capabilities, it is in warm, mostly tropical seas (> 18 °C) that corals are most diverse and common. The major calcifying organisms on coral reefs, such as corals, have developed an endosymbiotic relationship with unicellular plants. The most widely distributed endosymbiotic alga is the unicellular dinoflagellate, *Symbiodinium microadriaticum*, which in its various species resides in a large range of reef invertebrates including hydrozoan corals, scleractinian corals, and the various species of giant clam in the Pacific. The symbiotic algae, or zooxanthellae, within the tissues of most tropical corals provide their hosts with carbon products of photosynthesis and considerably enhance their rate of skeletal growth (calcification). Photosynthesis allows reef organisms to precipitate calcium carbonate faster than physical, chemical, and biological agencies can disperse it. The success of this symbiosis has allowed corals to form reefs, which are major geological structures on the earth's surface. Living reefs cover about 15% of the seabed in the 0–30 m depth range and they form about 0.2% of the world's ocean area. Geologists originally used the term "hermatypic" for corals that form reefs. Biologists have used this term to describe corals with zooxanthellae, which again emphasizes the importance of zooxanthellae to reef formation and maintenance. Light is considered to be the single most important environmental factor affecting coral growth. Light levels change most profoundly over the first 10–15 m of the water column. Over these depths, 60–75% of the surface light is being absorbed or scattered. The decrease in light intensity becomes more gradual deeper in the water column. The presence of zooxanthellate corals at 100 m depth indicates the ability of corals to live under very low light levels. Reduced growth rates with depth have been found for many corals. Increasing depth results in changes in growth form (mostly from rounded to flattened morphologies), changes in polyp and zooxanthellar densities, and changes in the types and concentrations of pigments associated with photosynthesis (Falkowski and Dubinsky 1981).

Corals acquire essential nutrients other than organic carbon by capturing zooplankton from the water column. The relative contribution of autotrophy (photosynthesis) and heterotrophy (particle feeding) may depend upon local availability from each source. Porter (1976) suggested that morphological variation might be related to this. In his model, corals with larger polyps are better able to feed heterotrophically while corals with high surface-to-volume ratios (e.g. branching species) depend more on autotrophy because of better light capturing capabilities.

Corals are marine animals of the phylum Cnidaria. Cnidarians are, after the sponges, the simplest form of metazoan life (see also Fig. 1.8). They differ from sponges in that their cells are organized into two distinct layers: the ectoderm and endoderm (literally outer and inner skin). The two layers share a common basal, mostly non-cellular, connective layer known as the mesoglea. In a coral polyp (see Fig. 2.33), which is the fundamental unit of a coral, the two layers of cells form a sack. Most corals are colonial and made up of many interconnected polyps, with no obvious diversity of function amongst the polyps. Each polyp is a cylindrical sack whose upper end is closed by a disk which bears six tentacles and has a centrally located mouth. The polyp can expand; this normally happens during the night when corals expand their tentacles to allow feeding whereas they are contracted during the day.

The tissue which closes the lower end of the cylindrical polyp is strongly invaginated into pockets. The skeleton is formed within these pockets and around the outside of the polyp. The result is a calcium carbonate cup, the corallite, which is divided radially by a series of walls, the septa. The skeleton is composed of nearly pure calcium carbonate such as aragonite. The shape of the skeleton reflects the folding of the skeleton-secreting layers of the polyp. The polyp is carried upwards by growth of the skeleton and vacates the lower regions. These unoccupied regions of skeleton are separated from occupied regions by dissepiments, which form as very thin bulkheads between the vertical skeletal elements.

Skeletal growth involves three process: (1) upward or outward extension of the vertical skeletal elements, (2) thickening of these vertical elements throughout the depth of the skeleton occupied by tissue and (3) periodic uplift of the lower regions of tissue with sealing off of the vacated regions by dissepiments (see Barnes and Lough 1992; Taylor et al. 1993). The skeletal growth of many corals can be visualized by sectioning the colony. If a slab is taken from such a section and x-rayed, it is possible to trace the growth process morphologically. In Fig. 2.34 a longitudinal section is made through the colony. The annual growth is visible as dark and light density bands in x-radiographs (see Graus and Macintyre 1982) and it is possible to distinguish growth lines.

Corals cannot grow tissue without skeleton and they cannot grow skeleton without tissue. Tissue growth and skeletal growth are affected by different environmental factors. Accommodation of the two forms of growth is thought to give rise to variations in growth form (see Barnes and Lough 1992).

Variation is not only restricted to over all colony morphology but is also found at the corallite level, where intra-colonial variation sometimes exceeds inter-specific or environmental variation. A dimorphism between radial and axial corallites in the genus *Acropora* is an extreme example of this variation (Veron and Wallace 1984). Furthermore geographic distance (say one to thousands of kilometers) can result in morphological variation in response to changed environmental conditions. This adaptation may be evolutionary and result in sub-species or species or it may simply reflect a plasticity that allows accommodation to altered conditions. Examples of such broad-scale variation in environmental factors are: increasing cloud cover resulting in lower irradiation, decreasing temperatures with increasing latitude, and decreasing nutrient concentrations and increasing water clarity with increasing distance from large landmasses or river deltas.

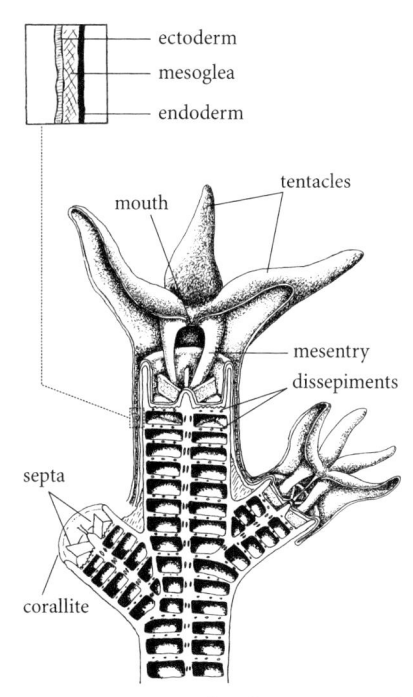

Fig. 2.33. Diagram of a scleractinian coral polyp (after Schumacher 1976)

Fig. 2.34. X-ray photograph of a longitudinal section through a column-shaped colony of the stony coral *Montastrea annularis*

The many factors now proven to affect coral growth partly explain the huge variation found in corals on the reef. Phenotypic plasticity within presumed morphological species boundaries questions the use of a morphological species concept for the Scleractinia. Accurate morphological analysis and understanding of the underlying processes affecting a coral's shape will be an important next step in coral research.

2.2.4 Biologically Inherent Regulation of Morphogenesis

In addition to the environmental influences on morphology discussed above, there is always some level of inherent, biological control of morphology. This allows us to identify species by their morphological characteristics. The expression of genetically determined morphological variants is the subject of modern developmental biology. At present, knowledge of the molecular mechanisms controlling development in seaweeds, sponges, and corals is still not available or fragmentary. In sponges some results have become available very recently on the genetic regulation of development; these will be discussed in this section. Although it is not confirmed that we can extrapolate these results to scleractinians, for some model organisms, for example the cnidarian *Hydra* (a fresh water, solitary polyp), there is a relatively extensive literature available on the molecular mechanisms controlling development (Wolpert et al. 1998). A simulation model of diffusive patterning mechanisms, based on the observations of *Hydra*, has been proposed by Meinhardt (1998). In this section the discussion will focus on the relatively well-studied stony coral *Stylophora pistillata*.

SEAWEEDS: BIOLOGICAL REGULATION OF MORPHOGENESIS. Very little is known, at the molecular level, about what controls the early development of seaweeds beyond the establishment of polarity of the zygote (Love et al. 1997). There are at least two examples of hormone-like molecules which are involved in the maintenance of the modular form of seaweeds.

The first example is a hormonal attractant called rhodomorphin, which is expressed by wounded tissues of red seaweeds attracting the surrounding cells to fuse (Waaland and Cleland 1974, Kim et al. 1995). Rhodomorphins may also be involved in other secondary fusions of red algal cells including the connections between filaments that maintain the integrity of the pseudoparenchymatous forms.

A second example is the presence of a molecular signal, secreted by growing cells into the cell walls of *Ulva sp.*, which inhibits the transformation of sheets of cells into free-living, flagellated cells (Stratman et al. 1996). When conditions are such that the concentration of the inhibitor drops below a critical threshold, the individual cells of the thallus separate from each other, become motile, and eventually form new colonies. It has long been known that at least some green algal species develop disorganized clumps of filaments, rather than the typical two-layered, sheet-like morphology, in the absence of extrinsic bacteria or bacterial growth factors (Provasoli and Pintner 1980, Nakanishi et al. 1996).

SPONGES: GENETIC REGULATION. The basic strategies that control individual cells in such a way that a highly organized pattern can emerge, can be studied best on the molecular level in species belonging to the simplest and evolutionary oldest still-extant metazoan phylum, the Porifera (sponges, see also Fig. 1.8). Hence sponges can be considered as living fossils which are witnesses of an evolutionary step that occurred during the maturation of the metazoans prior to the Cambrian explosion and during which the big bang of metazoan radiation took place (Müller 1998). This view, which classifies sponges as model organisms for the understanding of basic pattern formation in metazoans, was strengthened by the finding that their cells are provided with key molecules, receptors (even neuron-like receptors), and their interacting ligands or signaling proteins, allowing the construction of an integrated body plan. The existence of the same or similar molecules in sponges as in higher metazoans testifies to monophyletic origin of all (Eu)Metazoa (Müller 1995) but does not prove that these molecules have the same functions throughout the metazoan kingdom.

Sponges, like any other metazoan have a defined body plan as has been artistically illustrated by Haeckel (1872). In contrast to other metazoans, adult sponges are considered to have no pronounced anterior-posterior polarity; surely a dorsal ventral axis is absent. In higher metazoans the famous family of homeobox genes is involved in the patterning along the anterior-posterior axis. However, the related genes which have been identified in sponges display a more general function as transcription factors acting in all sponge cells (Seimiya et al. 1998); since they are also found in plants and fungi their existence in sponges can hardly be taken as evidence for the monophyly of metazoans.

The following groups of molecules and their corresponding genes which are involved in the establishment and maintenance of the body plan in sponges have been discovered: (i) molecules involved in the recognition of self/self and self/non-self, (ii) morphogens and (iii) enzyme(s) causing skeletal formation.

Recently it has been found that sponges are provided with molecules which ensure protection of their individuality against non-self by rejecting foreign tissue and recognizing self by fusion (Müller et al. 1999a). Sequence analyses revealed in some examples a closer similarity of sponge polypeptides to mammalian (Deuterostomia) molecules than to those found in higher invertebrates, for example, in nematode worms (Protostomia). The most prominent molecules involved in the recognition processes are the integrins, the putative aggregation receptor with the most complex composition of SRCR domains and the receptor tyrosine kinases with their polymorphic immunoglobulin-like domains, as well as the cytokine-like molecules, including the allograft inflammatory factor, pre-B-cell colony-enhancing factor, or the endothelial-monocyte-activating polypeptide. In Fig. 2.35 experiments are shown with the demosponge *Suberites domuncula*. This species is found in nature in a blue (Fig. 2.35a) and a red type (Fig. 2.35b). If tissue slices are attached and the parabionts (Fig. 2.35d) fixed with nylon fibers, autografts (Fig. 2.35c, tissue from the same specimen) fuse, while allografts (Fig. 2.35e, tissue from different specimens) reject each other and form a visible cleft. Since in allografts the rejection process is also preceded by a short fusion period of immune molecules, cell surface receptors – besides cytokines – are likely to be involved in immune recognition.

While the mentioned immune molecules display (very likely) no role in morphogenesis, i.e. no function which results in an arrangement of the cells into intricate tissue assemblies, special morphogens are present which do cause pattern formation. These are molecules present in different concentrations within the body that determine different structures. The first

Fig. 2.35a–f. Immune molecules in sponges, here the demosponge *Suberites domuncula*, which establish the integrity of the individual. This species is found in nature in a blue (*a*) and a red type (*b*). (*c*) autograft (tissue from the same specimen). (*d*) parabiont fixed with nylon fiber. (*e*) Allografts (tissue from different specimens). (*f*) Species-specific aggregates (aggregates from the sponge *Thoosa istriaca* (lilac-purple) and *Thoosa mollis* (white-yellow))

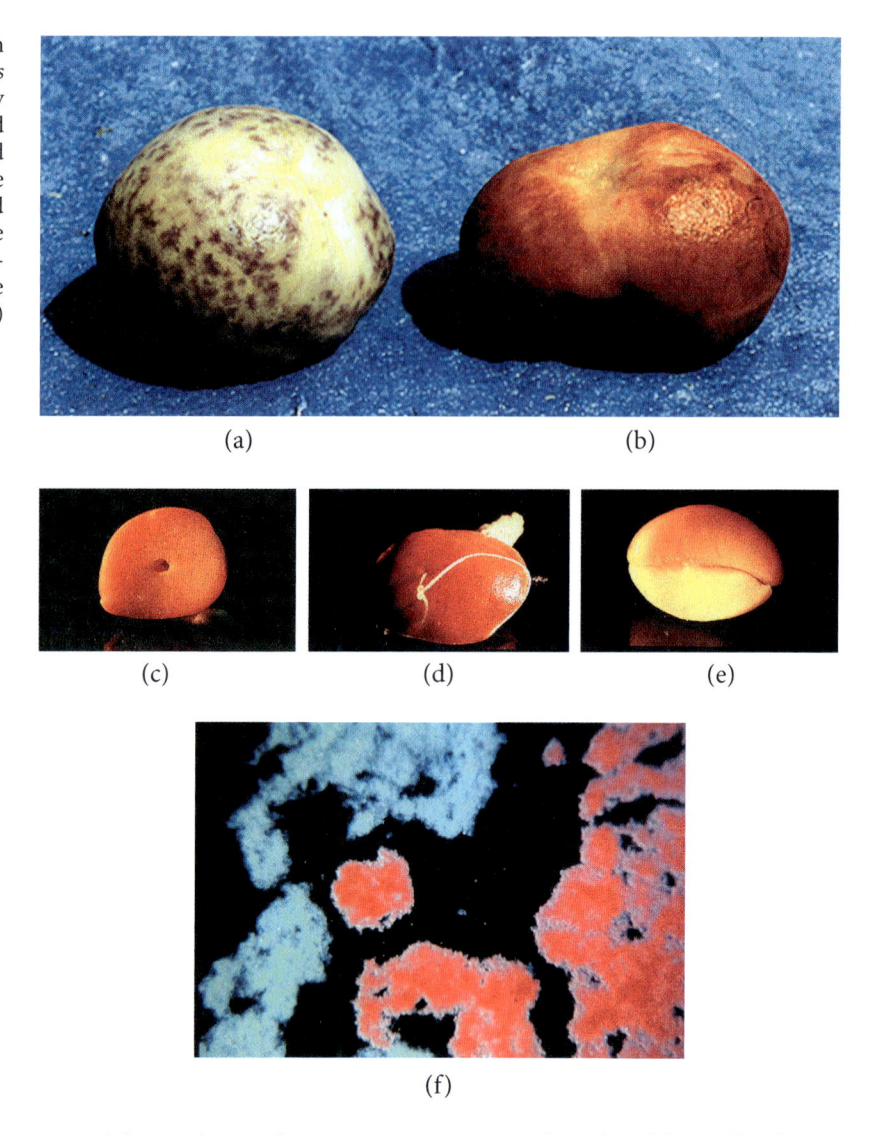

potential morphogen from sponges was recently isolated from the demosponge *Suberites domuncula* (see Fig. 2.36A; Schröder et al. 2000a). The gene was cloned from *S. domuncula* and the recombinant protein prepared. The sponge myotrophin polypeptide (MYOL_SUBDO) shares sequence similarity (and perhaps also homology) with the myotrophins from *Caenorhabditis elegans* (109a_CAEEL), chicken (V1P_CHICK), and mouse (V1P_MOUSE). It was found that the expression of the gene coding for myotrophin was highest around the oscule region (see Fig. 2.36b, c, and d). At this part of the body the excurrent canals are joined together and eject the water current into the surrounding milieu. In Fig. 2.36c and d an example is given of the calcareous sponges *Leucosolenia primordialis* and *Sycon raphanus*, where the aquiferous system consists of one osculum and inhalant pores. An example of the aquiferous system in a branching demosponge is shown in Fig. 2.20. Comparably lower is the expression of this potential morphogen in other parts of the sponge body. Therefore, there is reason to assume that a gradient of the expressed protein also exists – highest at the oscule and lowest at the

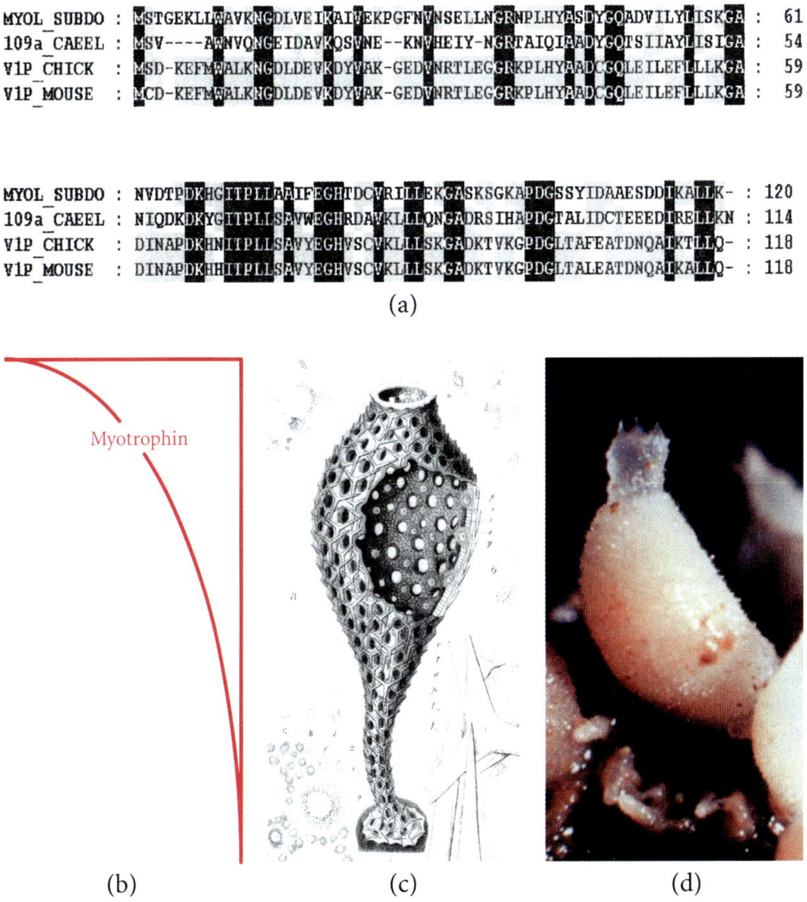

```
MYOL_SUBDO  : MSTGEKLLWAVKNGDLVEIKAIWEKPGFNVNSELLNGRNPLHYASDYGQADVILYLISKGA :  61
109a_CAEEL  : MSV----AWNVQNGEIDAVKQSWNE--KNWHEIY-NGRTAIQIAADYGQTSIIAYLISIGA :  54
V1P_CHICK   : MSD-KEFMWALKNGDLDEVKDYVAK-GEDVNRTLEGGRKPLHYAADCGQLEILEFLLLKGA :  59
V1P_MOUSE   : MCD-KEFMWALKNGDLDEVKDYVAK-GEDVNRTLEGGRKPLHYAADCGQLEILEFLLLKGA :  59


MYOL_SUBDO  : NVDTPDKHGITPLLAAIFEGHTDCVRILLEKGASKSGKAPDGSSYIDAAESDDIKALLK-  : 120
109a_CAEEL  : NIQDKDKYGITPLISAVWEGHRDAWKLLQNGADRSIHAPDGTALIDCTEEEDIRELLKN  : 114
V1P_CHICK   : DINAPDKHNIITPLLSAVYEGHVSCWKLLLSKGADKTVKGPDGLTAFEATDNQAIKTLLQ- : 118
V1P_MOUSE   : DINAPDKHHIITPLLSAVYEGHVSCWKLLLSKGADKTVKGPDGLTALEATDNQAIKALLQ- : 118
```

(a)

(b)   (c)   (d)

Fig. 2.36a–d. Myotrophin, the first identified potential morphogen from a sponge. (*a*) The sponge myotrophin polypeptide (MYOL_SUBDO) shares sequence similarity (and perhaps also homology) with the myotrophins from *Caenorhabditis elegans* (109a_CAEEL), chicken (V1P_CHICK), and mouse (V1P_MOUSE). (*b*) The expression of the gene is highest around the oscule and lowest at the basis of the sponge (the concentration of myotrophin is schematically indicated in a plot). (*c*) A drawing (Haeckel 1872) representing the calcareous sponge *Olynthus (Leucosolenia) primordialis* with a well-formed oscule (*d*) The calcareous sponge *Sycon raphanus*. The spicules are arranged around the many small incurrent canals in the drawing.

basis of the sponge – which might differentially activate the cells. It should be stressed here that from the oscule region a primordial organizer activity originates (Korotkova 1970). A threshold concentration of myotrophin exists, above which the cells respond to this factor in vitro. It remains to be studied if this gradient is also built up in vivo. One endpoint marker of the myotrophin response is the increased expression of the gene coding for collagen.

With the formation of the inorganic skeleton the body plan of metazoans is defined and its orientation fixed. In most sponges their solid support, the spicules, is composed of hydrated, amorphous, and noncrystalline silica (SiO_2/H_2O) as in the classes Demospongiae and Hexactinellida or of calcium carbonate ($CaCO_3$) as in the class Calcarea. The secretion of spicules occurs in specialized cells, the sclerocytes. While in Demospongiae silica is deposited around an organic filament, no organic axial structure is found in the spicules from Calcarea. Another difference is seen: the formation of spicules in Demospongiae starts intracellularly, in contrast to Calcarea which produce them extracellularly around a number of sclerocytes. It is surprising that two taxa within the kingdom of metazoans, the sponge groups Demospongiae and Hexactinellida, use silica instead of calcium in their mineral skeleton; calcium is otherwise the dominant inorganic skeletal component. This is especially interesting in view of the fact that the concentration in the seawater, the milieu in which most of the sponges live, is much higher for calcium ions (10 mM) than for silicon (25 µM). One possible reason why the Demospongiae use the more energy-consuming pathway, the formation of spicules from silica, can be seen in the occurrence of a large concentration of polyphosphate (10 g/g wet weight, Lorenz et al. 1995). Polyphosphate is known to chelate calcium and therefore counteracts the precipitation of calcium deposits. The origin of the polyphosphate in sponges is unclear; since Demospongiae harbor large amounts of (symbiotic) bacteria (Althoff et al. 1998) while Calcarea usually do not, it appears that the bacteria produce this polymer.

Based on the molecular biological/biochemical data available, it appears likely that the formation of the skeleton in sponges proceeds according to the following steps which are supported experimentally (Fig. 2.37). First, dissociated sponge cells form primmorphs (Fig. 2.37a and b), assemblies composed of dividing cells (Müller et al. 1999b), via an already highly complex interaction of cell surface receptors and their ligands, which allow a controlled cell-cell and cell-matrix interaction (Müller 1998). These primmorphs do not contain spicules. In the second step, the enzyme present in the axial filament, silicatein (Fig. 2.37c), recently also cloned (Shimzu et al. 1998, Krasko et al. 2000), is activated and promotes the silicification of the sponge spicules (Fig. 2.37d). In this phase polyphosphate suppresses the formation of calcium precipitates and likely attaches non-covalently to the silica spicules. Although sponges are rich in polyphosphate, they also contain large quantities of its catabolic enzymes; exo-polyphosphatases and alkaline phosphatases. Therefore, it can be hypothesized that the arrangement of spicules in the skeleton (Fig. 2.37e), i.e. their pattern formation, is (partially) directed by a gradient of these catabolic enzymes: Dissolution of polyphosphate by these enzymes removes the attached polyphosphate chains from the silica spicules and antagonizes their growing. This view is partially supported by the finding that the polyphosphate content in gemmules, the asexual propagative bodies, and in the following hatching stages from the gemmules is much higher than in an "adult" specimen (Imsiecke et al. 1996) which contains the sophisticated spicule skeleton. No enzymic data on the formation of spicules in Calcarea are known.

Another interesting result came from molecular biological and modern cell biological studies. Sponge cells have the capacity for indefinite proliferation due to the presence of high telomerase activity. However, since all sponge species have a characteristic body plan, it was compelling to postulate a developmental mechanism which is based on a balance between an (almost) unlimited production of immortal cells, a controlled elimination of cells by programmed cell death or apoptosis, and the existence of telomerase-negative, differentially developed somatic cells. For protostomian and deuterostomian metazoan animals it has been established that only immortal cells, those which are present in the reproductive lineage or which are cancer cells, show high levels of telomerase activity that maintains the telomeres. In contrast to these immortal cells, the somatic, mortal cells lack telomerase activity resulting in a shortening of telomeres with each cell division. In consequence, telomerase-negative cells show a limited number of cell divisions that range from 50 to 100, the Hayflick limit. If the cells reach the Hayflick limit, which is also called mortality phase 1, the critical telomere loss on chromosomes initiates a signaling event which results in

2.2. THE CASE STUDIES

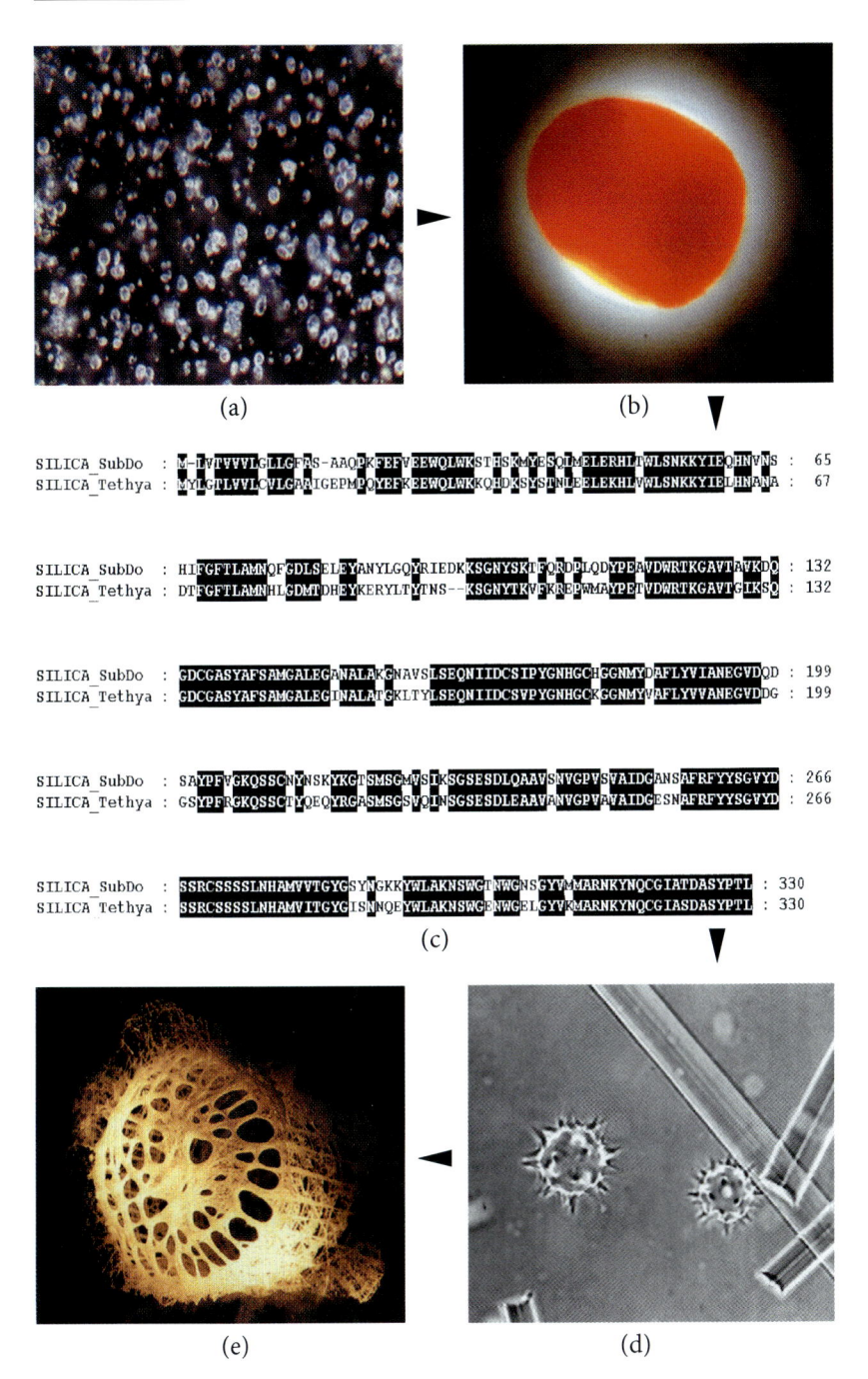

Fig. 2.37a–e. Postulated stages of skeleton-driven body plan formation in sponges. Dissociated sponge cells (*a*) form primmorphs (*b*), aggregates of dividing cells, a process which is mediated by characteristic metazoan cell surface receptors and their ligands. An enzyme present in the axial filament, silicatein, recently cloned from *S. domuncula* (SILICA_SubDo) and *Tethya aurantia* (SILICA_Tethya), is activated (*c*) which promotes the silicification of the sponge spicules (*d*). (*e*) The arrangement of spicules within the skeleton is (partially) directed by a gradient of polyphosphate-metabolizing enzymes. The dissociated cells and the primmorphs are from *S. domuncula* (see Fig. 2.35a and b); the spicules are from *Tethya lyncurium*; the skeleton is from *Euplectella aspergillum*.

an irreversible cell cycle arrest. In some instances spontaneous immortalization, due to transforming events or viral oncoproteins, allows metazoan somatic cells to bypass this mortality phase 1 without activating telomerase and to reach mortality phase 2 ("crisis", M-2 in Fig. 2.38). In M-2 cells often undergo apoptosis or necrosis as a consequence of high frequency of genomic instability. Rare clones of cells can again activate telomerase and acquire "secondary" indefinite growth capacity. Extension of cell life span can be achieved after transformation with viral oncoproteins, agents that

Fig. 2.38. Hypothetical determinants of immortality in species from higher metazoans and in sponges. The lack of telomerase activity, and in consequence telomere loss, in somatic cells of higher metazoans determines their fate to senescence (circles) via two phases: "Mortality Phase 1" (M-1) – cell cycle arrest – and, after transformation, "Mortality Phase 2" (M-2). Cells of the germ lineage from higher metazoans remain telomerase-positive and are immortal. In sponges (squares), experimental evidence suggests that the switch from immortal "somatic" cells, present in tissue and tissue-like assemblies (primmorphs), to mortal cells (in the single cell stage) is triggered by external as well as by internal programs. The mortal cells are eliminated by the process of apoptosis which is controlled by both pro- and anti-apoptotic programs; molecules presumably involved in this process (MA-3 protein, the potential anti-apoptotic Bcl-2, and the potential pro-apoptotic molecule which includes two death domains) have been identified. During the process of apoptosis in sponge cells, the expression of the gene SDLAGL, encoding the putative longevity-assurance-like polypeptide, is down-regulated.

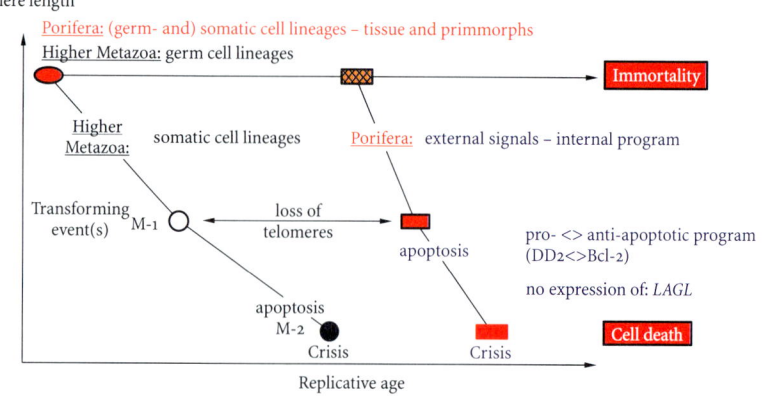

typically do not fully immortalize the cells. Without transformation by viral oncoproteins or acquisition of other mutational events the transition from the immortal to the mortal state is very rare.

Experimental studies revealed that the sponge telomerase-positive cells can be triggered to become mortal, telomerase-negative cells by dissociating them into single cells (Schröder et al. 2000b). In this state they lose the proliferation capacity and undergo apoptosis. Major apoptosis-controlling molecules in the Demospongiae *Geodia cydonium* and *S. domuncula* are the polypeptide encoded by the MA-3 gene, the potential anti-apoptotic Bcl-2 homologous proteins, and the potential pro-apoptotic molecule which includes two death domains (Wiens et al. 2000a and 2000b). In addition, it was found that *S. domuncula* contains a putative longevity-assurance-like protein whose expression is tightly connected with the proliferation and/or apoptotic state of the cells. It is concluded that this protein is involved in the shift of the immortal telomerase-positive cells to the telomerase-negative mortal sponge cells (Fig. 2.38).

In this section a short outline of present knowledge of factors and processes in sponges is given which will help to solve the problem of pattern formation in sponges by using mechanistic, mathematical analyses. The fantastic and unexpected discovery of the last few years, which is the result of molecular biological studies, is the fact that sponges can no longer be considered as organisms composed of "unspecialized flagellates held together by a glycoprotein extracellular matrix" or as animals which "originate from biofilms which were associated with choanoflagellates". Solid data are now available which show that sponge cells include the universal features of metazoan cells. Their characteristic metazoan molecules allow the integration of individual cells into a spatial organization, guaranteeing the formation of a species-specific body plan.

GENETIC REGULATION IN THE BRANCHING STONY CORAL *STYLOPHORA PISTILLATA*. Different approaches for modeling the colonial organization in stony corals can be arranged in order within a continuum connected between two extreme points. At one extreme is the idea that pattern formation of a colony is a morphologically rigid, an intrinsic process and genetically controlled. It is a centralized phenomenon working on the colony level. As a result, the outcome responses to environmental factors are achieved

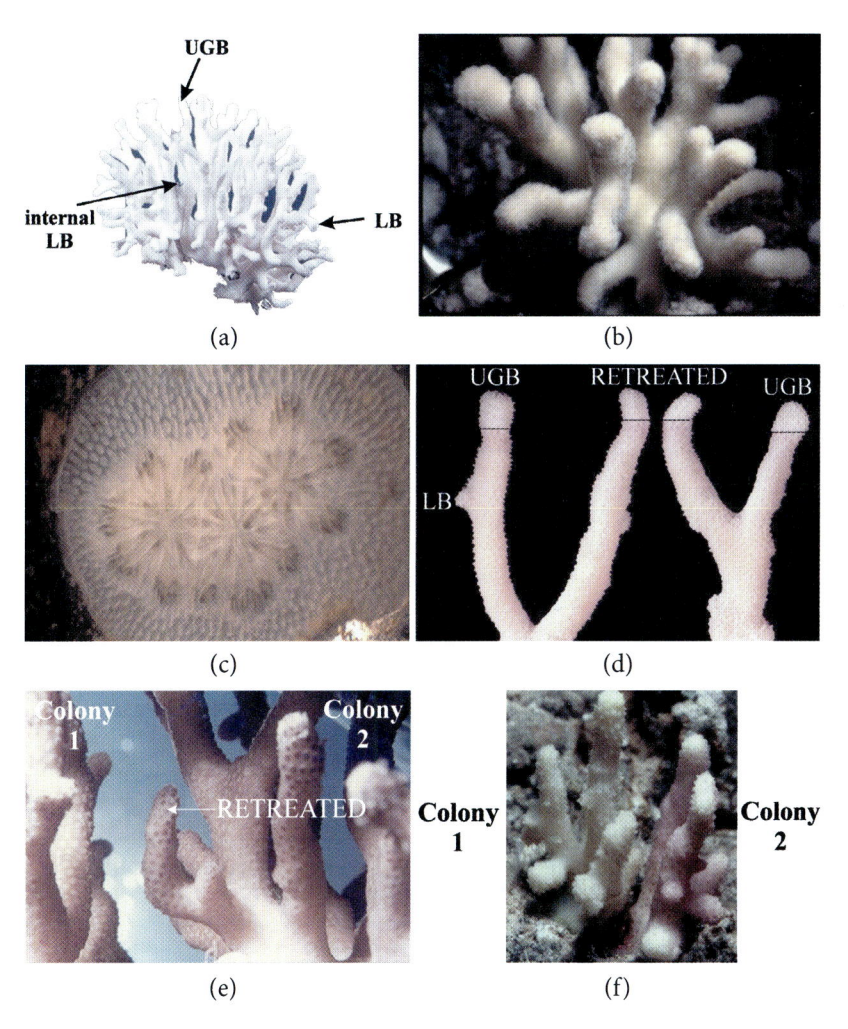

Fig. 2.39a–f. Transitions in *S. pistillata* colonial architecture. (*a*) A typical unharmed colony with numerous dichotomically splitting up-growing branches (UGBs) and lateral branches (LB). (*b*) A view from above a medium-sized colony with UGBs originating from a limited "stem" area. At this stage, a limited number of internal LBs are developed. (*c*) A one month old chimera produced by the fusion of three allogeneic conspecifics. The three polyps are clearly distinguished (growing on glass slides, photographed from behind). (*d*) Isogeneic retreat growth of closely situated UGBs as compared with normal growth of other UGBs. The lines depict the alizarin red labeling areas, done a few months ago (Rinkevich and Loya 1985a). (*e*) An allogeneic retreat growth of a branch closely situated to another colony in situ (6 m depth, Eilat's reef). (*f*) Two closely situated allogeneic small colonies, developing together the characteristic sphere-like structure of an unharmed *S. pistillata* colony. Neither of the colonies is growing towards the other (3 m depth, Eilat's reef).

through the establishment of ecological races, i.e. genetically different ecotypes. At the other extreme is the idea that colonial form is strictly flexible, shaped by trade-offs between different traits. Therefore, defining architecture is the science of defining a suite of characters that respond discretely to the environment. According to this notion "variations" is the conserved trait. Surprisingly enough, recent observations have documented that even such phenotypic plasticity phenomena are probably also controlled by genetics, the plasticity genes (Pigliucci 1996, Callahan et al. 1997). In this section the branching scleractinian coral *Stylophora pistillata* (see Fig. 2.39) will be used as a model case study, depicting the architectural beauty of this system.

Genetics is a key factor in shaping the colony landscape. *Stylophora pistillata* is a very abundant Indo-Pacific branching coral species. In the Gulf of Eilat, Red Sea, it is abundant in the lagoon, rear-reef, and reef flats, and common in the fore-reef (Loya 1976) found down to 60 m depth. This species is characterized by a rapid growth rate. Colonies of this species exhibit an axially rod-like growth form (Fig. 2.39a and b) and are represented by a variety of color morphs including dark brown, purple, yellow, and pale pink. Each branch in a colony consists of numerous minute polyps (0.8–1.0 mm in diameter).

Several morphological rules govern the formation of a typical *S. pistillata* colony. Primary polyps start to deposit calcareous skeletons about one day following metamorphosis. One week thereafter additional polyps are added extra-tentacularly from the peripheral tissue, forming a sheet-like structure. Growth rates in term of new polyps over time are highly variable among young colonies (Frank et al. 1997). As in *Pocillopora* (see Fig. 1.1, Stephenson 1931) the first established polyp "buds" simultaneously, usually forming six polyps that are arranged in a circle around the primary polyp. This kind of lateral expansion continues until at some, as yet unidentified, stage branches develop by apical growth. Establishing a chimera of young (< four months old) *S. pistillata* colonies before alloimmune maturation was achieved (Frank et al. 1997) not only resulted in tissue fusion, marked by a continuous layer of tissue and skeleton across the contact area, but also led to an unusual center of more than one polyp (Fig. 2.39c).

A single apical ramified structure is developed from each "base plate", produced either from a single genotype (founder polyp) or several fused genotypes, in the case of a chimera. New structures are then added, and develop in conformity with the basic architectural rules of this species. These structures are reiterated complexes. Looking at a mature *S. pistillata* colony (Fig. 2.39b) from above, it is clear that the available space left between the branches developed (up-growing and side-growing, see below) tends to be filled by the reiterated complexes (Dauget 1991) in such a way that branches will not come into tissue-to-tissue contact (Fig. 2.39a, d–f).

The resulting symmetry of a typical *S. pistillata* colony approximates a sphere (Loya 1976, Fig. 2.39a and b). Within the volume of the sphere, up-growing branches (UGBs; Fig. 2.39a and d, Fig. 2.40) are primarily added by dichotomous fission at the tip of a branch (Rinkevich and Loya 1985a). The apex of each axis (UGB) consists of several contiguous polyps. Apical ramification can produce two equally sized new axes, but usually forms unequally sized axes (as a result of fast growth of one of the new formed branches; Rinkevich 2000). This is also reflected by the measured high (70%) variation in mean growth rates of all UGB tips within single *S. pistillata* colonies. Tip

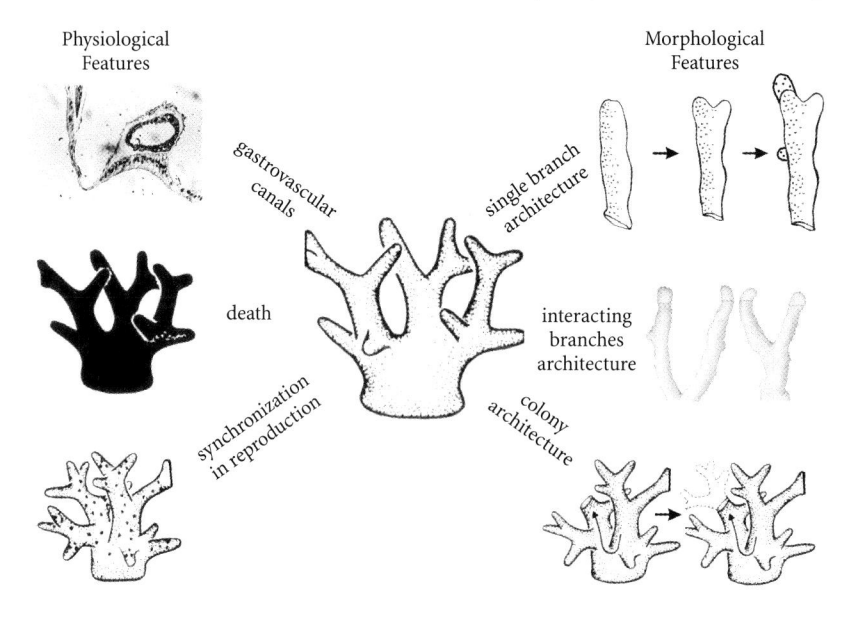

Fig. 2.40. Morphological and physiological features of a typical *S. pistillata* colony representing the holistic notion and some of the central coordinating patterns. Movement of photosynthates through the gastrovascular canals, synchronized death of all polyps (young and old) in a colony, and synchronization in breeding between different branches are some of the key physiological features. Architectural rules for up-growing and lateral growing branches, for isogeneic interacting branches and regenerating parts of the colony are some of the morphological features.

growth ratios within newly formed dichotomous UGBs differ significantly from older branches, further emphasizing the within-colony genetic background for spatial configuration (Rinkevich 2000). In addition to the UGBs many lateral, inside, and outside branches (LBs; Figs. 2.39a and d, Fig. 2.40) are formed. LBs facing out of the colony elongate similarly to UGBs, further adding lateral volume to the sphere-like structure (Loya 1976) of the colony. LBs facing the internal volume of the colony are developed from different zones along UGBs and it would be predicted that after prolonged elongation, LBs would encounter and fuse with UGBs. However, such fusing (anastomosis) was never observed in unharmed colonies growing either in the field (Rinkevich and Loya 1985a) or in the laboratory. The decrease in growth rates of internal LBs, the change of growth directions of isolated branches that are put in close contact with each other (Rinkevich and Loya 1985a, Fig. 2.39d), the lack of fusion between branches of a typical *S. pistillata* colony, and the retreat growth occasionally recorded between closely growing branches of allogeneic colonies (Rinkevich and Loya 1985b, Figs. 2.39e and f, Fig. 2.40) indicated the possible appearance of chemical signals ("isomone") carrying biological activities that control growth patterns (Rinkevich and Loya 1985a).

Several physiological characteristics of *S. pistillata* colonies also add to the "holism" claim. A short-term follow-up study of photosynthetic products within the coral tissue revealed that the area of the branch tip (the upper 0.5 cm) contains significantly less photosynthates than lower parts. In contrast to general belief, long-term observations have further excluded the possibility of a regular significant translocation of photosynthates along a branch, from branch bases to the tips (Rinkevich and Loya 1983a). This conclusion is also demonstrated by the fact that illumination of a branch base by an optic glass fiber resulted in accumulation of photosynthates in the illuminated zone and in non-upward translocation of materials, at least for 29 h post-incubation (Rinkevich and Loya 1984).

Translocation of photosynthates was however recorded when *S. pistillata* colonies were grafted with allogeneic ^{14}C-labelled branches. The host colonies translocated the labelled photosynthates towards the regenerating portions (Rinkevich and Weissman 1987), probably through the gastrovascular canals that connect different polyps (Fig. 2.40). Some of the daily fixed metabolites are stored for future use. For example, planula-larvae collected 1–7 months after coral tissue was labelled with ^{14}C were found to contain significant amounts of labelled photosynthates (Rinkevich 1989). In the same way photosynthetically fixed products of a specific single day were found to contribute to newly formed tissues months after labelling (Rinkevich 1991), or were contributed towards reproductive activities and the use of symbiotic invertebrates residing between the branches of *S. pistillata* colonies (Rinkevich et al. 1991).

Reproductive activities are also developed and shaped at the colony level. Onset of reproduction in *S. pistillata* colonies is at the age of about 2–3 years. In the first year of reproduction the vast majority of the population contains male gonads only, and with the increase in size, there is a tendency to an increase in the percentage of hermaphroditic colonies within the population (Rinkevich and Loya 1979). A long-term study on mature colonies in the field revealed that sexuality (reproductive state) and/or fecundity could be completely altered from one reproductive season to the next (Rinkevich and Loya 1987). In all cases, however, synchronization in breeding was recorded over all branches of a specific colony, independent of the age of the polyps (Rinkevich and Loya 1979; Fig. 2.40). The only case where this synchronization in reproductive activities was distorted has been documented in regenerative colonies (Rinkevich and Loya 1989). In old, senescent colonies, reproductive activity was decreased (as calcification rates), synchronically, in all branches. New and old polyps exhibited senescence simultaneously leading to complete tissue mortality (Rinkevich and Loya 1986, Fig. 2.40).

Although the above *S. pistillata*'s life history characterizations (Figs. 2.39 and 2.40) are probably genetically controlled, they were clearly affected by environmental and a variety of biological challenges. Colony architecture is probably a character shaped by selection, so trade-offs between that trait and other traits may define a suite of morphological responses, some influenced by genetics, others by epigenetic impacts. In *S. pistillata*, a developing colony responds to perturbations by "canalysing" (Waddington 1942) growth pattern to the typical species morphology. This was critically illuminated by Loya (1976) who demonstrated that in harmed *S. pistillata* colonies, where the spherical-like structure is lost through partial branch breakage, the pattern formation is maintained by fast growth in regenerating parts as opposed to reduction in growth in the intact branches, until the colony regains its former shape (Fig. 2.40). Photosynthetic products are channeled to the regenerating parts (Rinkevich and Weissman 1987) and as a result of the whole colony investment, the synchronization in reproductive activities (Rinkevich and Loya 1979) is lost (Rinkevich and Loya 1989). Changes in the colony morphology however, may be induced by intraspecific interactions (Rinkevich and Loya 1985a, Frank et al. 1997), by the employment of intracolonial morphological rules, such as the "no fusions between branches" concept (Rinkevich and Loya 1985a), or by biological interferences, such as symbiotic, mutualistic and parasitic interactions (Abelson et al. 1991).

Even a single separated polyp from a *S. pistillata* colony has the capacity to survive and develop a new colony. Therefore, as implied by the outcomes presented in this section, a single colony may be regarded as a "whole", where intrinsic orders of a branch's growth and related physiological parameters (such as sharing of resources between different branches, simultaneous aging processes, and reproductive activities) "produce" the structure we are familiar with.

More than two decades of studies on different life history traits of *S. pistillata* have resulted in much biological information on a variety of aspects that can be of help when analyzing colony architecture. Flexibility and variations in colony formation can be controlled by specific genetic rules, as can the general sphere-like structure of the colony and the interactive processes of branching. Although there is as yet no direct evidence for that proposal, phenotypic plasticity in the colonial organism, which may also be constructed by genetic rules for morphologies (Zilberberg and Edmunds 1999), should be seriously taken into consideration when studying marine invertebrate branching forms.

3. Measuring Growth and Form

A morphological analysis of the growth forms of marine sessile organisms is an important prerequisite for research on these organisms. A quantitative morphological analysis is required in studies on the morphological plasticity in marine sessile organisms. To verify simulation models a morphological comparison between actual and simulated forms is essential. In this chapter we will discuss a number of methods suitable for the quantitative morphological comparison of (branching) forms. In the first two sections methods will be discussed for computing ratios and fractal dimensions in two-dimensional images of branching structures. The third section focuses on the morphological analysis of two-dimensional images of growth forms sampled along environmental gradients. In the fourth section some preliminary results are presented for the morphological analysis of three-dimensional images of simulated and actual growth forms.

3.1 Metrics for Branching Networks

Networks are a universal feature of complex systems and structures in the world around us. The task of a network is often crucial: cardiovascular systems supply nutrients, central nervous systems carry biological information, social and business networks link people, and river networks transport water and chemicals, patterning the earth's topography. In the design of biological organisms we find networks in both body shapes and internal systems. In seeking to understand networks, we evidently need a way of quantifying network structure and in the present section we will discuss some fundamental network metrics.

A significant portion of important networks are purely branching networks, i.e. they contain no closed circuits and there is only one path along the network between any two points. The archetypal example of a branching network is a tree: a trunk rooted on a surface that repeatedly shoots off branches as it extends upwards with each branch possessing a similar form to the overall structure.

Of course, we find an abundant source of branching networks in the forms of seaweeds, sponges, and corals. A natural question is how do we begin to quantitatively compare and distinguish the branching patterns of these organisms? What are the relevant metrics? To this end, we consider a simple method of describing branching networks and then two complementary metrics for such networks.

3.1.1 Branch Ordering

Branching networks have a hierarchical structure that we would like to capture in a mathematical description. A reasonable scheme would assign the trunk or stem of a plant to be one end of a spectrum and the leaves or tips to be the other end with a range of intermediate levels in between.

One method that has these qualities was first developed in the study of river networks by Horton (1945). A later improvement by Strahler (1957) led to a method known as "Horton–Strahler stream ordering" or often simply as "stream ordering." Here, we will use the more general term "branch ordering" since the method applies to any branching network. Indeed, much use of this ordering technique has been made outside of the study of river networks, a good example being the study of venous and arterial blood networks in biology (Fung 1990, Zamir 1999).

The basic idea is to assign indices of significance to branches, affording a means of comparing branch lengths, branch cross-sections, numbers of branches, and so on. The process of ordering branches involves an iterative pruning of a tree which is depicted in Fig. 3.1. The first step is to identify all "first order branch segments." These are represented by dashed lines in Fig. 3.1a and are the outermost branches of the network, i.e. the leaves or tips of a plant. Imagine that we then prune our plant, removing all of its leaves. This gives Fig. 3.1b, where we observe a new set of outermost branch segments. These are then classified as being "second order" and are themselves removed from the network. This gives the sole stem of Fig. 3.1c, which is itself identified as a "third order" branch segment. For this particular example, we would say that the entire network is an order-three network. In general, the ordering process is iterated until we have labelled all branch segments. Because branch ordering proceeds rapidly through a network, network order typically does not exceed ten and rarely fifteen.

The definition of branch ordering given above can also be defined algorithmically. The junction of two "offspring" branches of order ω_1 and ω_2 will

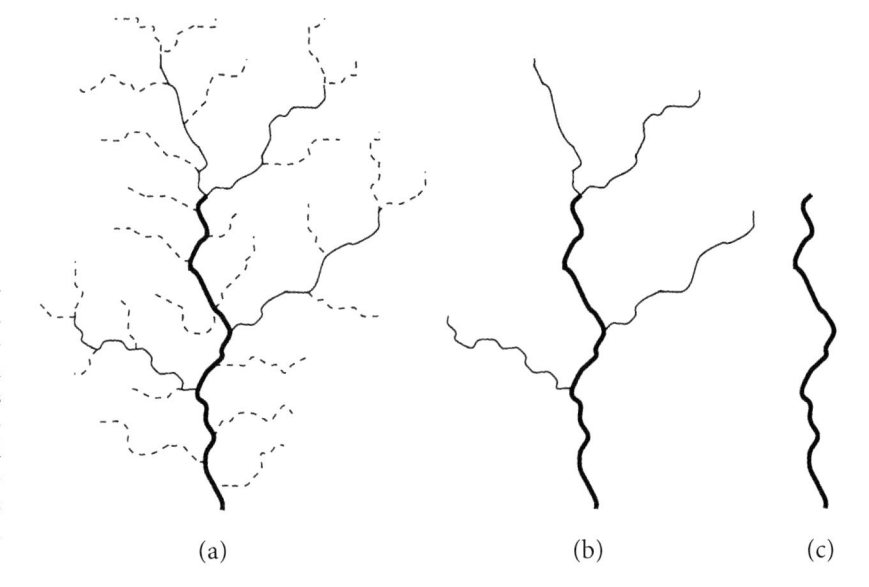

Fig. 3.1a–c. Horton–Strahler "branch ordering." (*a*) shows a simple network. (*b*) is created by removing all outermost branches (i.e. leaves or tips) from the network in (*a*), these same branches being denoted as "first order branch segments". The new "leaves" in the pruned network of (*b*) are labelled as second order branch segments and are themselves removed to give (*c*), a third order branch segment.

(a)   (b)   (c)

form a "parent" branch of order ω according to the rule

$$\omega = \max(\omega_1, \omega_2) + \delta(\omega_1, \omega_2) \tag{3.1}$$

where $\delta(\omega_1, \omega_2) = 1$ if $\omega_1 = \omega_2$ and 0 otherwise. Note that in all of these definitions we are moving from the outermost parts on the network inwards, counter to the direction of growth. Thus, a parent branch of order ω is only produced when two offspring branches of order $\omega - 1$ come together. In all cases where offspring branches of different orders join, the parent will have the same order as the maximum of the two offspring. The pruning method is more useful when a specimen is examined manually while the algorithmic method is preferable when digitized data of a branching structure are available.

Now that we have a way to break a network into logical pieces, we may begin to analyze the way that these pieces fit together.

3.1.2 Horton Statistics

On a network with branch ordering, various natural quantities to measure arise. Some principal ones are n_ω, the number of branch segments for a given order ω; $\bar{l}_\omega$, the average branch segment length; $\bar{a}_\omega$, the average cross-sectional branch area; and the variation in these numbers from order to order. In some cases, the volume or area surrounding a branch segment can be measured. For example, $\bar{a}_\omega$ for river networks would refer to the average area of land from which water drains into stream segments of order ω. Networks internal to bodies such as blood networks have a typical volume surrounding each branch. For marine organisms, a similar volume may be determined if a particular species "fills space," i.e. its branches are separated by some characteristic scale. However, for any tree-like organism, the fact that there is no body surrounding the structure means that it may grow "loosely" and such a measure would be irrelevant.

Horton (1945) and later Schumm (1956) observed for river networks that network quantities such as those above approximately change by the same ratio from order to order. For the present situation this would amount to the follow statements:

$$\frac{n_\omega}{n_{\omega+1}} = R_n, \qquad \frac{\bar{l}_{\omega+1}}{\bar{l}_\omega} = R_l, \qquad \text{and} \qquad \frac{\bar{a}_{\omega+1}}{\bar{a}_\omega} = R_a. \tag{3.2}$$

All of these "Horton ratios" are defined so that they are greater than unity. The numbers of branch segments decrease with order while branch segment length and cross-sectional area usually increase. Typical values of the Horton ratios are $R_n \cong 4$, $R_l \cong 2$ and $R_a \cong 4$.

It is to be expected that these relations hold over a range of orders but not all orders and some care must be taken in measuring these Horton ratios. Note that if the relations do not hold for a network then order-dependent ratios should be used. So instead of a single value of R_n we would have $R_n(\omega) = n_\omega/n_{\omega-1}$.

As an example, we can readily calculate the n_ω for the example network in Fig. 3.1. We find $n_1 = 26$, $n_2 = 4$, and $n_3 = 1$. Note that for an order Ω network, we always have $n_\Omega = 1$. Now, $n_2/n_1 = 4$ and $n_3/n_2 = 6.5$ but for such a small network we should not expect these to be the same. Care must be taken to find ranges of orders where ratios are relatively constant. Typically, networks of at least six orders are necessary for this.

Another quantity we can use is the distance from the end of an order ω sub-network out along branches to the furthest tip. Writing this length as l_ω we see that it can be broken down into branch segments and we have that $\bar{l}_\omega = \sum_{k=1}^{\omega} \bar{l}_k$. A simple calculation shows that the corresponding ratio R_l must be equivalent to the ratio for branch segment lengths R_l (Dodds and Rothmann 1999). We therefore have in principle three independent Horton ratios R_n, R_l, and R_a. For networks that are constrained to two dimensions and fill space (e.g. river networks), only two of these ratios are independent since $R_n \equiv R_a$ (Dodds and Rothmann 1999). An important point to remember is that the relations in (3.2) are only for mean values of branch length and cross-sectional area. In general, similar relationships hold for higher order moments (Peckham and Gupta 1999, Dodds and Rothman in prep.). The above should be seen as a first measurement after which a more careful examination of frequency distributions of quantities such as l_ω and a_ω may be carried out.

3.1.3 Tokunaga Statistics

Not surprisingly, branch ordering allows for other interesting metrics and eventually Tokunaga introduced the idea of measuring side branch statistics (Tokunaga 1966, 1978, 1984). As with Horton's work, the first setting for Tokunaga's method was river networks. This technique arguably provides the most useful measurement based on branch ordering but has only recently received much attention (Turcotte et al. 1998, Dodds and Rothmann 1999). The idea is simply, for a given network, to count the average number of order ν "side branches" attached to order μ branch segments. This gives $\langle T_{\mu,\nu} \rangle$, a set of double-indexed parameters for a network which we will call "Tokunaga ratios." Specifically, only side branches that attach along a branch segment and not at its beginning or end are counted. Note that $\Omega \geq \mu > \nu \geq 1$, so we can view the Tokunaga ratios as a lower (or upper) triangular matrix of numbers.

Referring back to Fig. 3.1, we have three Tokunaga ratios to measure: $T_{2,1}$, $T_{3,1}$ and $T_{3,2}$. From Fig. 3.1b we see that there are two $\omega = 2$ side branches for the single $\omega = 3$ branch segment. Note that the other two $\omega = 2$ branches are not side branches since they meet the $\omega = 3$ branch segment at one end. Therefore, we have $T_{3,2} = 2$. It is left as an exercise for the reader to check that $T_{2,1} = 2.25$ and $T_{3,1} = 9$.

Tokunaga made several key observations about these side branch ratios and while this was done in the context of river networks, the same notions hold in general. The first is that because of the self-similar nature of branching networks, the $\langle T_{\mu,\nu} \rangle$ should not depend absolutely on either of μ or ν but only on the relative difference, i.e. $k = \mu - \nu$. The second, which also follows from considerations of self-similarity, is that in changing the value of $k = \mu - \nu$, the $\langle T_{\mu,\nu} \rangle$ must themselves change by a systematic ratio. These statements lead to "Tokunaga's law":

$$\langle T_{\mu,\nu} \rangle = \langle T_k \rangle = \langle T_1 \rangle (R_T)^{k-1} \tag{3.3}$$

Thus, for a strictly self-similar network only two parameters are necessary to characterize the set of $T_{\mu,\nu}$: T_1 and R_T.

Now, it turns out that in theory $R_T \equiv R_l$ (Peckham 1995, Dodds and Rothmann 1999) so we can understand R_T as a ratio of length scales. The parameter $T_1 > 0$ is the average number of side branches of one order lower than the branch they are attached to, typically on the order of 1.0–1.5. In general, larger values of $\langle T_1 \rangle$ correspond to wider structures while smaller values are in keeping with structures with relatively thinner profiles.

It can be shown that Horton's laws of branch numbers and branch segment length follow from what we have called Tokunaga's law, (3.3). The required observation is that n_ω, the number of order ω branches, is related to the number of higher order branches that have order ω side branches by

$$n_\omega = 2n_{\omega+1} + \sum_{\omega'=\omega+1}^{\Omega} \langle T_{\omega'-\omega} \rangle n_{\omega'} \tag{3.4}$$

The extra $2n_{\omega+1}$ accounts for those order ω branches that are not side branches but rather generators of order $\omega+1$ branches. A difference equation for R_n is obtained by dividing through by $n_{\omega+1}$ and writing $n_\omega/n_{\omega+1} = R_n$ and $n_{\omega'}/n_{\omega+1} = (R_n)^{\omega+1-\omega'}$. For an infinitely large network, the solution is

$$R_n = A_T + [A_T^2 - 2R_T]^{1/2} \tag{3.5}$$

where $A_T = (2 + R_T + T_1)/2$. For a finite network, an exact solution can still be obtained but is more complicated (Tokunaga 1978, Dodds and Rothmann 1999).

We see that the Horton ratios (3.2), although indicative of the network structure, do not give the full picture. One cannot picture a network using the Horton ratios alone since we do not know which branch segments connect with which. The network perhaps suggested by the Horton ratios is one where all branch segments of order ω join branch segments of order $\omega+1$, a true hierarchy. But this is misleading since branch segments of a certain order have side branches of all lower orders.

Nevertheless, one can further argue that Horton's laws also lead to Tokunaga's law (Dodds and Rothmann 1999) An invertible transformation between the remaining pairs of parameters may be deduced to be (Tokunaga 1978)

$$R_n = 1/2(2 + R_T + T_1) + \left[(2 + R_T + T_1)^2 - 8R_T\right]^{1/2} \tag{3.6}$$

$$R_l = R_T \tag{3.7}$$

This equivalence between the two descriptions means we have a useful cross-check between measurements. Note that $(R_n, R_l) = (4, 2)$ matches exactly with $(T_1, R_T) = (1, 2)$. Of course, this sort of equivalence applies only to exactly self-similar networks. In practice deviations from self-similarity occur and Tokunaga's statistics then carry more raw information (Cui et al. 1999, Dodds and Rothman in prep.).

We have thus defined a simple method for quantifying network structure in branch ordering. Based on this ordering scheme, the two measures of Horton and Tokunaga provide a good basis for comparison of network structure. It is important to keep in mind that for small networks, strict self-similarity is unlikely and that tables of numbers rather than poorly estimated ratios are more useful (e.g. it would be wise to keep all of the n_ω along with any estimate of R_n). Finally, wherever it is possible, statistics should always be improved by averaging over many samples of a species.

3.2 Morphological Analysis of a Branching Sponge

In Sect. 2.2.2 the biology of the branching sponge *Raspailia inaequalis* was described. Because of its planar form, accurate measurements can be made from an analysis of photographs, making it a convenient species for the study of sponge growth. In this section a detailed analysis is made of the branching pattern of *Raspailia inaequalis*, as described in Abraham (2001).

A prediction of models of sponge growth is that the branching pattern should be self-similar. To test whether the branching structure of *Raspailia inaequalis* is fractal, 45 specimens were harvested and photographed in the laboratory. The specimens were taken from a site, known as Leigh Reef, where *Raspailia inaequalis* is abundant. The sponges were growing at 25 m depth on a flat reef and, in order to minimize variation in environmental conditions, all the specimens were collected from within a circle of 10 m radius. The sponges were chosen to represent the range of morphologies which was present at the site, although the smallest sponges ($<$ 5 cm high) were not collected. After being harvested the sponges were laid flat on a black background and photographed. The digitized images were used for the morphological analysis.

3.2.1 Image Processing

Some finger sponges have an amorphous lobed morphology, so the branching pattern is difficult to define unambiguously. In contrast, the skeleton of *Raspailia inaequalis* is axially condensed, each finger having a spicule bundle at its core. Because of this structure, the photographic images may be used to obtain a reliable representation of the branching. The method used is a common procedure in digital analysis. The image is first cleaned by removing any stray light pixels from the dark background. Some of the sponges were anastomosed: branches that touched while the sponge was growing had fused together. In order to define the branching, the anastomoses were removed by inserting a line of zero-valued pixels between fused branches. A few sponges appeared to have been damaged while growing and had a confused three-dimensional structure. These specimens were not analyzed. After this preprocessing the images of the sponges were stripped to single lines of pixels, or skeletonized, using the following steps (see also Fig. 3.2):

Fig. 3.2a–c. Stages in the image processing of a photograph: (*a*) The original digitized photograph after preliminary processing to remove stray light pixels from the dark background and to separate branches which were touching. (*b*) Each pixel on the sponge is given a value which is its distance from the background. This distance is smoothed so that the processing avoids producing spurious branches. (*c*) The pixel skeleton from which the branching pattern is abstracted

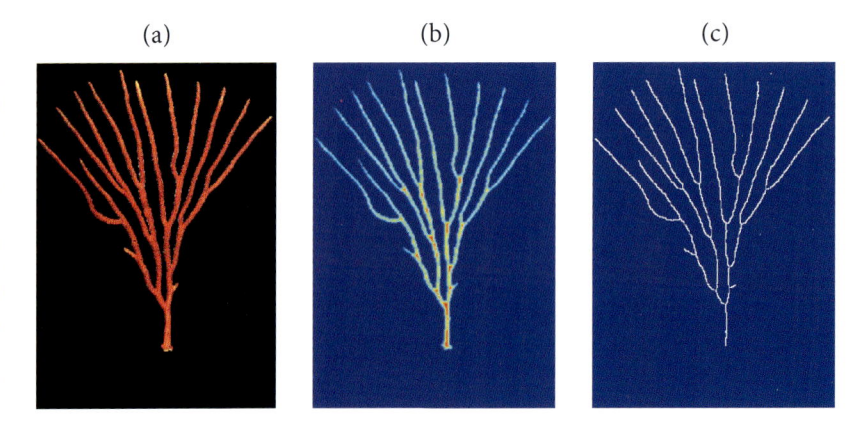

```
Set the value of the background pixels to zero
Set the value of other pixels to be the distance from the
nearest background pixel
Replace the value of each non-zero pixel with the average
value of its neighbors
for (each non-zero pixel, proceeding from the
lowest to the highest valued){
    if ((any of the pixel's non-zero neighbors are
    neighbors of each other) or
    (a pixel has more than three non-zero neighbors)){
    set the pixel to zero
    }
}
```

The remaining non-zero pixels form the digital skeleton. They may be classified according to the number of their non-zero neighbors. Branch-tips have only one neighbor, branches have two neighbors, and branch-vertices have three neighbors. In order to abstract the branch structure a root pixel is chosen at the base of each sponge's stipe and then, starting with the root, a linked list of pixels is formed. Each member of the list contains the elements (`x`-position, `y`-position, pointer to parent pixel, pointer to first child pixel, pointer to second child pixel). The root pixel has a null parent, tips have two null children, and points along the branches have one null child. The `x`-position and `y`-position data are converted to centimeters using a scale-bar included in each photograph, and the data is smoothed while holding the positions of the tips, root, and vertices fixed. From this linked list the branch structure may readily be determined, and its scaling properties explored, using Horton and fractal analyses.

3.2.2 Horton Analysis

The methods of Sect. 3.1.2 are used to calculate the Horton–Strahler order of the branch networks for each specimen (Fig. 3.3). Once these orders have been assigned, the bifurcation ratios R_n, the length ratios R_l, and the Tokunaga ratios can be calculated. These ratios capture the principal features of the branching pattern in a simple way. Most of the sponge specimens (29) are order 4, with seven order 3, and eight order 5 specimens. There was one order 2 sponge which was something of a morphological outlier and was not included in the calculation of the branching ratios. The sponges have a well-defined bifurcation ratio, with a value of $R_n = 2.71 \pm 0.046$ (Fig. 3.4a). By comparison, the regular branching pattern shown in Fig. 3.5 has a bifurcation ratio of

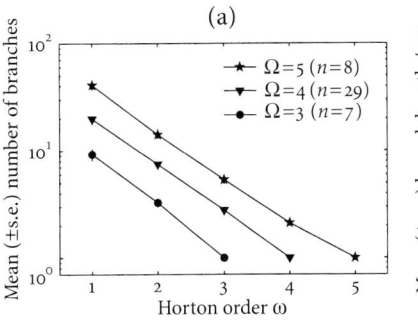

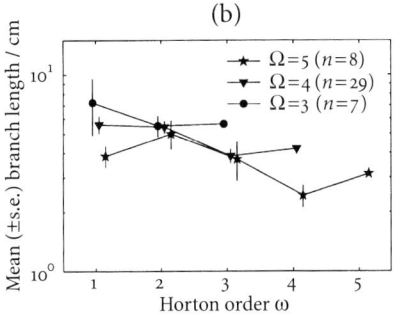

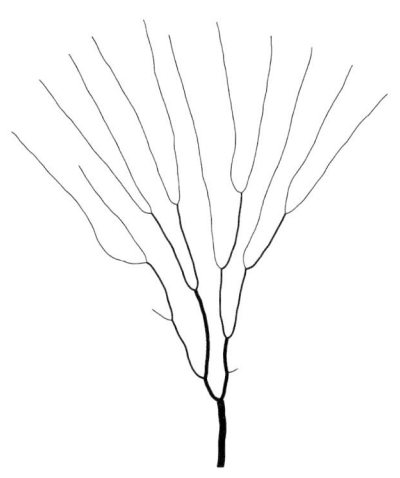

Fig. 3.3. The branches with a higher Horton order are drawn with thicker lines. This specimen has 14 first order branches, 5 second order branches, 2 third order branches, and a fourth order stipe.

Fig. 3.4a,b. Summary of the Horton analysis of *Raspailia inaequalis*: (*a*) The mean number of branches of each order. Where they cannot be seen the error-bars are smaller than the symbols. The number of branches decreases exponentially as the order increases, so the bifurcation ratio is well-defined. (*b*) The average branch length. The mid-order branches show a consistent decrease in their length as the order increases.

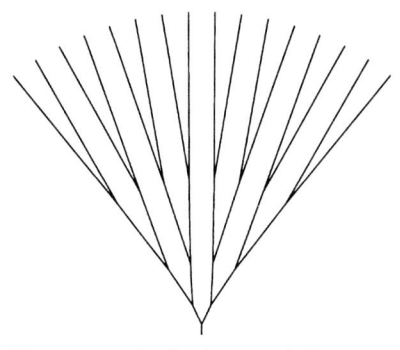

Fig. 3.5. An idealized network for comparison with the branching patterns of *Raspailia inaequalis*. The branches in this network double in length between each bifurcation, the distance between the branches remaining constant as the network grows. This branching structure has the Horton ratios $R_n = 2$ and $R_l = 0.5$.

Table 3.1. Average Tokunaga ratios for the order 4 and 5 specimens of *Raspailia inaequalis*. The equivalence of the ratios along the diagonals supports the hypothesis that the sponge's branching pattern is self-similar.

$T_{1,2} = 0.58 \pm 0.04$ $T_{1,3} = 0.19 \pm 0.04$ $T_{1,4} = 0.05 \pm 0.04$	$T_{2,3} = 0.67 \pm 0.10$ $T_{2,4} = 0.19 \pm 0.08$	$T_{3,4} = 0.73 \pm 0.12$

$R_n = 2$, the minimum possible for a dichotomous network. The Tokunaga ratios (Table 3.1) are consistent with a self-similar network which has $T_1 \cong T_{1,2} \cong 0.6$ and $R_T \cong 0.3$. The low value of R_T means that the sponge branching networks are nearly hierarchical: most branches stem from a branch of the next highest order. If the network was strictly hierarchical then the relation $R_n = 2 + T_1$ would hold, and this is nearly satisfied by the sponge networks.

The branching lengths tell a more complex story (Fig. 3.4b). The lengths decrease as the branch order increases. The length ratio calculated from the mid-order branches of the fourth and fifth order sponges is $R_l = 0.75 \pm 0.043$. The sponges were still growing when harvested so the first order branches would not have reached their full length, consequently the first order branches are shorter than would be expected from self-similarity. The highest order branch, which includes the stipes, appears to be longer than would be expected. This is interesting in light of the theory that the sponge growth is being organized by the flow. The boundary layer over the substrate will be thicker than the boundary layer around the sponge itself and this may be why the stipes are longer than the higher-order branches of the fan.

Horton analysis was first used to characterize the branching pattern of river networks. Natural river systems have bifurcation ratios in the range $3 < R_n < 5$ and length ratios in the range $1.5 < R_l < 3.5$ (Marani et al. 1991). In contrast to the sponges, the lower order streams are shorter than the higher order rivers. River networks also have a bushier branching pattern, with many side-branches or tributaries, and this is reflected in the much higher value of the ratio $R_T \cong 2$. So, although they are consistent with self-similarity, the sponges do not look at all like rivers. Does Horton analysis distinguish the branching of *Raspailia inaequalis* from the branching of other marine organisms? The branching patterns of several gorgonian species (see also Sect. 2.2.3) have been quantified using Horton analysis (Brazeau and Lasker 1988, Mitchell et al. 1993). The five species (from the genera *Plexaura* and *Leptogorgia*) have bifurcation ratios that are between 3 and 4, depending on the species and the site that the specimens were sampled from. These gorgonians have a form which is intermediate between *Raspailia inaequalis* and the river networks, with more side-branching than the sponge specimens.

It has been suggested that the self-similar branching structure of rivers arises as an optimal pattern for the network, minimizing the energy dissipated by the flow (Rinaldo et al. 1992). Other fractal branching structures, such as the foraging trails of ants (Ganeshaiah and Veena 1991), have also been proposed as optimizations of harvesting and transport problems (West et al. 1997). Without understanding more of the sponges, interactions with their fluid environment it is not possible to convincingly argue the adaptive merits of the branching pattern. A challenge for any complete organismal modeling is to reach an understanding of not only how an individual develops, but also why a particular morphology has been favored by evolution. In this light, it is interesting to note the variety of morphologies that are found at a site such as the Sponge Garden. At least five different species from the family *Axinellidae* grow there and *Raspailia inaequalis* is the only one which has a planar branched form. Two others have a three-dimensional branching structure, one grows as a single unbranched finger on a short stalk, and one forms a solid vertical sheet which is set perpendicular to the flow. Explaining the diversity of sponge growth patterns remains a great challenge.

3.2.3 Fractal Analysis

It appears from the Horton analysis that the sponge branching network is consistent with self-similarity, although the order of the sponge specimens is too small to make a strong statement. An interesting question to ask, from a functional viewpoint, is how the spacing between the branches changes with distance from the base of the sponge fan. It might be expected that the sponge would maintain a constant distance between its branches, in order to efficiently filter the water. However, the skeleton of *Raspailia inaequalis* is flexible and the whole sponge sways in the current. As the sponge bends the outer branch tips are swept towards one another, so an *a priori* statement of what the branch spacing should be is difficult to make without a better understanding of the interaction between the flow, the mechanical flexing, and the feeding of the sponge. In order to explore this aspect of the branching structure, the length of branch, l, within a distance r of the primary vertex is determined. The length is measured along the branches, so that it is not influenced by the way the specimens have been arranged before being photographed. For an ideal radial dichotomous branching pattern (Fig. 3.5) the distance between the branches is constant and

$$l(r) \sim r^2 \qquad (3.8)$$

If the sponge has a fractal structure then

$$l(r) \sim r^d \qquad (3.9)$$

where $d \neq 2$ is the fractal dimension. The scaling of the branch length is shown in Fig. 3.6. For small r, l is proportional to r and l is limited above by the total length of the sponge's branches. This means that fractal scaling can be established over only a short range, after the branching has reached sufficient complexity but before the finite size of the specimen becomes important. In order to account for this, the scaling analysis was begun at a radius $r_{\min}$, where the specimen first had four branches, and stopped at $\bar{r}$, the average radius of the fan. Only the 26 sponges for which $\bar{r} > 2r_{\min}$ were included in the analysis. By a linear regression of $\log(l)$ against $\log(r)$ over this range was found that $d = 1.63 \pm 0.047$, suggesting that, on average, the branching of *Raspailia inaequalis* tends to become more open as the distance from the primary vertex increases.

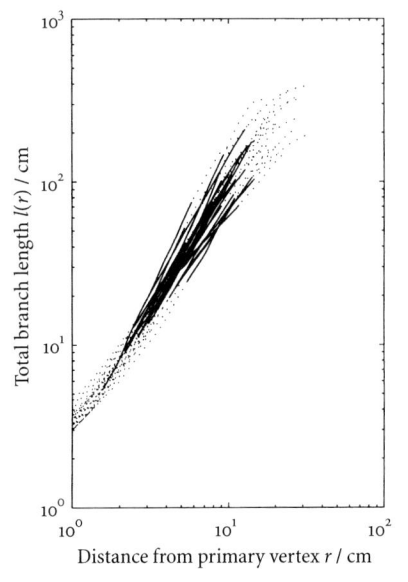

Fig. 3.6. Scaling of branch length with distance from the primary vertex, for each sponge. The solid lines show the range over which the scaling relation was assumed to hold. The average rate of increase of the branch length is slower than the r^2 relation expected if the distance between the branches remained constant.

3.3 Two-dimensional Morphological Analysis of Ranges of Growth Forms

As was discussed in two previous chapters, many marine sessile organisms from various taxonomical groups exhibit considerable morphological plasticity; in many cases this is related to the impact of the physical environment.

The local availability of light and the exposure to water movement are the dominant environmental influences; sedimentation and transport of food particles are closely related to the hydrodynamic parameter. Typical examples of this morphological plasticity and the relation to the physical environment are: Fig. 2.31 showing a range of growth forms of the scleractinian *Montastrea annularis* and local light intensities (Barnes 1973, Graus and Macintyre 1982); growth forms of the hydrozoan *Millepora spp.* and exposure to water movement (Stearn and Riding 1973, de Weerdt 1981); the variations in morphology due to differences in exposure to water movement in the scleractinian *Pocillopora damicornis* shown in Fig. 1.1 (see also Lesser et al. 1994), *Madracis mirabilis* (see Fig. 2.5 and Sebens et al. 1997), and *Agaricia agaricites* (Helmuth and Sebens 1993); the shape of coralline algae (see Fig. 1.3d and e) and the effect of exposure to water movement (Bosence 1976); and the growth forms of the sponge *Haliclona oculata* shown in Fig. 2.16 (Kaandorp 1991, Kaandorp and de Kluijver 1992) and the exposure to water movement.

Since the physical environment has a strong impact on the growth process, distinct growth forms of marine sessile organisms can often be associated with an environmental gradient, where the amount of water movement is usually the most dominant parameter. In Veron and Pichon (1976) several series of growth forms of scleractinians (for example *Pocillopora damicornis* and *Seriatopora hysterix*) are presented, which are arranged along a gradient of the amount of water movement. Among studied specimens of both scleractinians the growth forms showed a gradual transformation from a compact shape, under exposed conditions, to a thin-branching shape under sheltered conditions. In Fig. 1.1 a range of growth forms of *P. damicornis* are shown. Form (a) originated from the most exposed site, form (f) originated from the most sheltered site, and in the range (a) to (f) the exposure to water movement gradually decreases. In the growth forms of the sponge *Haliclona oculata* shown in Fig. 2.16 a plate-like, more compact shape was found at exposed sites; this shape was gradually replaced by a thin-branching form at less exposed sites (Kaandorp 1991, 1994b). A similar trend was observed in growth forms of the hydrozoan *Millepora alcicornis* (de Weerdt 1981). In this species the shape changed from plate-like forms at shallow and exposed sites to thin-branching forms at deeper and sheltered locations.

In this section a range of growth forms for each of three species – the sponge *Haliclona oculata*, the scleractinian coral *Pocillopora damicornis*, and the hydrozoan *Millepora alcicornis* – was morphologically analyzed. Samples of the species were collected along a gradient of exposure to water movement. The idea was that by analyzing growth forms from very different taxonomical groups, it might be possible to identify some generic effects of the impact of water movement on the overall growth form of organisms with accretive growth, and to distinguish these from species-specific morphological features, which are presumably controlled by genetic regulation (Kaandorp 1999). An example of a morphological feature which might be regulated by biological mechanism is the average distance between branch tips and neighboring branches ("branch spacing"). In studies on particle capture in the branching scleractinian *Madracis mirabilis* (see Fig. 2.5) and the influence of hydrodynamics (see also Sect. 2.1.1 and Sebens et al. 1997), it was demonstrated that branch spacing is a crucial morphological property. Branch spacing is variable and may be controlled by a chemical agent (see Sect. 2.2.4 and Rinkevich and Loya 1985a). Sebens et al. (1997) argue that through modifications of its branch structure and branch spacing, *M. mirabilis* can function efficiently as a passive suspension feeder over a wide range of different degrees of exposure to water movement.

3.3.1 Sampling Growth Forms Along a Gradient of Increasing Water Movement

For the measurements a series of sponge samples (*Haliclona oculata*) was used; these samples were collected at different sites in the Oosterschelde, the Netherlands. For each site the exposure to water movement was estimated quantitatively by relating the exposure to water movement to the erosion of gypsum blocks. The erosion value (g h^{-1}) is expressed as the weight loss of the gypsum blocks during a lunar day (24 h 45 min) (see de Kluijver 1989, Kaandorp 1991). The series of samples of the hydrozoan *Millepora alcicornis* was collected at different depths in the Caribbean area (see de Weerdt 1981). The exact exposure to water movement was not known for these samples, but it was assumed that exposure decreases uniformly with depth. For the scleractinian coral *Pocillopora damicornis* measurements were based on photographs published in Veron and Pichon (1976) and shown in Fig. 1.1. In their book Veron and Pichon arranged samples in 11 classes of growth forms originating from sites with different values of exposure to water

Fig. 3.7a–f. *Pocillopora damicornis* (*a,b*), *Millepora alcicornis* (*c,d*), *Haliclona oculata* (*e,f*). Contour images of the extreme growth forms of the three species. (*a*) *P. damicornis* from an exposed site (Class 11) and (*b*) collected at a sheltered site (Class 1). (*c*) *M. alcicornis* from an exposed site (depth 2 m) and (*d*) the sheltered extreme (depth 25 m). (*e*) *H. oculata* collected at an exposed site (erosion value 0.12 g h^{-1}) and (*f*) at a sheltered site (erosion value 0.05 g h^{-1})

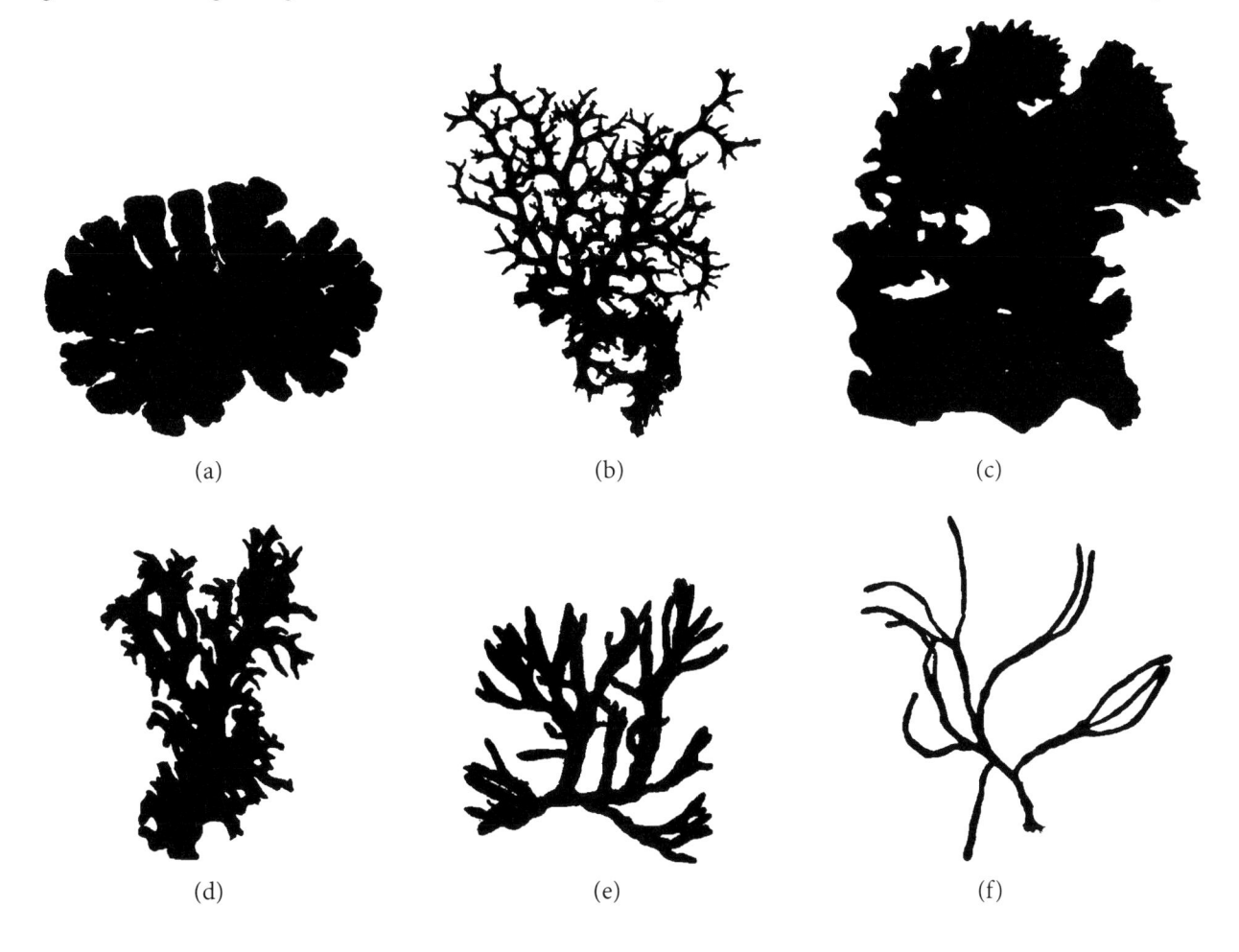

movement. For this species only a qualitative description of the amount of exposure to water movement was available, where Class 1 indicates the lowest and 11 the highest exposure to water movement.

For *Haliclona oculata* and *Millepora alcicornis* photographs of the samples were made using a high-contrast film in order to obtain sharp contours (Kaandorp 1999). In the case of *Pocillopora damicornis* the original film provided by Veron and Pichon was used, and sharp contour prints were developed. The photographs were scanned using a 600 dpi scanner and converted into pbm-files: a lattice representation of the photograph where parts of the object are in state "1" and the environment is in state "0". In Fig. 3.7 representative examples are shown of the resulting contour images of the extreme growth forms of each of the three species.

3.3.2 Morphological Measurements in a Range of Growth Forms

The measurements are based primarily on the morphological skeleton of the contour images. The morphological skeleton was obtained by applying the thinning algorithm developed by Zhang and Suen (1984). The skeleton is defined by connecting the center points of the maximum discs which fit exactly within the contour (Rosenfeld and Kak 1982). In Fig. 3.8 an example is shown of the morphological skeleton constructed within the contour shown in Fig. 3.7f. In the construction of the morphological skeleton several artefacts may be generated. By occlusion effects some branches of the growth form may overlap other branches and produce junctions that are not actually present; furthermore, contamination and damage at the object (for example at the holdfast of the organism) may produce "false" junctions. In many marine sessile organisms (for example in many sponges) real junctions are also present; these are formed by fusion of branches (anastomosis). In the morphological measurements, both "false" junctions and junctions formed by anastomosis were detected by comparing the images to the actual objects and are not used in the measurements.

Diameter was measured for two types of maximum discs: the diameter da of the disc a with a center point at a junction of the morphological skeleton and the diameter db of the disc adjacent to disc a. Disc b was measured in an area of the contour which represents a younger part of the organism compared with the area where disc a was located. Both types of discs are shown in Fig. 3.9. The diameter da represents an estimation of the maximal thickness of a branch, while db is an estimation of the minimal thickness. The measurement rb is the branching rate, and is defined by the length of an edge connecting the centers of two successive a discs (see Fig. 3.9). A low value of rb indicates a high branching rate, while high values indicate relatively slow formation of branches during the growth process. Two types of angles were measured using the skeleton: b_angle is the branching angle, and is defined by the intersection points of the skeleton and the outer circle of disc a (see Fig. 3.9); g_angle is the geotropy angle, and is defined by the angle between the vector connecting two successive a discs and the positive y-axis. The y-axis in the construction of g_angle corresponds to the original growth position, and is directed away from the substrate. The original growth position of the object was approximated using the shape and position of the holdfast. The angle g_angle expresses the degree of substrate tropism; low values of g_angle reflect a high degree of negative substrate tropism, with branches

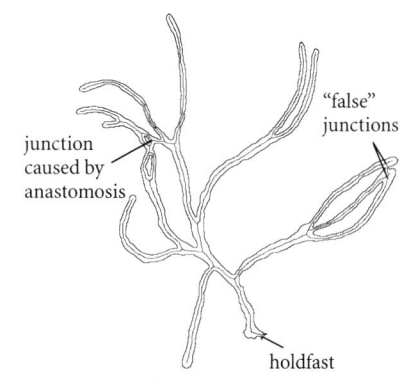

Fig. 3.8. *Haliclona oculata*. Morphological skeleton constructed within the contour shown in Fig. 3.7f

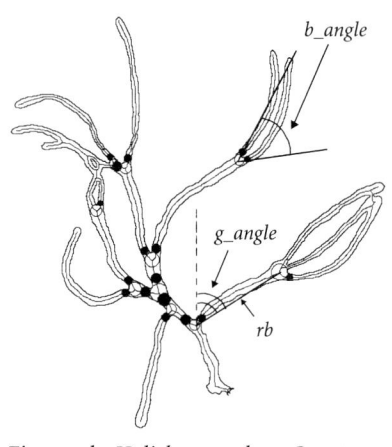

Fig. 3.9a,b. *Haliclona oculata*. Construction of the maximum discs (*a*) (*solid discs*) and (*b*) (*open discs*), the branching rate rb, the branching angle b_angle, and the geotropy angle g_angle within the contour shown in Fig. 3.7f

of the organism showing a tendency to develop away from the substrate. In Fig. 3.10 the measurements of the diameter *br_spacing* are depicted. The diameter *br_spacing* is defined as the radius of a maximum disc which can be constructed using the tip of the skeleton, located at the tip of a branch, and construction stops when the outer circle of the disc intersects with a part of the skeleton which is not connected to a part of the skeleton situated within the maximum disc. The standard deviation of *br_spacing* expresses the degree to which branches tend to fuse; a relatively low value indicates that there is a low degree of self-intersection between branches, while a high value shows that there is no mechanism present preventing self-intersection, and anastomosis of branches may occur. In a number of cases the value of *br_spacing* is undefined, for example, when the tip of the root of the object shown in Fig. 3.10 is used as the center of construction. In morphological measurements these non-valid centers were not used.

The algorithm with which *da*, *db*, *b_angle*, *g_angle*, *rb*, and *br_spacing* were measured can be subdivided into five distinctive stages:

```
1.  Determination of the morphological skeleton.
2.  Determination of the junctions in the skeleton.
3.  Determination of the a discs at the junctions of the
    morphological skeleton, where "false" junctions and
    junctions formed by anastomosis are detected by visual
    inspection of the data set and eliminated.
4.  Measurement of b discs, rb lengths, and the angles
    b_angle and g_angle in successive order by retracing the
    morphological skeleton. The endpoints of the skeleton,
    located within the tips of the branches, are determined.
    Non-valid endpoints are detected by visual inspection of
    the data set and are eliminated. Retracing starts at the
    root of the skeleton, which is close to the holdfast of
    the object.
5.  Construction of the br_spacing discs, using the endpoints
    of the skeleton as centers of construction.
```

After this procedure an arrangement of the *a* discs is possible. Since the skeleton is retraced starting near the holdfast, arrangement of the *a* discs in the order of emergence during the growth process can be ensured; i.e. an *a* disc which is directly connected via the skeleton to a previous *a* disc will be younger than its predecessor. Furthermore this procedure ensures that the *b* disc is positioned within the contour (immediately after an *a* disc) representing a part of the organism which was added to the organism after the formation of the part in which the previous disc *a* is located. The branching angle, *b_angle*, is measured in branches formed after the part represented by disc *a*, while in the measurement of the geotropy angle, *g_angle*, the direction of the vector connecting two successive *a* discs can be unambiguously determined and corresponds to the growth direction. A complication in the algorithm is the occurrence of loops (in Fig. 3.9 several loops can be observed); in the simple version of the algorithm used in this section loops prevented the detection of two more *b* discs near the *a* disc positioned at the very right of Fig. 3.9. The loops were caused by "false" junctions and fusion of branches. As soon as a loop is detected in the algorithm, successive measurements are aborted in that branch, since the loop may cause the *a* discs to be arranged in reverse order to their emergence during the growth process. In Stage 5 the *br_spacing* discs are constructed at valid tips of the skeleton, which are located in the youngest parts of the growth form.

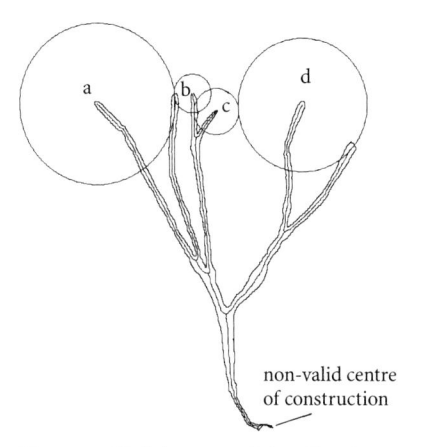

Fig. 3.10. *Haliclona oculata*. Construction of four *br_spacing* maximum discs, using the tips of the skeleton situated in the branches of a contour of *H. oculata* as centers of construction. The four centers of construction are marked a to d.

Fig. 3.11a–d. *Pocillopora damicornis*. Construction of the morphological skeleton within the contours shown in Fig. 3.7a and b. (*a*) sample from an exposed site; (*c*) from a sheltered site. (*b*), (*d*) The corresponding valid *a*-discs which can be constructed at the junctions of the skeleton

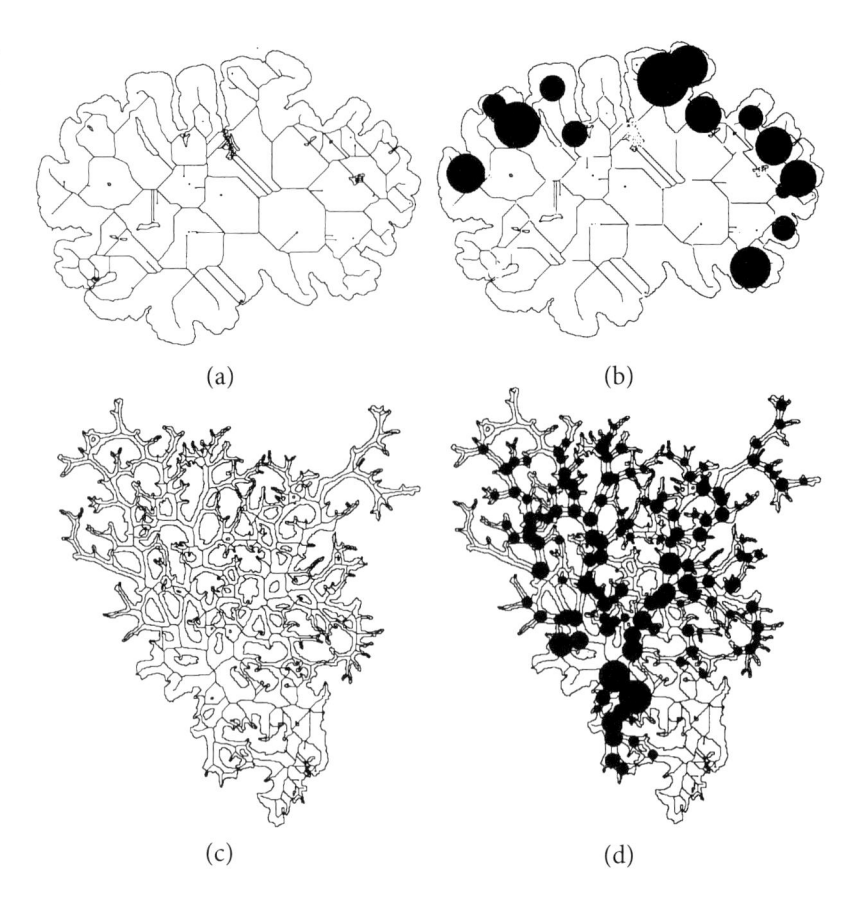

In Fig. 3.11a and c the morphological skeleton was constructed using the contour photographs of both extremes of the growth forms of *Pocillopora damicornis* shown in Fig. 3.7. Fig. 3.11b and d show the valid *a* discs which remained after removal of all "false" junctions from the data set. The object shown in Fig. 3.11a and b represents the "worst case" of the contour images studied in this section, where the highest number of junctions due to occlusions had to be discarded.

Apart from the measurements based on the morphological skeleton, the plane-filling properties of the outline of the contour images were determined by estimating the box dimension, D_{box} (Feder 1988). To measure D_{box} a grid is constructed within the lattice containing the image. The grid divides the lattice into a number of square boxes, where the length of one edge of the box is equal to the grid spacing ε. In the box counting method the number of boxes $N(\varepsilon)$, containing a lattice site which is a part of the outline of the contour, for grid spacing ε, is determined. D_{box} can be determined by the relation:

$$N(\varepsilon) \sim \varepsilon^{D_{box}}. \tag{3.10}$$

Thus, for the contour in Fig. 3.7f D_{box} was estimated by calculating the slope of the line in the log–log plot in Fig. 3.12. In this graph estimates are presented for the entire length of the outline of the contour for various values of ε. Within a certain range of observations the contour behaves as a typical fractal object, where the total length $N(\varepsilon)$ does not stabilize but increases steadily

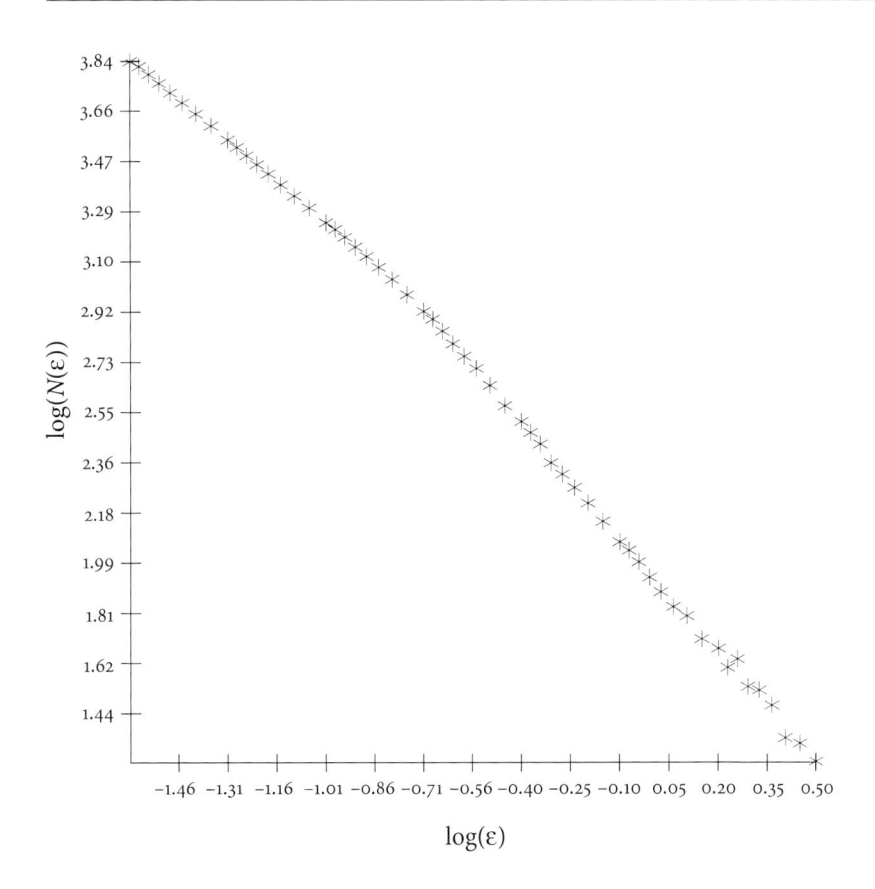

Fig. 3.12. *Haliclona oculata*. Log–log plot of D_{box}, estimated by the box counting method using the outline of the contour shown in Fig. 3.7f

the lower the values of ε. D_{box} values indicate the plane-filling properties of the outline of the contour, and can vary between 1, for "normal" Euclidean curves in the plane, and 2, for complete plane-filling curves. The range of ε within which a meaningful value of D_{box} can be determined depends on the size and resolution of the image, for values of ε close to the resolution of the image and values near the total image size, the outline of the contour behaves as a "normal" Euclidean curve. To determine the range within which the outline is characterized by a fractal dimension, for a given image size and resolution, the D_{box} method was calibrated using some examples from the literature on fractal geometry (Mandelbrot 1983), for which the fractal dimension is analytically known. Here the range of 0.18 to 1.15 cm for ε is used to estimate D_{box}. When applied to projections of three-dimensional objects onto a plane, artefacts in this method created by occlusion are disregarded.

The measurements carried out on the *Millepora alcicornis* and *Pocillopora damicornis* samples are shown in Figs. 3.13, 3.14, and 3.15. The morphological measurements are plotted against the depth at which the *M. alcicornis* were collected (in this case the depth is used as an estimate of the exposure to water movement). For the *P. damicornis* samples, the measurements are plotted against the "exposure to water movement" classes of Veron and Pichon (1976), while in *Haliclona oculata* they are plotted against an estimate of the amount of exposure to water movement which was expressed in the erosion rate of gypsum blocks (g h^{-1}). The plots for *M. alcicornis* (Fig. 3.13) of *da*, *db*, *b_angle*, *g_angle*, *rb*, and D_{box} are based respectively on 360, 485, 135, 460, 385, and 10 measurements. The measure-

Fig. 3.13. *Millepora alcicornis*. Plots of the morphological measurements *da*, *db*, *b_angle*, *g_angle*, *rb*, and D_{box} against depth. Mean values are indicated with an asterisk and the standard deviation with a dashed line. Values of the regression coefficient *rc* are listed in Table 3.2.

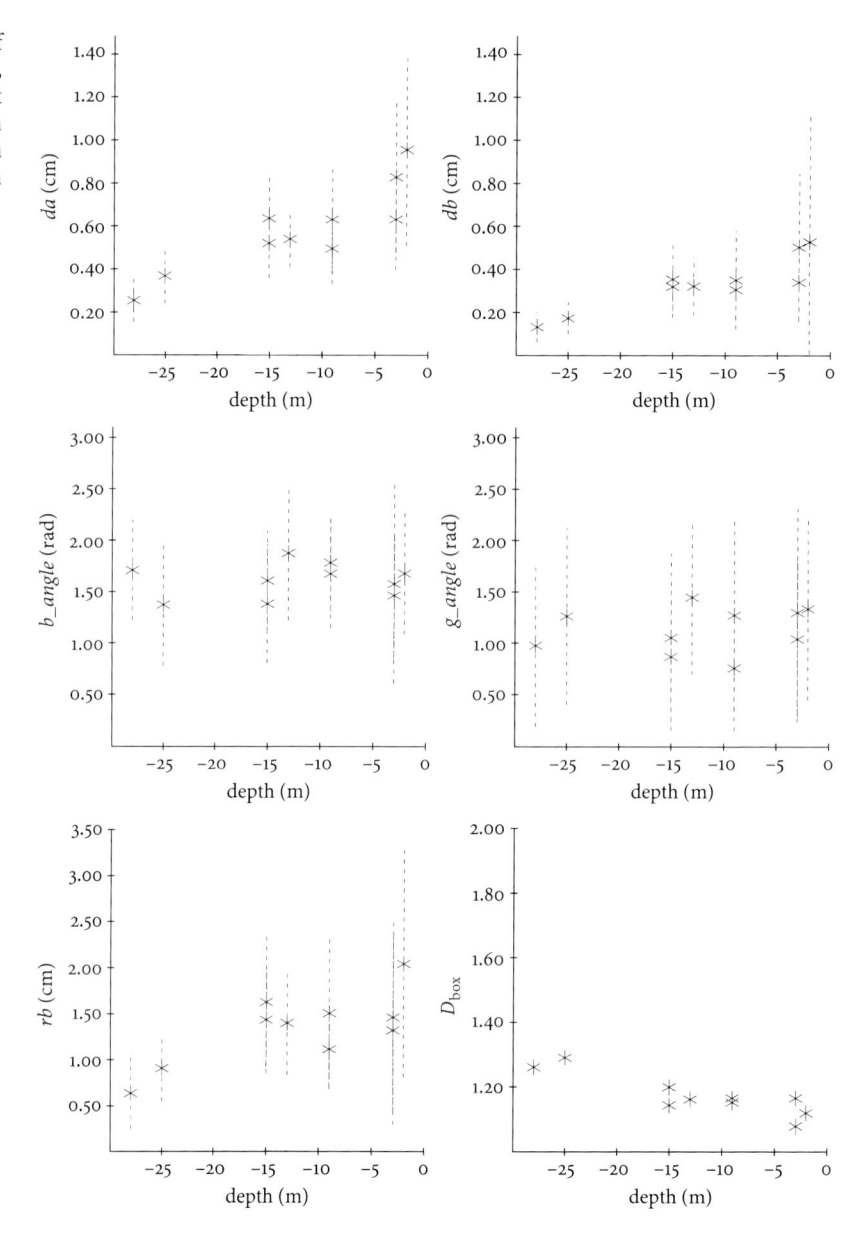

Table 3.2. Values of the regression coefficient *rc*, obtained from morphological measurements versus "exposure to water movement estimate" plots, carried out in the three investigated species. The hypothesis that *rc* = 0 was tested against the alternatives that *rc* < 0 and *rc* > 0 (*rc* = 0.0 indicates that the hypothesis is accepted).

	da	db	b_angle	g_angle	rb	br_spacing	D_{box}
Millepora alcicornis	0.02	0.01	0.0	0.0	0.03	0.02	−0.006
Pocillopora damicornis	0.07	0.04	0.0	0.0	0.11	0.11	−0.037
Haliclona oculata	0.94	0.65	0.0	0.87	−1.17	−2.55	−0.472

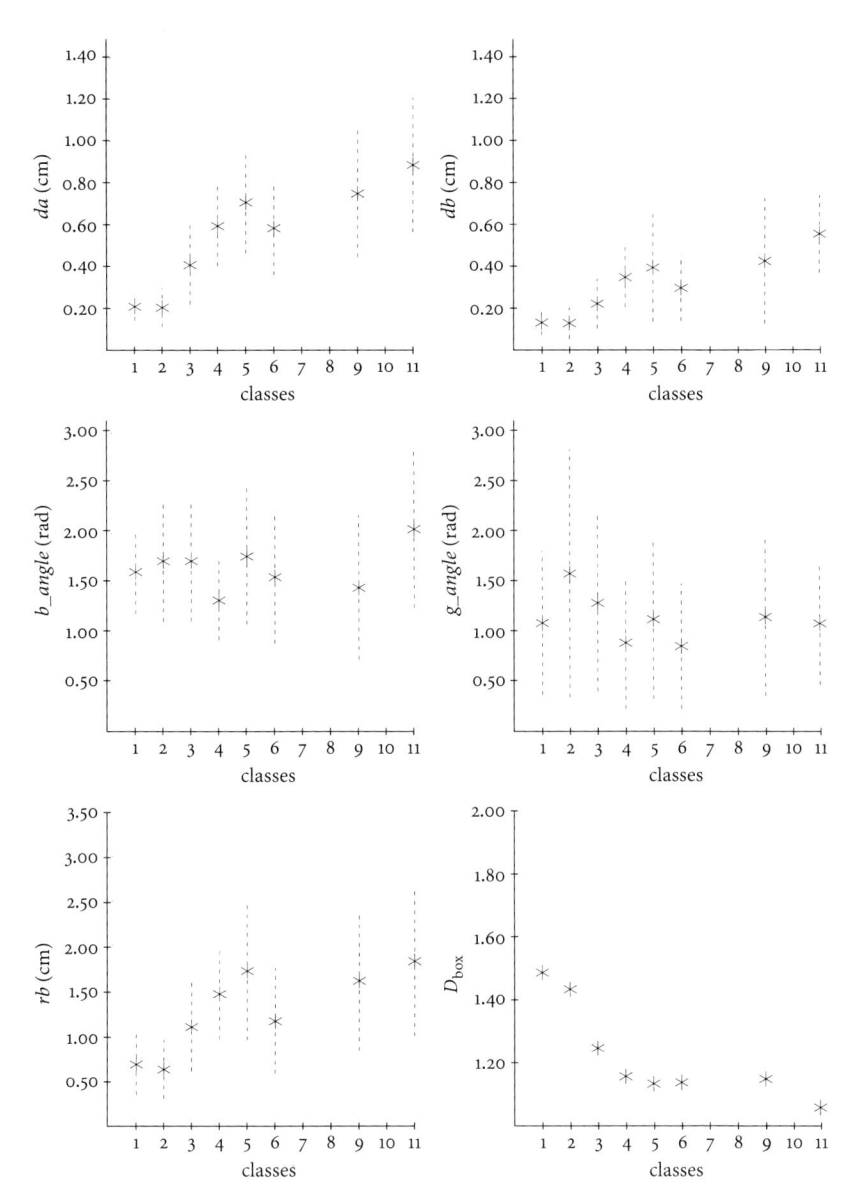

Fig. 3.14. *Pocillopora damicornis*. Plots of the morphological measurements *da*, *db*, *b_angle*, *g_angle*, *rb*, and D_{box} against "exposure to water movement" classes of Veron and Pichon (1976). Mean values are indicated with an asterisk and the standard deviation with a dashed line. Values of the regression coefficient *rc* are listed in Table 3.2.

ments for *P. damicornis* (Fig. 3.14) were based respectively on 443, 627, 197, 544, 461, and 8 measurements. The measurements for *H. oculata* (Fig. 3.15) were based respectively on 276, 359, 105, 359, 275, and 12 measurements. The *br_spacing* measurements, carried out in all three species, are shown in Fig. 3.16; values for *M. alcicornis*, *P. damicornis*, and *H. oculata* are based respectively on 170, 298, and 134 measurements. In all plots the regression coefficient *rc* was tested with a significance of 5%; the hypothesis that *rc* = 0 was tested against the alternatives that *rc* < 0 and *rc* > 0. The data on the three species are summarized in Table 3.2. In this table the values of *rc*, obtained from the morphological measurements versus estimates of "exposure to water movement", are shown.

Fig. 3.15. *Haliclona oculata*. Plots of the morphological measurements *da*, *db*, *b_angle*, *g_angle*, *rb*, and D_{box} against the corresponding erosion values. Mean values are indicated with an asterisk and the standard deviation with a dashed line. Values of the regression coefficient *rc* are listed in Table 3.2.

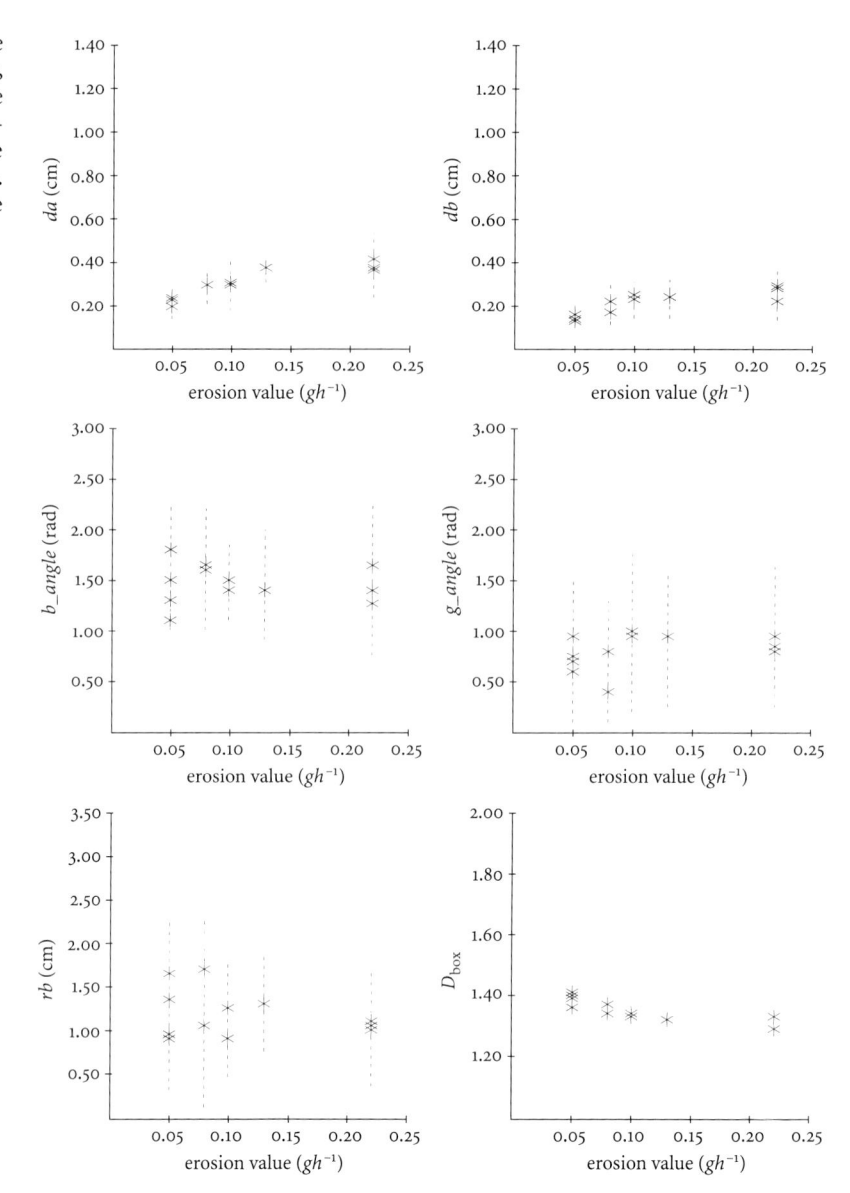

3.3.3 A Comparison of the Morphological Measurements in a Range of Growth Forms of the Three Species

Table 3.2 shows that all three species exhibited a similar trend in the measurements pertaining to the thickness of branches (*da* and *db*). The growth forms gradually transformed from a thin-branching shape to a compact shape along a gradient of increasing exposure to water movement. In Figs. 3.13 and 3.14 it can be seen that the variance of *da* and *db* increased with exposure to water movement. A possible explanation is that with the increase in exposure to water movement the probability also increases that the growth form may be damaged, adding to the variance. The results in Table 3.2 also demonstrate,

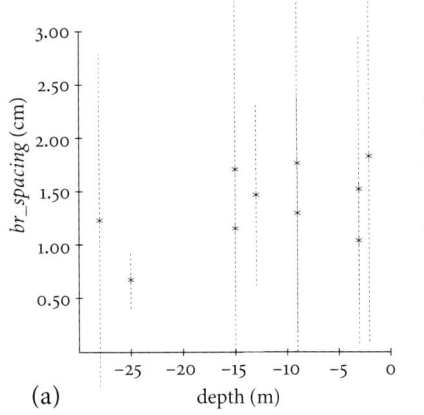

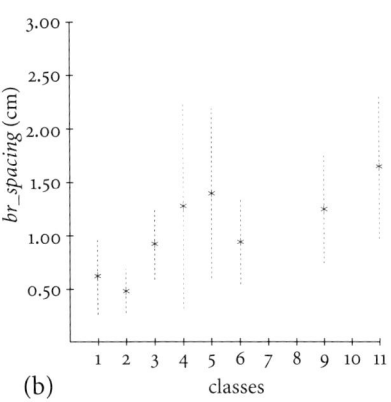

Fig. 3.16a–c. *Millepora alcicornis* (*a*), *Pocillopora damicornis* (*b*), *Haliclona oculata* (*c*). Plots of *br_spacing* against depth, "exposure to water movement" classes of Veron and Pichon (1976), and erosion rate of gypsum blocks (g h^{-1}). Mean values are indicated with an asterisk and the standard deviation with a dashed line. Values of the regression coefficient *rc* are listed in Table 3.2.

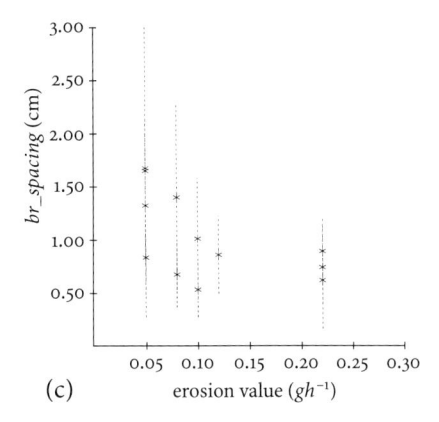

for the growth forms of all three species, that the plane-filling properties of the contour outlines decrease with increasing exposure to water movement. The D_{box} values thus express an overall increase in compactness of the contour line. Although the three species are taxonomically very different, the impact of water movement on the growth process showed a similar trend. Although for *Millepora alcicornis* and *Pocillopora damicornis* local light intensities are another major environmental influence, the amount of water movement seems to be the dominant parameter causing morphological plasticity.

In all three species the branching angle *b_angle* (see Table 3.2; Figs. 3.13, 3.14) was invariant to the change in exposure to water movement in the growth forms. In all three species *b_angle* fluctuated around the value of approximately 1.50 rad, which seems to indicate that this parameter is not a very species-specific property either. In both *Millepora alcicornis* and *Pocillopora damicornis*, *rb* increased with the amount of water movement; this observation contradicts measurements on *Haliclona oculata* and on simulated growth forms of *H. oculata* (Kaandorp 1995). There is no possible explanation for this observation, except that the formation of branches may follow a different pattern in *M. alcicornis* and *P. damicornis* than in *H. oculata*. In all three species negative substrate tropism was detected. The value of *g_angle* in Figs. 3.13 and 3.14 was $< 1/2\pi$; this was also observed in *H. oculata*. In contrast with the observations on *H. oculata* (Table 3.2) the degree of negative substrate tropism does not seem to be affected by the change in exposure to water movement in the two other species.

In Fig. 3.16 notable differences can be seen in the standard deviation of *br_spacing* between *Pocillopora damicornis* and the two other species. In *P. damicornis* the standard deviation was relatively low compared to the two other species, fusion of branches never occurred, and there seemed to be a mechanism which suppressed growth of branches in the immediate vicinity of other branches. In the study by Rinkevich and Loya (1985a) it was proposed for the branching scleractinian *Stylopora pistillata* (see also Sect. 2.2.4) that there is a chemical signal mechanism which suppresses the growth of branches when a certain distance to another branch is reached. Their experiments indicated that a chemical signal was possibly secreted into the water column and worked as a repellent, growth suppressing agent. A similar mechanism might be present in *P. damicornis*. In Fig. 3.16b it can be seen that the average *br_spacing* values decreased at increasing exposures to water movement. A possible explanation might be that the repelling agent was also dispersed by the hydrodynamic action, which may have diluted the overall concentration of the chemical signal. In Fig. 3.16a it can be seen that there was a high standard deviation in branch spacing in *Millepora alcicornis*; in this species there does not seem to be an active chemical agent controlling growth of branches as these tend to grow in proximity to one another and fusion of branches frequently occurs. This observation corresponds to the observations of Rinkevich and Loya (1985a) in experiments with the branching hydrozoan *Millepora dichotoma*. In Fig. 3.16c it can be seen that the variance in branch spacing and also the probability of fusion of branches increased with increasing exposure to water movement in *Haliclona oculata*.

Morphological measurements on projected images are a substantial simplification of reality, and this method was used only for practical reasons. However, several morphological properties of the organisms may be adequately represented by this approach. The branches, for example in *Haliclona oculata*, are formed more or less in one plane. In more complex growth forms, especially the more compact growth forms of *Pocillopora damicornis* where branches often overlap, information is lost by projection. In Fig. 3.11b, the "worst case", it can be observed that due to occlusion effects the individual branches cannot be distinguished in the inner part of the object. Valid measurements can be made only at the periphery. In Fig. 3.11b it can also be seen that the morphological skeleton does not give meaningful results in the inner part of the image; in this area artificial straight lines and junctions are formed in the thinning algorithm. Fig. 3.11 shows that the method works only partially for objects with a complex three-dimensional geometry; the number of meaningful results decreases with an increasing degree of compactness of the objects. For these more complex forms a full three-dimensional analysis of the growth form is ultimately the only solution. A three-dimensional analysis would also allow determination of the degree of anastomosis, which is another relevant morphological property of these organisms.

3.4 Three-Dimensional Morphological Analysis of Growth Forms of *Madracis Mirabilis* (Preliminary Results)

In Sect. 3.3 the morphological analysis of two-dimensional images of growth forms of the three species, collected along environmental gradients, was discussed in detail. The morphological analysis only partly works using two-dimensional images; for complex-shaped growth forms with many overlapping branches, as shown in Fig. 3.11, a two-dimensional analysis fails and a three-dimensional analysis is required. In this section we will present some preliminary results on how similar methods, as discussed in Sect. 3.3, can be extended towards a three-dimensional analysis. Since a complete set of measurements is not yet available (will be described in Vermeij et al. in prep.), we will focus on the three-dimensional data acquisition and give a short description of the thinning procedure, the construction of morphological skeletons, and the measurements based on these skeletons, for the three-dimensional case.

For the morphological analysis we have used growth forms of the stony coral *Madracis mirabilis* (see Fig. 2.5c and a). For the three-dimensional data acquisition we have used X-ray CT (Computed Tomography) scanning techniques. The CT scan data was stored in DICOM format (a general data format used for medical images). The CT scan data consists of $512 \times 512 \times z$ ($20 \leq z \leq 50$) three-dimensional pixels, the so-called "voxels". The slice thickness of the CT scan data is 2.5 mm in the xy direction. Each voxel represents a density value between 0 and 2^{12}, where 0 is the lowest den-

Fig. 3.17a,b. Volume rendered images of CT scans of *Madracis mirabilis*: in (*a*) the growth form of Fig. 2.5c, and (*b*) the growth form of Fig. 2.5a

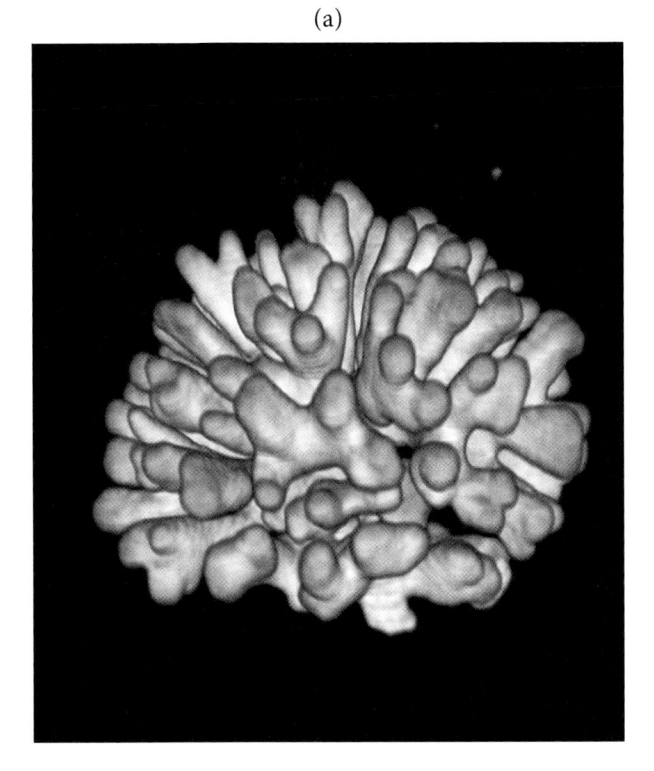

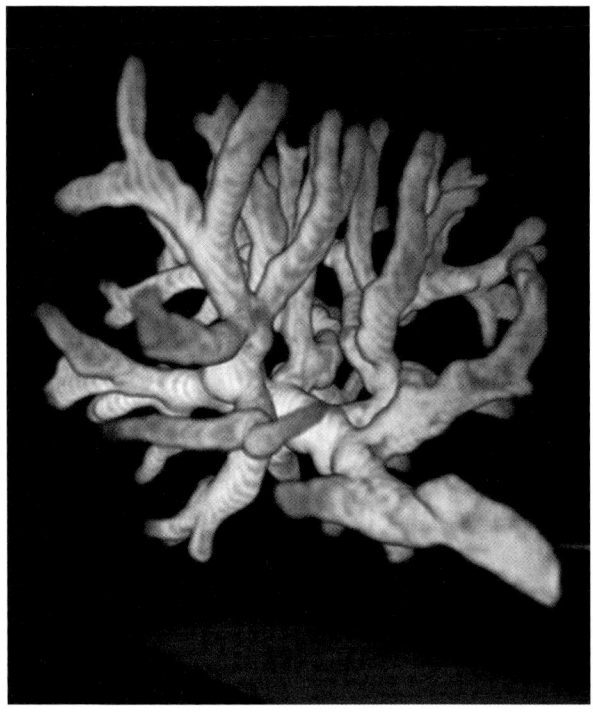

(a) (b)

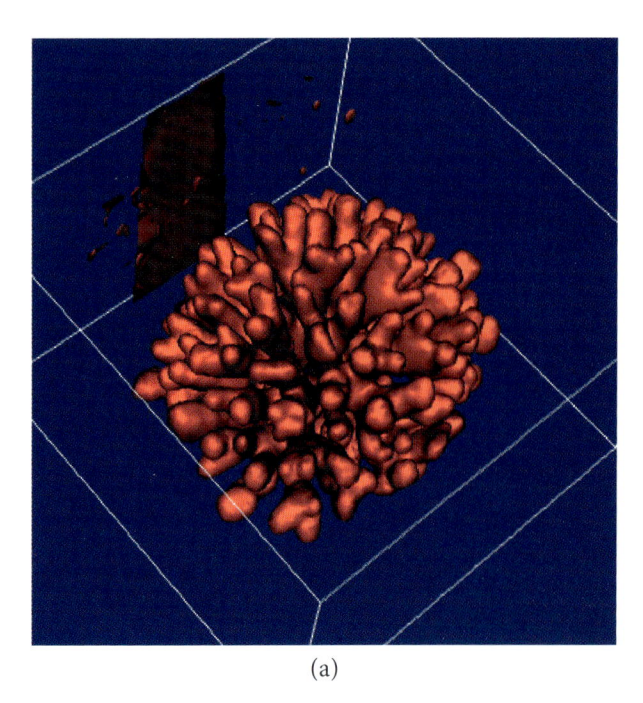

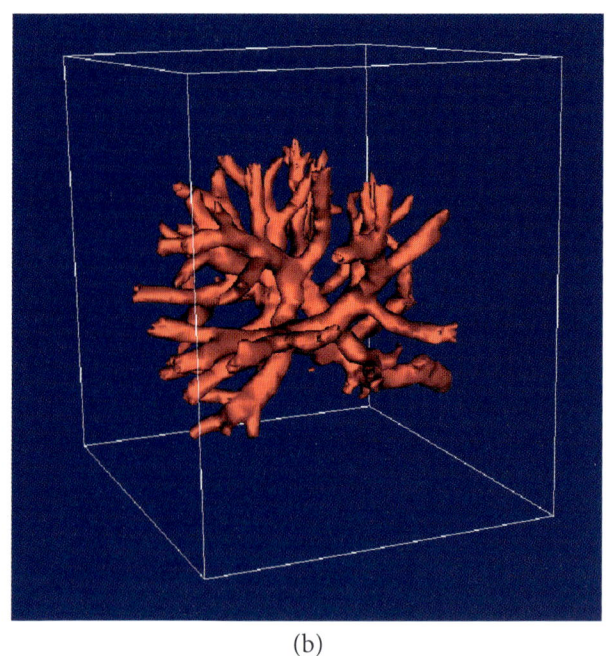

Fig. 3.18a,b. Images of CT scans of *Madracis mirabilis* generated with a surface rendering technique: (a) the growth form of Fig. 2.5c, and (b) the growth form of Fig. 2.5a

sity (the air around the coral skeleton), while high values indicate the calcium carbonate of the coral skeleton. Within the coral skeleton these density values vary (see also Fig. 2.34); in a number of cases, growth layers within the skeleton can be distinguished using these density variations. In Fig. 3.17 two images are shown of CT scans of the *Madracis mirabilis* growth forms shown in Fig. 2.5c and a. In this figure the data sets are visualized using a volume rendering technique (see Upson, 1991). In this figure all density values, including those for some of the surface structures (for example corallites), are visualized. The same technique can be used to visualize density variations in the skeleton, showing growth layers in sections through the data set. In Fig. 3.18 the same data set is displayed with a surface rendering technique (the surface was constructed with the marching cube technique discussed in Lorensen and Cline (1987). The surface is constructed, approximately, at the boundary between air and the calcium carbonate skeleton of the coral. With this technique, using the original data set of $512 \times 512 \times z$ voxels, an image is reconstructed with an equal resolution in x, y, and z directions (see for details Schroeder et al. 1997). In these images only the surface of the coral is visualized, without any surface structures such as corallites. On the voxels representing the surface of the corals a triangulated mesh was constructed using this surface rendering technique. This triangulated surface representation can be used to map the form onto a lattice of 512^3 voxels, with an equal resolution in the x, y, and z directions.

In Fig. 3.19a an example is shown of a simple branching object, which is visualized using a surface rendering technique where the surface is tessellated with a triangular mesh. Fig. 3.19b is obtained by mapping the object shown in Fig. 3.19a onto a 144^3 lattice. The original surface is now converted into a solid and discrete lattice representation. This three-dimensional lattice representation of the branching object corresponds to the two-dimensional discrete images used in Sect. 3.3, where a voxel in the state '1' represents

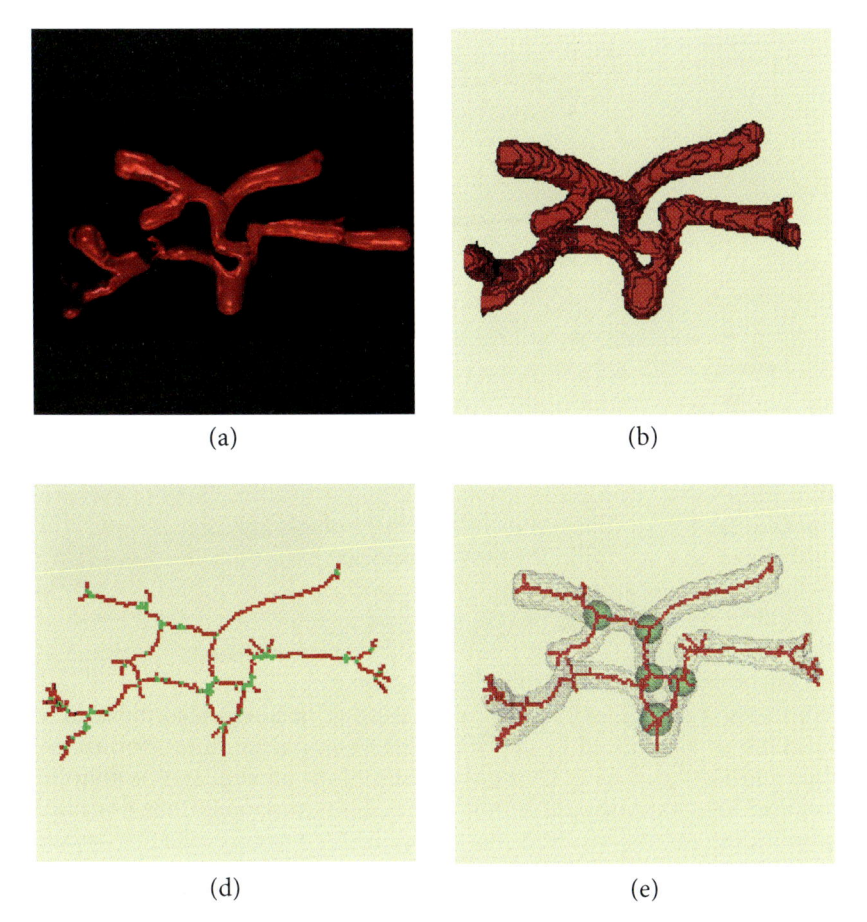

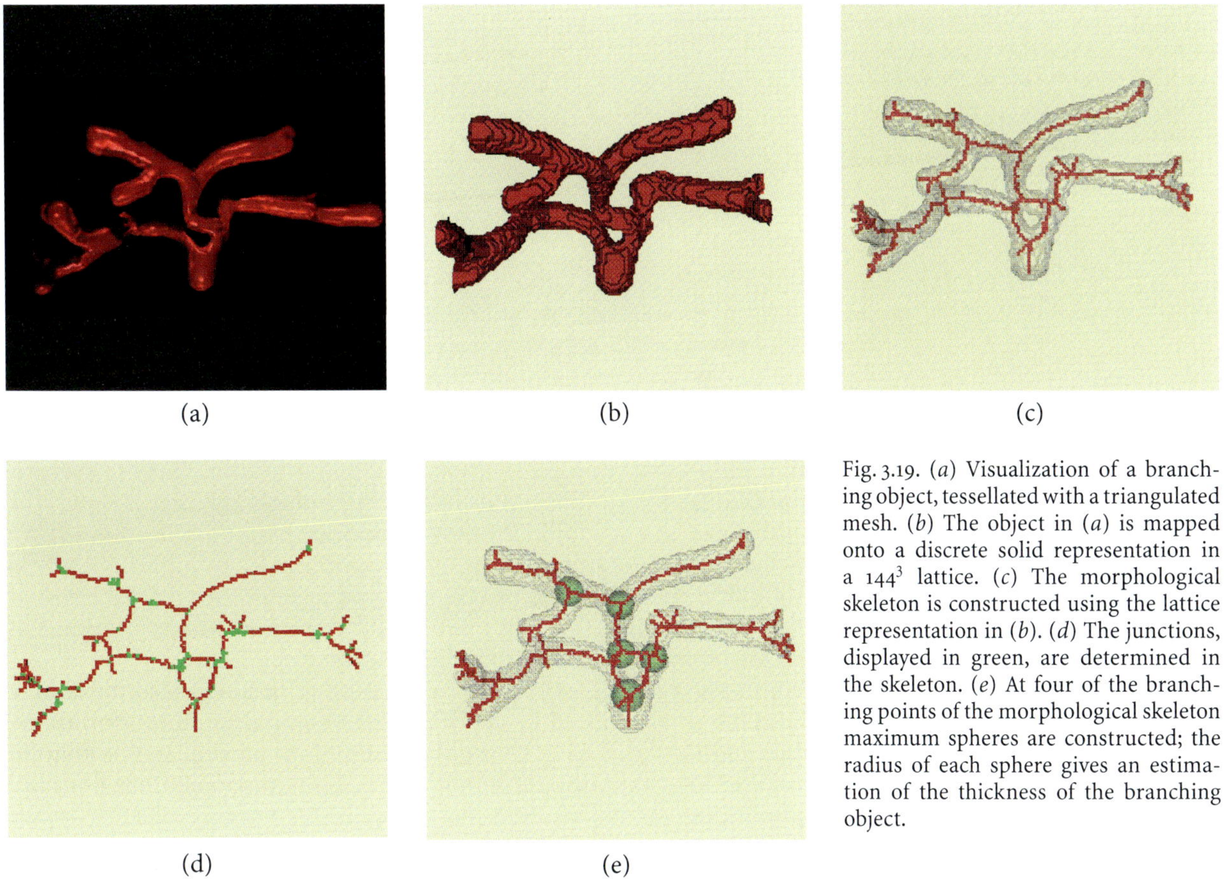

3.4. THREE-DIMENSIONAL MORPHOLOGICAL ANALYSIS OF GROWTH FORMS OF *MADRACIS MIRABILIS*

(a)　　　　　　　　　(b)　　　　　　　　　(c)

(d)　　　　　　　　　(e)

Fig. 3.19. (*a*) Visualization of a branching object, tessellated with a triangulated mesh. (*b*) The object in (*a*) is mapped onto a discrete solid representation in a 144^3 lattice. (*c*) The morphological skeleton is constructed using the lattice representation in (*b*). (*d*) The junctions, displayed in green, are determined in the skeleton. (*e*) At four of the branching points of the morphological skeleton maximum spheres are constructed; the radius of each sphere gives an estimation of the thickness of the branching object.

the object, while a voxel in state 'o' represents the environment. This lattice representation can be used as the input data set for a three-dimensional version of the thinning algorithm applied in Sect. 3.3. An overview of three-dimensional thinning techniques can be found in Jonker and Vossepoel (1995). In theory these techniques construct the medial axis or medial plane in a three-dimensional object. Fig. 3.20 shows a slice made through a lattice

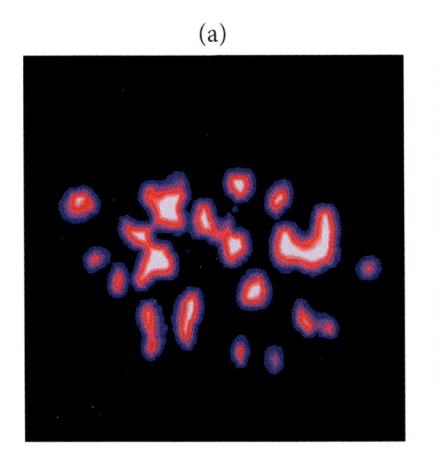

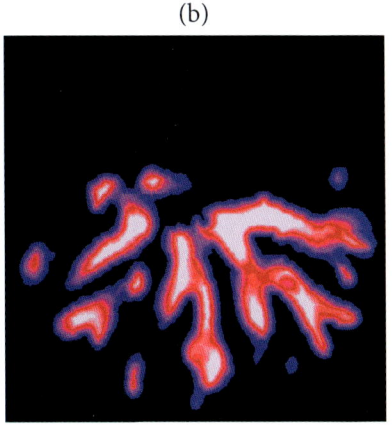

(a)　　　　　　　　　(b)

Fig. 3.20a,b. Slices of the discrete representation, obtained by mapping the surface shown in Fig. 3.18a onto a 512^3 lattice at respectively the level $j = 151$ (*a*) and $j = 251$ (*b*), approximately through the middle of the colony. The color gradient indicates, for every point in the coral, the shortest distance to the environment. Blue indicates the points which are relatively close to the environment, while the white points are more remote from the environment.

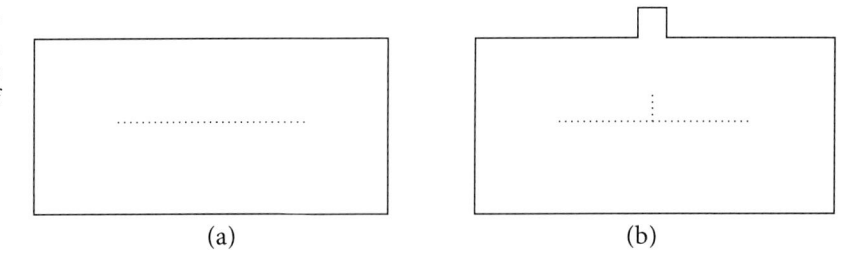

Fig. 3.21. (*a*) Construction of a morphological skeleton in a cylindrical object. (*b*) Formation of a new side branch due to a small perturbation on the surface of the object

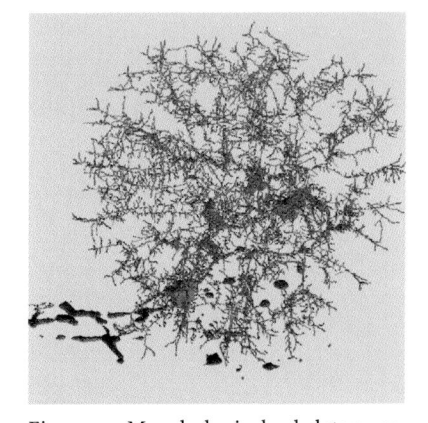

Fig. 3.22. Morphological skeleton resulting from applying the thinning procedure to the three-dimensional lattice representation of the coral shown in Fig. 3.18a

obtained by mapping the surface shown in Fig. 3.18a onto a 512^2 lattice. The color gradient indicates, for every point in the coral, the shortest distance to the environment, measured by constructing a sphere in every voxel in the coral, which is extended until the surface of the sphere reaches the exterior of the coral. The radius of this sphere is used to estimate the shortest distance from every coral voxel to the exterior. The points relatively close to the environment are colored blue, while more remote points are colored white. The medial axis can now, as in the two-dimensional case, be defined as a curve connecting the local maxima in the three-dimensional "distance map". In three dimensions the ridge of local maxima might be arranged along an axis (for example when thinning a cylindrical object) or consist of a plane of maxima (for example when thinning a flattened object).

In Fig. 3.19c the morphological skeleton of the solid object shown in Fig. 3.19b is constructed by applying the thinning algorithm described by Tsao and Fu (1981). As in the measurements done in Sect. 3.3, this skeleton can used to measure several morphological properties, such as the thickness of branches and the determination of branching points, branching angles, branch spacing etc. In the next stage, shown in Fig. 3.19d, the branching points (junctions) are determined in the skeleton. In Fig. 3.19e the result of Fig. 3.19d, is used to construct maximum spheres, where the centers of the spheres are located at the branching points in the morphological skeleton. The radius of these spheres gives an estimation of the maximal thickness of the branches and corresponds to the *da* measure discussed in Sect. 3.3. The thinning algorithm described by Tsao and Fu (1981) gives reasonable results for relatively simple branching objects as shown in Fig. 3.19a. One of the major pitfalls in many of the thinning techniques, for example in the algorithm applied above, is the generation of new branches due to small perturbations at the surface of the object. This phenomenon is demonstrated in Fig. 3.21. In Fig. 3.21a the morphological skeleton consists of one medial axis, and in Fig. 3.21b the addition of a small perturbation results in the formation of a new side branch. Especially for large three-dimensional objects, such as the three-dimensional lattice representation of the object shown in Fig. 3.18a, the result is a highly complex ("bushy") skeleton. The result of the thinning procedure applied to the lattice representation of Fig. 3.18a is shown in Fig. 3.22. For practical measurements this complicated structure is virtually useless. In a forthcoming paper (Vermeij et al. in prep.), we are planning to analyze three-dimensional data sets obtained from CT scans of various *Madracis* species by using an improved version of the thinning technique, producing less "bushy" skeletons which are more suitable for carrying out measurements such as those shown in Fig. 3.19d.

4. Simulating Growth and Form

In this section we will discuss a number of methods which have applied in modeling growth and form of marine sessile organisms. The models have been, more or less, been arranged according to level of abstraction with respect to the actual growth process, starting at the highest level of abstraction. The first section on L-system models, focuses on a method capturing the iterative structure of a modular organism into an algorithmic form, using formal languages. In the second section an alternative method for modeling the iterative structure in seaweeds is presented, based on iterative geometric constructions. In the examples shown of the L-system models and the example using iterative geometric constructions, the level of abstraction is relatively high and no model of the influence of the physical environment is present. In the third section a method, stemming from computational physics, is presented which is very suitable for modeling hydrodynamics in three-dimensional irregular geometries, as frequently found in marine sessile organisms. In the fourth section on Laplacian growth models a two-dimensional model is discussed which is in some ways intermediate between approaches such as L-systems and models which have a representation of the fluid environment. In the fifth section a simple three-dimensional model of the growth process, the aggregation model, is discussed, while the model of the fluid environment is extended to a full three-dimensional model of hydrodynamics, using the results from the second section on the lattice Boltzmann method. In the sixth section, the three-dimensional growth model of marine sessile organisms is further extended to a model of accretive growth, including the influence of light, hydrodynamics, and biological regulation mechanisms. In the last section a model with the relatively lowest level of abstraction is presented, which simulates one specific internal component of the growth process, the fluid transport in the gastrovascular system of a hydrozoan.

4.1 L-systems

4.1.1 Introduction to Modeling Using L-systems

L-systems were introduced by A. Lindenmayer as a mathematical model of multicellular organisms that form linear or branching filaments (Lindenmayer 1968, 1971). The models are inherently *dynamic*, which means that the form of an organism is considered "an event in space-time, and not merely a configuration in space" (D'Arcy Thompson 1917). The whole organism is treated as an assembly of discrete units, called *modules*. Although the nature of the modules is not predefined by the formalism, they often represent individual cells. Each module is characterized by its type and, possibly, one or more numerical parameters (Prusinkiewicz and Lindenmayer 1990), which collectively determine the module's *state*. The development of a structure is described in terms of *rewriting rules* or *productions* that replace a module by zero, one, or more new modules, or change the module's state. Productions are applied in parallel in order to capture the simultaneous development of different parts of the organism.

One of the simplest biologically relevant examples of L-systems is the model of the filamentous bacteria *Anabaena catenula* (Lindenmayer 1987, Lindenmayer and Jürgensen 1992, Prusinkiewicz and Lindenmayer 1990), which formalizes the developmental rules first formulated by Mitchison and Wilcox (1972). The model describes the development of a so-called vegetative segment of *Anabaena* using productions that operate on two types of cells: large cells L and small cells S. Each cell is assumed to have one of two possible polarities, indicated by superscript arrows: $\vec{L}, \overleftarrow{L}$, and $\vec{S}, \overleftarrow{S}$. During the development, cells S elongate and change their state to L, while cells L divide, producing a cell L and a cell S. Taking the polarities into account, this process is captured by the following productions:

$$\vec{L} \longrightarrow \overleftarrow{L}\vec{S} \qquad \overleftarrow{L} \longrightarrow \overleftarrow{S}\vec{L} \qquad \vec{S} \longrightarrow \vec{L} \qquad \overleftarrow{S} \longrightarrow \overleftarrow{L}$$

The development of the filament is simulated as a sequence of stages, with the next stage obtained by applying appropriate productions simultaneously to all cells of the previous stage (Fig. 4.1).

In spite of its simplicity, this example reflects the essential features of the L-system formalism. The model is discrete in three senses: "the state transformations are defined on discrete subunits (cells); each subunit may be present in one of a finite set of states; and the transformations are performed in discrete time steps" (Lindenmayer and Jürgensen 1992). The arrangement of cells in the filament is determined by the pattern of cell division. No mechanism exists to rearrange a set of existing cells, since in algae and plants the cells are tightly cemented together.

The above model implies that the time between the formation and division of a small cell is twice as long as the time between the formation and division of a large cell. In reality, a large cell takes only about 20% longer to divide than a small cell. This behavior can be incorporated into the model by assuming that the cells undergo sequences of state transitions before they divide, as represented by the following L-system (Lindenmayer 1978).

$$\vec{S} \longrightarrow \vec{L} \qquad \vec{L} \longrightarrow \vec{A} \qquad \vec{A} \longrightarrow \vec{B} \qquad \vec{B} \longrightarrow \vec{C} \qquad \vec{C} \longrightarrow \overleftarrow{L}\vec{S}$$
$$\overleftarrow{S} \longrightarrow \overleftarrow{L} \qquad \overleftarrow{L} \longrightarrow \overleftarrow{A} \qquad \overleftarrow{A} \longrightarrow \overleftarrow{B} \qquad \overleftarrow{B} \longrightarrow \overleftarrow{C} \qquad \overleftarrow{C} \longrightarrow \overleftarrow{S}\vec{L}$$

The same model can be described more concisely using a parametric L-system (Prusinkiewicz and Hanan 1990, Prusinkiewicz and Lindenmayer 1990). In that case, modules are associated with numerically-valued parameters, and productions have the form:

predecessor : condition $\longrightarrow$ successor.

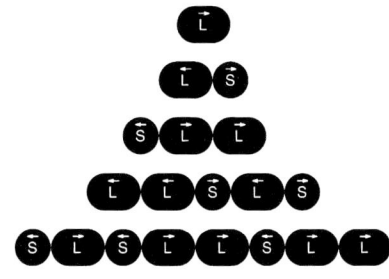

Fig. 4.1. Visualization of the L-system model of *Anabaena catenula*. The model captures the arrangement of shorter and longer cells in a vegetative segment of the filament.

Specifically,

$$\vec{M}(s): s < 5 \longrightarrow \vec{M}(s+1)$$
$$\vec{M}(s): s == 5 \longrightarrow \vec{M}(2)\overleftarrow{M}(1)$$
$$\overleftarrow{M}(s): s < 0 \longrightarrow \overleftarrow{M}(s+1)$$
$$\overleftarrow{M}(s): s == 5 \longrightarrow \overleftarrow{M}(1)\vec{M}(2)$$

The logical expression *condition* determines whether or not the production can be applied to a given module. The non-parametric and parametric L-systems describing the development of *Anabaena* are equivalent due to the one-to-one correspondence between module types in the non-parametric L-systems and parameter values in its parametric counterpart: $S \leftrightarrow 1$, $L \leftrightarrow 2$, $A \leftrightarrow 3$, $B \leftrightarrow 4$, and $C \leftrightarrow 5$, as illustrated in Fig. 4.2a and b.

Both L-systems can be easily modified by changing the number of steps between cell divisions and the polarities of the cells resulting from the division. Lück and Lück (1976) showed (see also Lindenmayer 1978) that some of these patterns describe cell division patterns in green algae. For example, filaments of *Chaetomorpha linum* grown under natural conditions have division rules:

$$\vec{M}(s): s == 5 \longrightarrow \vec{M}(3)\overleftarrow{M}(1)$$
$$\overleftarrow{M}(s): s == 0 \longrightarrow \vec{M}(1)\overleftarrow{M}(3)$$

Under optimal laboratory growth conditions the pattern of cell divisions in *Chaetomorpha linum* changes to:

$$\vec{M}(s): s == 5 \longrightarrow \vec{M}(3)\vec{M}(1)$$
$$\overleftarrow{M}(s): s == 5 \longrightarrow \overleftarrow{M}(1)\overleftarrow{M}(3)$$

These patterns are compared in Fig. 4.2c and d.

An important feature of L-systems is their ability to capture the development of branching structures. Notationally, branches are represented by substrings enclosed in brackets. For example, one of the original models proposed by Lindenmayer to illustrate the concept of L-systems was a model of the development of a red alga *Callithamnion roseum* (Lindenmayer, 1968; see also Fig. 1.2). The model postulates that cells can be in one of nine possible states, denoted by digits from 1 to 9. The development is captured by the following rules (productions):

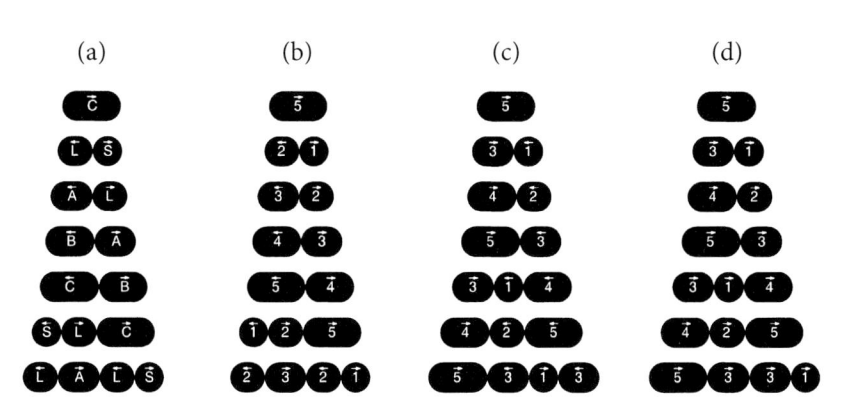

$$1 \longrightarrow 23 \quad 2 \longrightarrow 2 \quad 3 \longrightarrow 24 \quad 4 \longrightarrow 25 \quad 5 \longrightarrow 65$$
$$6 \longrightarrow 7 \quad 7 \longrightarrow 8 \quad 8 \longrightarrow 9[3] \quad 9 \longrightarrow 9 \quad [\longrightarrow [\quad] \longrightarrow]$$

Fig. 4.2a–d. L-system models of filamentous bacteria and algae: (*a*) Improved model of the *Anabeana catenula* development, taking into account timing of the small and large cell divisions. (*b*) Parametric version of the same model. (*c*) Model of *Chaetomorpha linum* developing under normal conditions. (*d*) Model of *Chaetomorpha linum* developing under optimal conditions

Fig. 4.3. Developmental stages w_0, w_4, w_7, w_9, w_{11}, w_{13}, and w_{15} of the model of *Callithamnion roseum*. Reproduced from Prusinkiewicz and Kari (1996)

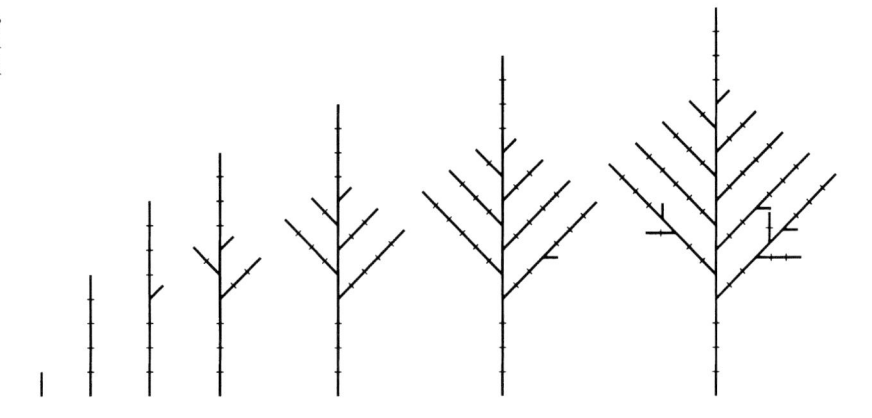

The initial stages of the development of this alga are represented by the following strings of symbols:

$w_0 = 1$
$w_1 = 23$
$w_2 = 224$
$w_3 = 2225$
$w_4 = 22265$
$w_5 = 222765$
$w_6 = 2228765$
$w_7 = 2229[3]8765$
$w_8 = 2229[24]9[3]8765$
$w_9 = 2229[225]9[24]9[3]8765$
$w_{10} = 2229[2265]9[225]9[24]9[3]8765$
$w_{11} = 2229[22765]9[2265]9[225]9[24]9[3]8765$
$w_{12} = 2229[228765]9[22765]9[2265]9[225]9[24]9[3]8765$
$w_{13} = 2229[229[3]8765]9[228765]9[22765]9[2265]9[225]9[24]9[3]8765$

Selected developmental stages obtained by this model are shown in Fig. 4.3.

In the above model it was assumed that all cells have the same size, and branches are issued alternately to the left and right. More explicit control over the size of cells and branching angles is also possible using reserved symbols with a geometric interpretation to specify the appearance of the models (Prusinkiewicz and Lindenmayer 1990).

L-system models of other algae have also been proposed. For example, Fig. 4.4 shows branching patterns of selected genera and species of red algae (see also Fig. 2.8) generated using L-systems presented by Schneider and Walde (1992).

4.1.2 Examples of L-systems for Modeling Seaweed

CLONAL ARCHITECTURE IN SEAWEEDS. Due to modular intrinsic way of growth, algae are constructed by the iteration of modules. This characteristic can be successfully used to create grammatical algorithms to simulate and compute growth based on species-specific rules. Lindenmayer Systems

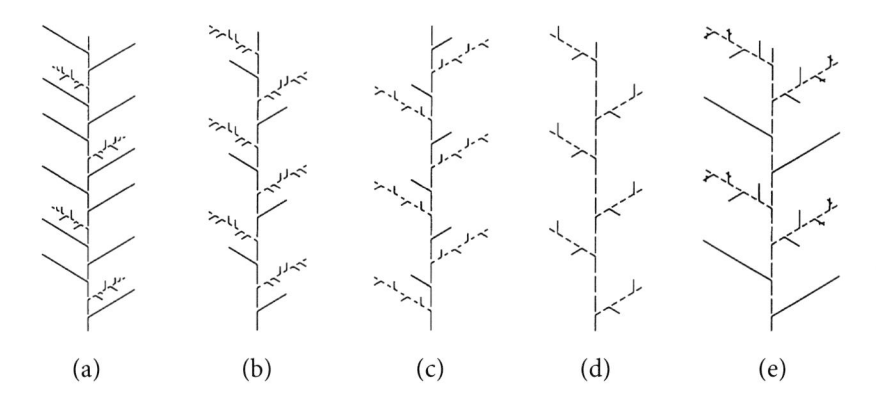

Fig. 4.4a–e. L-system models of selected members of the red algal tribe Polysiphoniae (Rhodomelaceae, Rhodophyta): (*a*) *Herposiphonia*, (*b*) *Dipterosiphonia*, (*c*) *Dipterosiphonia reversa*, (*d*) *Ditria zonaricola*, and (*e*) *Metamorphe*. Simulations show mature intercalary portions of the filaments. Recreated from Schneider and Walde (1992)

have been extensively used to simulate different structures, forms, and developmental models, taking advantage of branching patterns present in the majority of sessile organisms (Prusinkiewicz and Lindenmayer 1990).

Several models of filamentous marine algae using L-Systems have been created. Morelli et al. (1991) were interested in an evolutionary discussion of two similar species of *Dipterosiphonia*. Schneider and Walde (1992) analyzed the evolutionary relationship of two-dimensional branching topologies in some dorsiventral members of the Rhodomelaceae family. Garbary and Corbit (1992) created models of the morphology and development of several marine red algae. Schneider et al. (1994) used L-Systems for the generation of three-dimensional models to evaluate taxonomic relationships between two genera. Collado-Vides et al. (1997) discussed the strategy of growth and space utilization of *Bostrychia radicans*, based on the generation of a model of growth using L-Systems and rules of growth.

In the case of *Bostrychia radicans*, a filamentous branched algae (see Fig. 4.5), a theoretical experiment was done using L-Systems. In Table 4.1 the L-system for generating models of *Bostrychia radicans* is shown; the resulting simulated objects are depicted in Figs. 4.6 and 4.7.

Fig. 4.5. The filamentous red seaweed *Bostrychia radicans*

Angle 12	
Axiom *o*	
o = *fgd*	Unit of construction
d = *g*[+ + +*fgfgrg*][−*gf*]*gfga*	Apical Activity
a = *fgfgfgfg*[+ + +*fgr*][−*gf*]*gfgb*	Construction of lateral stolon
b = *fgfgfgfg*[+ + +*fgr*][−*gf*]*gfgc*	Call for 2nd ramet
c = *fgfgfgfg*[+ + +*fgr*][−*gf*]*gfgk*	Call for 3rd ramet
r = *fg*[+*fgfgu*]*gfgfge*	Construction of ramet axis
e = [−*fgfgx*]*gfgfgh*	2nd branch of 1st order
h = [+*fgfgw*]*gfgfgzfg*	3rd branch of 1st order
k = *fgfgfg*[+ + +*fgr*][−*fga*]*gfgl*	Call for right lateral stolon
l = *fgfgfg*[+ + +*fgr*][−*gf*]*gfgp*	Call for ramet of lateral stolon
p = *fgfgfg*[+ + +*fgr*][−*gf*]*gfgq*	Call for ramet of lateral stolon
q = *fgfgfgt*	Distance between ramets
t = *fgfgfg*[+ + +*fgr*][+*fga*]*gfga*	Call for left lateral stolon
u = [+*fgfgf*]*gfg*	2nd branching of 1st ramet branch
x = [−*fgfgf*]*gfg*	2nd branching of 2nd ramet branch
w = [+*fgfgf*]*gfg*	2nd branching of 3rd ramet branch
z = [−*fgfgf*]*gfg*	4th branch of 1st order

Table 4.1. Program of growth of *Bostrychia radicans* using L-Systems. *f*: draw forward one segment, *g*: move forward one unit, +: increase angle by a fixed amount, −: decrease angle by a fixed amount, [: push the drawing turtle onto a stack,]: pop the drawing turtle off a stack

Fig. 4.6. Module iteration and the architectural unit of construction. By iteration of the program 8 and 20 times the gradual use of space of *Bostrychia radicans* is shown.

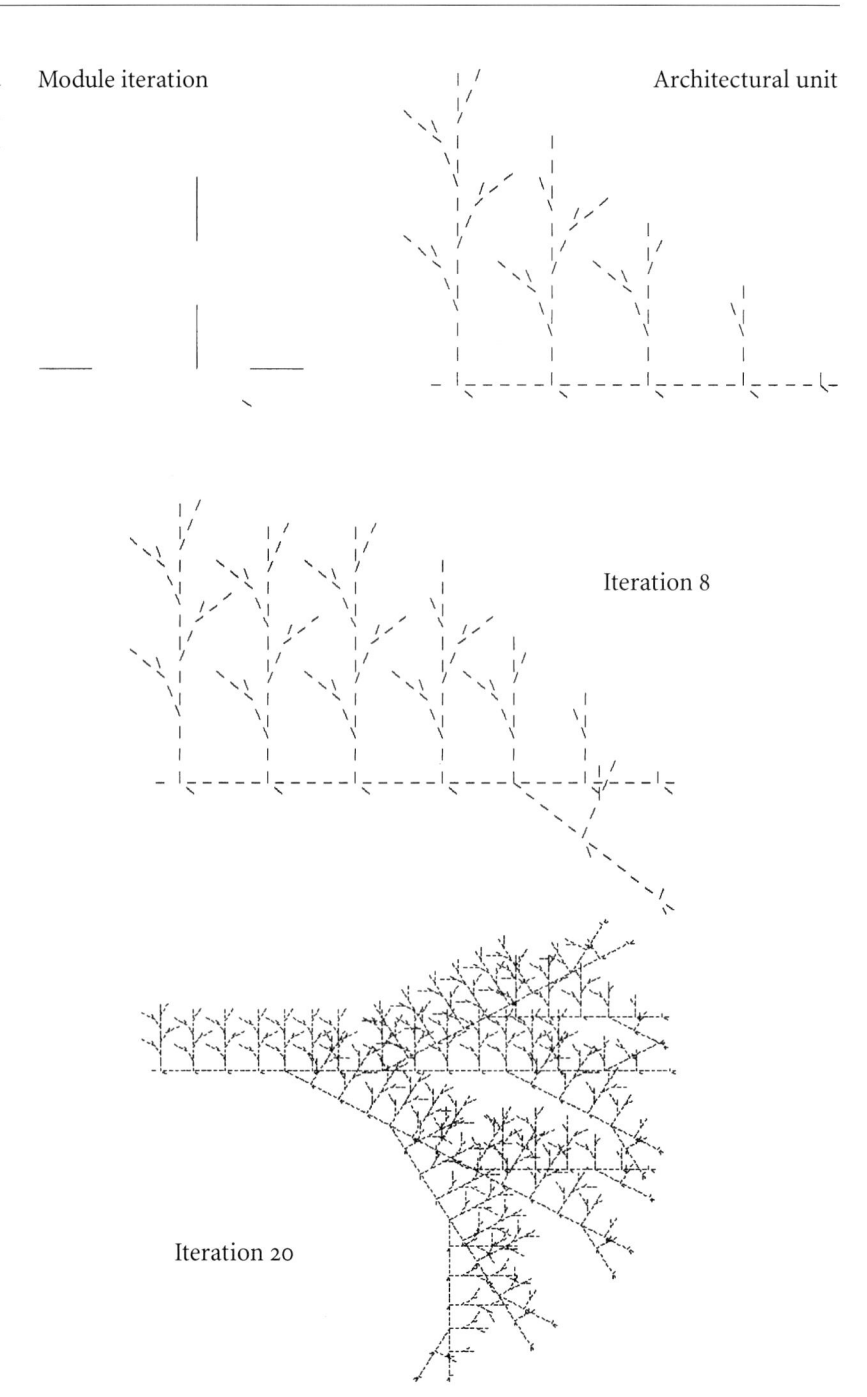

Based on the rules of growth, from the module to the architectural unit, and by the repeated iteration of this unit, it was possible to predict the way in which *Bostrychia radicans*, a modular clonal algae, uses the space. In the absence of neighbors, and in a theoretical homogeneous ground, the growth of this particular algae species is regular and in response to its architectural rules (Fig. 4.6). As the number of iterations is increased (the algae grows) the entire ground is covered filling up any available space

Iteration 40

Iteration 50

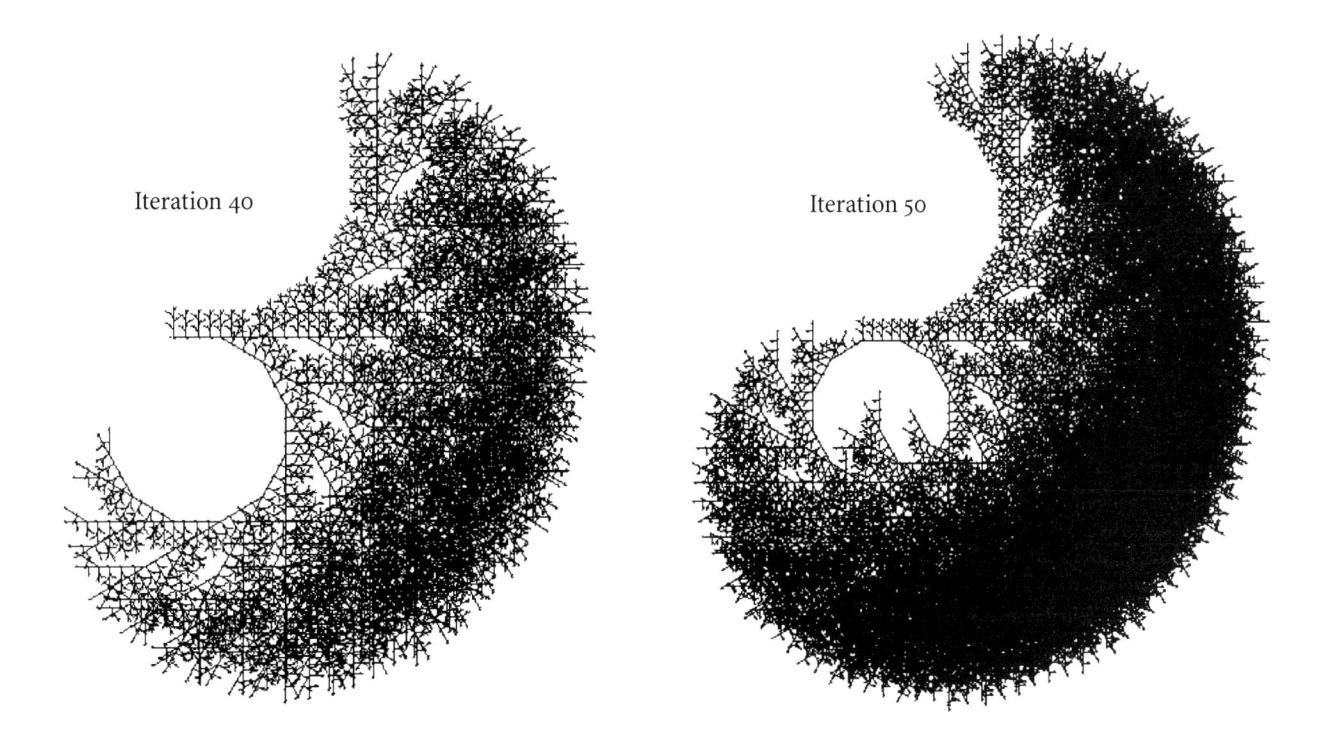

(Fig. 4.7). This way of growth is equivalent to the Phalanx strategy, a compact way of growth, proposed by Lovett Doust (1981). This strategy has several ecological consequences such as an intense use of resources. It offers a close barrier to neighbors increasing interactions at the intra-specific level such as intra-competition for light.

Bostrychia radicans grows in mangrove environments which consist, for the algae, primarily of prop roots and/or pneumatophores as substrata to attach to. These roots function as an island where space is limited, and physical conditions are difficult due to tidal changes and stream water flow conditions. In this context, the model described above can help to explain the domination of *Bostrychia radicans* in mangrove habitats (King and Puttock 1989). The phalanx strategy, or compact way of growth, allows *B. radicans* to use the available space intensively, and a highly branched morphology favors the accumulation of detritus which traps moisture and reduces desiccation stress, hence, such organisms well suited to resist the mangrove habitat conditions.

Fig. 4.7. Intense use of space after 40 and 50 iterations of the program. A phalanx strategy of growth is clearly demonstrated by iteration 50

4.2 Example of a Simple Model of Plasticity in Algal Morphology

The red seaweed *Chondrus crispus* is a foliose, branching form constructed of intertwined microscopic branching filaments. The macroscopic form of each frond is roughly planar and dichotomously branching (Figs. 1.3–6 and 4.8) but there is a great deal of variation in morphology both geographically and within populations. The dichotomous branching can also be observed

Fig. 4.8. The red seaweed *Chondrus crispus*

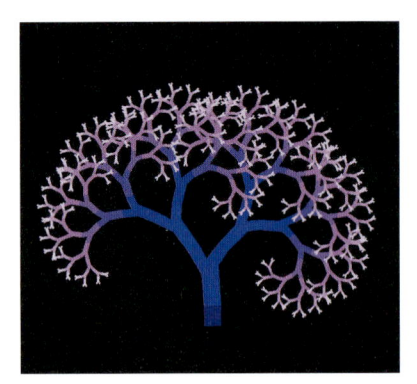

Fig. 4.9. Pythagorous tree model of the red seaweed *Chondrus crispus*

Fig. 4.10. Silhouettes of apices of *Chondrus crispus* grown at two different temperatures

in various other seaweeds, see for example Figs. 1.3b–d and 2.8. Some of the variation in morphology has been shown to be due to plasticity of the growth process in response to changing environmental conditions such as light supply, hydrodynamics, or temperature (Kübler and Dudgeon 1996). Such changes in morphology can compensate for lower nutrient availability, increased metabolic costs related to temperature, or decreased light intensity by increasing the fractal dimension of fronds and thereby the ratio of reactive surface available for exchanges with the environment to the volume of living tissue.

A simple, iterative fractal called the Pythagoras tree was used as the basis of a model of the growth of *Chondrus crispus* in two dimensions. In this species, macroscopic branches represent the continued growth of bunches of microscopic filaments until the next time that a "decision" is made to branch. So a model which introduces entire branches punctuated by branch points, while very simplistic, is reasonable for examining the effects of changes in the timing of that branching decision relative to other growth processes. The model starts with a box, adds the hypotenuse of a right triangle to the top of the box, then smaller boxes to each of the free sides of the triangle, and iterates this pattern as illustrated in Fig. 4.9. Each block in the model represents the pseudoparenchymatous bundle of filaments that make up one macroscopic branch, and the apex of each triangle represents the point where inhibition of growth of one filament divides the branch. Changing the dimensions of the boxes or the ratio of the sides of the triangles results in different overall shapes of the simulated objects after n iterations or after the accumulation of some total surface area. Built-in randomization was added to provide a population of related forms for each set of initial parameters. Data for the rates of linear extension and area-specific growth rates at a range of temperatures were used to set the initial parameters of the model.

The model was tested for two different cases: plasticity due to temperature change in a single population, and geographic variation in morphology between populations. The model produced shorter branches at high temperature and therefore, a branchier, higher fractal dimension structure in a given increment of linear growth. This is in agreement with the results of a growth experiment in which *Chondrus crispus* was grown at two temperatures. Apical segments of the branches are shown in Fig. 4.10, illustrating the increased branching frequency at the higher temperature.

When individuals from two distinct populations were simultaneously grown at two different temperatures, the above pattern held, overall, but the individuals from the northern population better fit a model with more scope for random variation. The comparison of this very simplistic model to actual growth experiments gives some insight into the potential mechanisms of two types of morphological variation in this seaweed. In the first case, plasticity within the lifetime of a simple organism is, in part at least, regulated by a temperature sensing and the next step in identifying the morphogenetic pathway should be focused on diffusive limitation. In the case of the geographic differences in morphology, greater random asymmetry in the branching process for one population relative to the other suggests an instability in the sequence of events that leads to branch formation or the physical integrity of the structure. A suitable hypothesis to follow up this result might be that the number of secondary connections which crosslink the filaments of this pseudoparenchymatous structure might be genetically fixed and different in the two populations. Although it would have been possible to determine if this were the case without using any kind of growth model, looking only at the microscopic internal structures from individuals of both populations, it is much easier to invest the effort once a hypothetical mechanism has been identified. In this case, comparison of actual growth dynamics to a simplistic model allowed us to distinguish between different hypothetical modes of morphogenesis.

4.3 Modeling Fluid Flow Using Lattice Gases and the Lattice Boltzmann Model

As discussed in Sect. 2.1.1 hydrodynamics has a strong impact on the growth process of marine sessile organisms. Traditionally hydrodynamics is modeled using the macroscopical equations, shown in (2.1) and (2.2), describing the conservation of mass and momentum. Solutions to these equations are usually approximated by using numerical solvers (see for example Roache 1976). When developing models of the impact of hydrodynamics on the growth process of marine sessile organisms we are faced with a number of problems, which cannot easily be solved using numerical approximations of the macroscopic equations. Some of the problems we want to address are varying the Reynolds number (Re, see (2.3)) and the Péclet number (Pe, see (2.4)) independently, and studying the influence of turbulence and the contribution of diffusion processes in the dispersion of nutrients. Marine sessile organisms usually have complex-shaped, fractal-like growth forms, which impose a number of problems in the specification of boundary conditions in numerical solvers. Similar problems are encountered in studies on flow and diffusion processes in porous media (Heijs and Lowe 1995). Furthermore, as discussed in Sect. 2.1.1, hydrodynamics influences the distribution of food particles, so it is required to include simulated food particles in the hydrodynamic model. In addition the hydrodynamic impact consists of the influence of hydrodynamic forces, erosion effects, and deposition processes. Another complication is that the simulation of hydrodynamic processes around complex-shaped, three-dimensional objects requires an enormous computational effort, which can be done only with solving techniques which are suitable for large-scale computing on parallel computers. In the next section an alternative method, stemming from the study of porous media, is presented in which the problems mentioned above can be solved to a certain extent. This alternative method is a particle-based technique in which the fluid and suspended material are described by microscopic rules and it seems to be in general very suitable for studying diffusion and flow processes in biology.

4.3.1 Cellular Automata as Models for Fluid Flow

INTRODUCTION. Here we introduce a very specific cellular automata (CA) which, as will become clear later on, can be used as a model of fluid flow. This class of CA is called the Lattice Gas Automata (LGA), and they are described in detail in two recent books (Rothman and Zaleski 1997, Chopard and Droz 1998).

Suppose that the state of a cell is determined by b_m surrounding cells. Usually, only the nearest and next-nearest neighbors are considered. For example, on a square lattice with only nearest-neighbor interactions $b_m = 4$, while if next-nearest neighbors are also included $b_m = 8$; and on a hexagonal lattice with nearest-neighbor interactions $b_m = 6$. Furthermore, suppose that the state of the cell is a vector $\boldsymbol{n} = (n_1, n_2, \ldots, n_b)$ of $b = b_m$ bits. Each element of the state vector is associated with a direction on the CA lattice. For example, in the case of a square grid with only nearest-neighbor interactions we may associate the first element of the state vector with the *north* direction, the second with *east*, the third with *south*, and the fourth with *west*. With these definitions we construct the following CA rule (called the LGA rule), which consists of two sequential steps:

1. Each bit in the state vector is moved in its associated direction (thus in the example, the bit in element 1 is moved to the neighboring cell in the north) and placed in the state vector of the associated neighboring cell, in the same position (so, the bit in element 1 is moved to element 1 of the state vector in the cell in the north direction). In this step each cell is in fact moving bits from its state vector in all directions, and at the same time is receiving bits from all directions, which are stored into the state vector.

2. Following some deterministic or stochastic procedure, the bits in the state vector are reshuffled. For instance, the state vector $(1, 0, 1, 0)$ is changed to $(0, 1, 0, 1)$.

As a refinement, one may also introduce b_r extra bits in the state vector which, as if they were residing on the cell itself, are not moved to another cell in step 1 of the LGA rule. In that case the length of the state vector $b = b_m + b_r$. These b_r residing bits do however participate in the reshuffling step 2.

It is clear that the class of LGA-CA that we have just defined is very large. We have the freedom to choose the CA lattice, the interaction list, the number of residing bits, and the reshuffling rule. Once all this is done, we may expect that the specific LGA-CA that we defined has a very rich dynamic behavior, depending on the initial conditions and the size of the grid. Except maybe for 1-dimensional lattices, a detailed study of the dynamics of such CA is probably not feasible. It was shown by Moore and Nordhal (1997) that the problem of LGA prediction is P-complete, and thus cannot be solved in parallel in polylogarithmic time. This implies that the only solution is a step-by-step explicit simulation. Our new CA therefore seems like a nice toy that may exhibit a very complex dynamic behavior, but no more than that. However, maybe surprisingly, if we associate physical quantities with our CA, enforce physical conservation laws on the bit-reshuffling rule of step 2, and use methods from theoretical physics to study the dynamics, we *are* in fact able to analyze the CA in terms of its *average* behavior, i.e. the average state vector of a cell and the average flow of bits between cells can be calculated.

Even better, it turns out, again within the correct physical picture, that this CA behaves like a real fluid (such as water) and therefore can be used as a model for hydrodynamics. Furthermore, as the LGA rule is intrinsically local (only nearest and next-nearest neighbor interactions) we can construct an inherently parallel model for fluid flow.

ASSOCIATING PHYSICS WITH THE LGA-CA. Our current image of the LGA-CA is that of bits that first move from a cell to a neighboring cell and are then reshuffled into another direction. Now we associate the bits in the state vector with *particles*; a one-bit code for the presence of a particle, and a zero-bit code for the absence of a particle. Assume that all particles are equal and have a mass of 1. Step 1 in the LGA-CA is now interpreted as a streaming of particles from one cell to another. If we also introduce a length scale, i.e. a *distance* between the cells (usually the distance between nearest-neighbors cells is taken as 1), and a time scale, i.e. a *time duration* for the streaming (i.e. step 1 in the LGA-CA rule, usually a time step of 1 is assumed), then we are able to define a *velocity* c_i for each particle in direction i (i.e. the direction associated with the i-th element of the state vector n). Step 1 of the LGA-CA is the streaming of particles with velocity c_i from one cell to a neighboring cell. The residing bits can be viewed as particles with a zero velocity, or rest particles. Now we may imagine, as the particles meet in a cell, that they collide. In this collision the velocity of the particles (i.e. both absolute speed and direction) can be changed. The reshuffling of bits in step 2 of the LGA-CA rule can be interpreted as a collision of particles.

In a real physical collision, mass, momentum, and energy are conserved. Therefore, if we formulate the reshuffling such that these three conservation laws are obeyed, we have constructed a true Lattice Gas Automaton, i.e. a gas of particles that can have a small set of discrete velocities c_i, moving in lock-step over the links of a lattice (space is discretized) and such that all collide with other particles arriving at a lattice point at the same time. In the collisions, particles may be sent in other directions, in such a way that the total mass and momentum in a lattice point is conserved.

We can now associate with each cell of the LGA-CA a density ρ and momentum ρu, with u the velocity of the gas:

$$\rho = \sum_{i=1}^{b} N_i \qquad (4.1)$$

$$\rho u = \sum_{i=1}^{b} c_i N_i \qquad (4.2)$$

where $N_i = \langle n_i \rangle$, i.e. a statistical average of the Boolean variables; N_i should be interpreted as a particle density.

If we now let the LGA evaluate and calculate the density and momentum as defined in (4.1) and (4.2), these quantities behave just as in a real fluid.

In Fig. 4.11a and b we show an example of an LGA simulation of flow around a cylinder. In Fig. 4.11a we show the results of a single iteration of the LGA, so in fact we have assumed that $N_i = n_i$. Clearly, the resulting flow field is very noisy. In order to arrive at smooth flow lines one should calculate $N_i = \langle n_i \rangle$. Because the flow is static, we calculate N_i by averaging the Boolean variables n_i over a large number of LGA iterations or a patch of surrounding cells. The resulting flow velocities are shown in Fig. 4.11b.

Fig. 4.11. (*a*) LGA simulation of flow around a cylinder; the result of a single iteration of the LGA is shown. The arrows are the flow velocities, and the length is proportional to the absolute velocity. The simulations were done with the so-called FHP-III model, on a 32 × 64 lattice, the cylinder has a diameter of 8 lattice spacings, only a 32 × 32 portion of the lattice is shown, and periodic boundary conditions are imposed in all directions. (*b*) As in (*a*), now the velocities are shown after averaging over 1000 LGA iterations

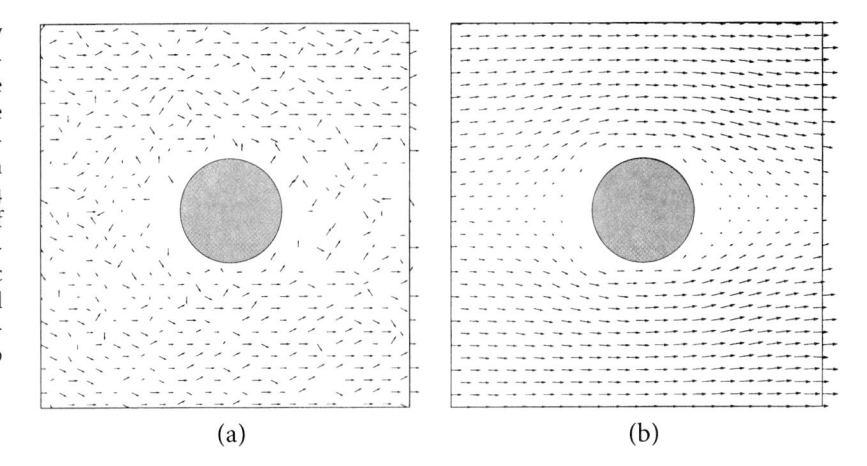

(a) (b)

The evolution of the boolean variables n_i can be expressed as

$$n_i(x + c_i, t+1) - n_i(x, t) = \Delta_i(n(x, t)) \tag{4.3}$$

where x denotes the position of a lattice point and Δ is the collision operator. The collision operator must obey mass, momentum, and energy conservation, i.e.

$$\sum_{i=1}^{b} \Delta_i(n) = 0 \tag{4.4}$$

$$\sum_{i=1}^{b} c_i \Delta_i(n) = 0 \tag{4.5}$$

$$\sum_{i=1}^{b} c_i^2 \Delta_i(n) = 0 \tag{4.6}$$

where $c_i = |c_i|$. One can ask if the evolution (4.3) is also valid for the averaged particle densities N_i. It turns out that this is possible, but only under the Boltzmann molecular chaos assumption which states that particles that collide are not correlated before the collision, or, in equations, that for any number of particles k, $\langle n_1 n_2 \ldots n_k \rangle = \langle n_1 \rangle \langle n_2 \rangle \ldots \langle n_k \rangle$. In that case one can show that $\langle \Delta_i(n) \rangle = \Delta_i(N)$. By averaging (4.3) and applying the molecular chaos assumption we find

$$N_i(x + c_i, t+1) - N_i(x, t) = \Delta_i(N(x, t)) \tag{4.7}$$

A first-order Taylor expansion of $N_i(x + c_i, t+1)$ substituted into (4.7) results in

$$\partial_t N_i(x, t) + \partial_\alpha c_{i\alpha} N_i(x, t) = \Delta_i(N(x, t)) \tag{4.8}$$

Note that the shorthand ∂_t means $\partial/\partial t$, the subscript α denotes the α-component of a D-dimensional vector, where D is the spatial dimension of the LGA lattice, and we assume the Einstein summation convention over repeated Greek indices (e.g. in two dimensions $\partial_\alpha c_{i\alpha} N_i = \partial_x c_{ix} N_i + \partial_y c_{iy} N_i$). Next we sum (4.8) over the index i and apply (4.1), (4.4), and (4.5), thus arriving at $\partial_t \rho + \partial_\alpha (\rho u_\alpha) = 0$, or

$$\frac{\partial \rho}{\partial t} + \nabla \cdot \rho \boldsymbol{u} = 0 \tag{4.9}$$

which is just the equation of continuity that expresses conservation of mass in a fluid. One can also first multiply (4.8) with c_i and then sum over the index i. In that case we arrive at

$$\partial_t \rho u_\alpha + \partial_\beta \Pi_{\alpha\beta} = 0 \tag{4.10}$$

with

$$\Pi_{\alpha\beta} = \sum_{i=1}^{b} c_{i\alpha} c_{i\beta} N_i \tag{4.11}$$

The quantity $\Pi_{\alpha\beta}$ is the momentum density flux tensor and must be interpreted as the flow of the α-component of the momentum into the β-direction. In order to proceed (i.e. express $\Pi_{\alpha\beta}$ in terms of ρ and $\boldsymbol{u}$), one must be able to find expressions for the particle densities N_i. This is a highly technical matter that is described in detail in e.g. Rothman and Zaleski (1997) and Chopard and Droz (1998). The bottom line is that one first calculates the particle densities for a LGA in equilibrium, N_i^o, and then substitutes them into (4.11). This results in an equation that is almost similar to the Euler equation, i.e. the expression of conservation of momentum for an inviscid fluid. Next, one proceeds by taking into account small deviations from equilibrium, resulting in viscous effects. Again, after a very technical and lengthy derivation one is able to derive the particle densities, substitute everything into (4.11) and derive the full expression for the momentum conservation of the LGA, which again very closely resembles the Navier–Stokes equations for an incompressible fluid. The viscosity and sound speed of the LGA are determined by its exact nature (i.e. the lattice, the interaction list and the number of residing particles, and the exact definition of the collision operator).

At first sight the average, macroscopic behavior of the LGA may come as a surprise. The LGA-CA is a model that reduces a real fluid to one that consists of particles with a very limited set of possible velocities, that live on the links of a lattice, and all stream and collide at the same time. Yet, theoretical analysis and a large body of simulation results show that, although the LGA is indeed a very simple model, it certainly is a realistic model of a real fluid. However, it is true that not all LGA behave as a real fluid. The underlying lattice must have enough symmetry such that the resulting macroscopic equations are isotropic, as in a real fluid. For instance, the first LGA, the so-called HPP model, which is defined on a two-dimensional square lattice with only nearest-neighbor interactions and no rest particles, is not isotropic. The FHP models, which have a two-dimensional hexagonal lattice, do possess enough symmetry and their momentum conservation laws have the desired isotropy property.

To end this section we stress once more that the LGA is an intrinsically local CA and therefore gives us an inherently parallel model for fluid flow simulations.

THE FHP MODEL. The FHP model, named after its discoverers Frisch, Hasslacher, and Pomeau, was the first LGA with the correct (isotropic) hydrodynamic behavior. The FHP model is based on a two-dimensional hexagonal

Fig. 4.12a–c. Lattice and update mechanism of the FHP-I LGA. A dot denotes a particle and the arrow its moving direction. In (a) to (c) the propagation and collision phases are shown for some initial configuration

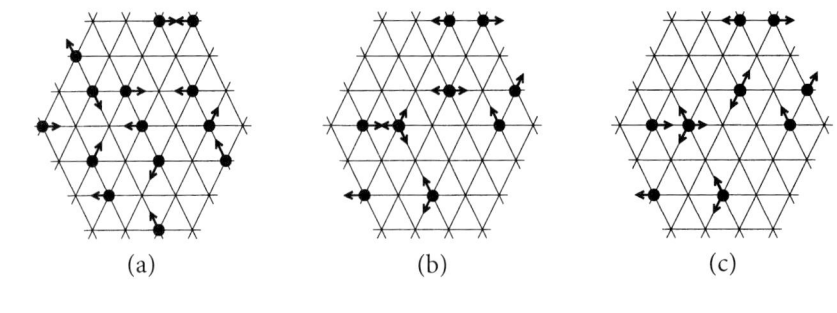

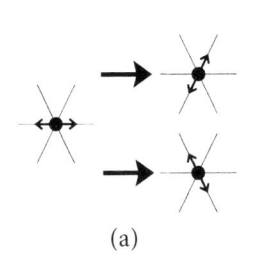

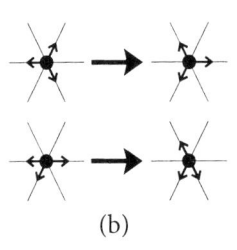

(a) (b) (c)

Fig. 4.13. Collision rules of FHP-I. A dot denotes a particle and the arrow its moving direction. The left figure shows the two-particle collisions, the right figure the three-particle collisions

lattice, as in Fig. 4.12. This figure also shows examples of streaming and collisions of particles in this model. In the FHP-I model, which has no rest particles (i.e. $b_r = 0$ and $b = b_m = 6$), only two-body and three-body collisions are possible, see Fig. 4.13. Note that all these collision configurations can of course be rotated over multiples of 60°.

For this model we can easily write an explicit expression for the collision operator, as

$$\Delta_i(\mathbf{n}) = \Delta_i^{(3)}(\mathbf{n}) + \Delta_i^{(2)}(\mathbf{n}) \tag{4.12}$$

The three-body collision operator is

$$\Delta_i^{(3)}(\mathbf{n}) = n_{i+1} n_{i+3} n_{i+5} \bar{n}_i \bar{n}_{i+2} \bar{n}_{i+4} - n_i n_{i+2} n_{i+4} \bar{n}_{i+1} \bar{n}_{i+3} \bar{n}_{i+5} \tag{4.13}$$

where $\bar{n}_i = 1 - n_i$ and the subscript should be understood as "modulo 6". A similar expression can be obtained for the two-body collisions (see e.g. Rothman and Zaleski 1997). It is clear that the implementation of this LGA, i.e. the evolution equation (4.3) with the FHP-I collision operator (4.12), using bit wise operations, is straightforward and can result in very fast simulations with low memory consumption. Furthermore, the inherent locality of the LGA rule makes parallelization trivial. Next, by averaging the Boolean variables $\mathbf{n}_i$, either in space or in time, one obtains the particles densities N_i and from that, using (4.1) and (4.2), the density and fluid velocity. Many people, especially those who are used to simulating flow patterns based on numerical schemes derived from the Navier–Stokes equations, find it hard to believe that such a simple Boolean scheme is able to produce realistic flow simulations. Yet the LGA, which in a sense originated from the original ideas of von Neumann who invented CA as a possible model for simulating life, is a very powerful and viable model for hydrodynamics.

THE LATTICE BOLTZMANN METHOD. Immediately after the discovery of LGA as a model for hydrodynamics, it was criticized on three points: noisy dynamics, lack of Galilean invariance, and exponential complexity of the collision operator. The noisy dynamics is clearly illustrated in Fig. 4.11a. The lack of Galilean invariance is a somewhat technical matter which results in small differences between the equation for conservation of momentum for LGA and real Navier–Stokes equations; for details see Rothman and Zaleski (1997). Adding more velocities in an LGA leads to increasingly more complex collision operators, exponentially in the number of particles. Therefore, another model, the Lattice Boltzmann Method (LBM), was introduced. This method is reviewed in detail in Chen et al. (1992).

The basic idea is that one should not model the individual particles n_i, but instead the particle densities N_i, i.e. one iterates the Lattice Boltzmann

Equation (4.7). This means that particle densities stream from cell to cell, and particle densities collide. This immediately solves the problem of noisy dynamics. In a strict sense we no longer have a CA with a Boolean state vector, but rather we can view LBM as a generalized CA. By a clever choice of the collision operator, the model becomes isotropic and Galilean invariant, thus solving the second problem of LGA. Actually, a very simple collision operator is introduced, namely the so-called BGK collision operator which models the collisions as a single-time relaxation towards a local equilibrium distribution N_i^o, i.e.

$$\Delta_i^{BGK}(N) = \frac{1}{\tau}(N_i^o - N_i) \qquad (4.14)$$

where τ is the parameter representing the relaxation time. The distributions N_i^o are given functions of ρ and u (see (4.1) and (4.2)) whose expression can be found in Rothman and Zaleski (1997) or in Chopard and Droz (1998). Equations (4.7) and (4.14) together with a definition of the equilibrium distributions result in the Lattice-BGK (L-BGK) model. The L-BGK model leads to correct hydrodynamic behavior. The viscosity of a two-dimensional L-BGK on a hexagonal lattice is given by:

$$\nu = \frac{1}{4}\left(\tau - \frac{1}{2}\right) \qquad (4.15)$$

Note that the limit of zero viscosity (i.e. $\tau \to 1/2$) is numerically unstable.

The L-BGK is also developed for other lattices, e.g. cubic lattices in two- or three-dimensions with nearest and next-nearest neighbor interactions. The LBM, and especially the L-BGK, has found widespread use in simulations of fluid flow.

4.3.2 Transport, Erosion, Deposition, and Hydrodynamic Forces

Transport processes and sedimentation problems in rivers or in coastal environments are important issues for the understanding of the influence of physical environments on the growth and form of marine sessile organisms. For instance, tidal movement is responsible for the large amount of sand that may deposit in places where it is locally screened by an obstacle or by an abrupt change of the ground profile. The prediction of sediment transport in water is also important for understanding the evolution of, for instance, river beds. The creation of meanders is an interesting example of a pattern that is generated in an erosion–deposition process. Similarly, it is well known that the presence of a dike or other human construction may severely affect the profile of a coastal line. Here again, an effective numerical simulation can be a very useful prediction tool.

In this section we model granular material (typically sand) by a multi-particle stochastic cellular automata (Chopard and Droz 1998) in which an arbitrary number of point particles may exist at each lattice site.

In addition to the LBM fluid introduced previously, the second important ingredient in the erosion model is the granular particles. Suspensions are represented by an integer $n(x, t) \geq 0$ indicating how many solid particles are present on site x at time t. Suspensions move on the same lattice as the fluid particles and interact with them.

It is important to remember that in the present mesoscopic approach we do not attempt to represent a specific granular material. Rather, we want to capture the generic features of the erosion-deposition process. The existence of universal behaviors in systems with many interacting particles is common in many areas of science and there are numerous examples where the macroscopic behavior depends very little on the microscopic details of the system. For this reason it is expected that, to first approximation, this dynamics of fictitious particles produces the same deposition patterns as real systems, even if not all the parameters are of the correct order of magnitude.

TRANSPORT RULE FOR SUSPENSIONS. We describe briefly the rule of motion for solid particles. After each time step, the particles jump to a nearest-neighbor site, under the action of the local fluid flow and gravity force. Gravity is taken into account by imposing a falling speed u_{fall} on the particles. Here, suspensions are passive particles since their presence does not modify the flow field, except when they form a solid deposit. However, it would be quite easy to modify the fluid properties so as to make the relaxation time τ vary according to the local density of transported particles since, in real systems, it is observed that the fluid viscosity depends on the local concentration of the suspension.

If the local fluid velocity at site x is $u(x)$, the particles located at that site will move to site $x + \tau_s(u + u_{\text{fall}})$ where τ_s is the time unit associated with the motion of the granular particles. Unfortunately, this new location usually does not coincide with a lattice site. The solution to this problem is then to consider a stochastic motion: each of the $n(x)$ particles jumps to neighbor $x + c_i$ with a probability proportional to the projection of $\tau_s(u + u_{\text{fall}})$ on lattice direction c_i. The quantity τ_s is adjusted so as to maximize the probability of motion, while ensuring that the jumps are always smaller than a lattice constant (see Chopard and Droz 1998, Masselot and Chopard 1998, for more information).

This stochastic cellular automata rule produces a particle motion with the correct average trajectory and a variance which can be interpreted as a local diffusive behavior (Masselot 2000). Note that in this model no specific rule is needed to split the transport among creeping, saltation, and suspension, as is usually done in traditional numerical models.

DEPOSITION RULE. The next aspect of the particle dynamics is the deposition rule. Under the effect of gravity, the particles keep moving downward until they land on a solid site (e.g. the bottom of the system or the top of the deposition layer). On such deposition sites, when motion is no longer possible, particles start piling up. In our model, up to N_{thres} particles can accumulate on a given site (N_{thres} gives a way to specify the space scale of the granular particles with respect to the fluid system). When this limit is reached, the site solidifies and new incoming particles pile up on the site directly above. The solid sites formed in this way represent obstacles over which the fluid particles bounce back from where they came. Thus, this solidification process implies a dynamically changing boundary condition for the fluid.

Note that before solidification, the fluid is not affected by the presence of the rest particles piling up on top of a solid site. Also, these rest particles are no longer subject to the suspension transport rule. Only the erosion mechanism discussed below can move them away.

In case of particles with high cohesion, such as snow under some conditions, one can keep piling particles without worrying about a possible

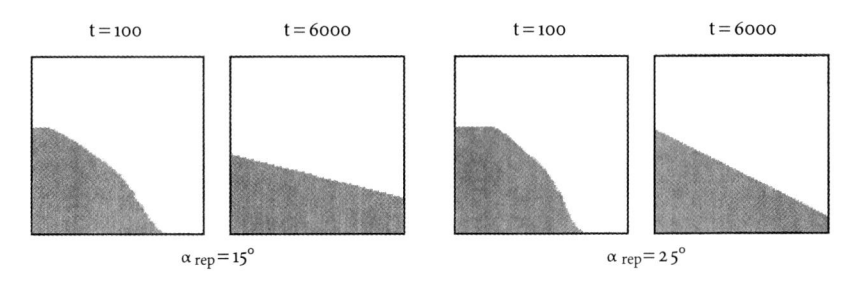

Fig. 4.14. Toppling evolution to a stable configuration for two different angles of repose α_{rep}, as produced by the deposition rule. The variable t indicates the number of iterations

toppling. However, for dry sand for instance, some toppling mechanism should be added. The rule we consider is the following: when a lattice site contains more than δN deposited particles with respect to its left or right neighbors (in 2D), toppling occurs. During this process, all unstable sites send half of their excess of grains to the less occupied neighbors. With this rule, the stable configuration may not be reached after one time step.

The quantities δN and N_{thres} give a simple way to adjust the angle of repose of the pile. Since, in the stable state, the model tolerates a maximum difference of δN particles between two adjacent sites, two solidified sites are separated by k sites, where $k = N_{\text{thres}}/\delta N$ and the angle of repose α_{rep} satisfies $\tan \alpha_{\text{rep}} = 1/k$. Fig. 4.14 illustrates the effect of changing α_{rep} on two toppling simulations.

EROSION. Erosion is a complicated phenomenon and many different explanations can be found in the literature. Here, the mechanism we propose is quite simple: with probability p_{erosion}, each of the first N_{thres} particles belonging to the top of the deposition layer is ejected vertically (i.e. it gets a vertical velocity). Usually, these candidates for erosion are distributed on both a solid site and the rest particles that have accumulated directly above it.

If the local fluid flow is fast enough, the particle will be picked up and moved further away due to the transport rule. Otherwise, if the flow is slow, the resulting motion will be to land on the same site where the particle started.

This rule captures the important effect that a strong flow will result in an important erosion process. It also implements naturally the idea that erosion starts only if the local speed is larger than some threshold. One could also make the probability parameter p_{erosion} depend on the number of particles already in suspension, as is often suggested in phenomenological models. However, in several cases this does not turn out to be necessary and so p_{erosion} is a constant that is modified only from one simulation to the next, when representing different qualities of the suspensions.

HYDRODYNAMIC FORCES. At the surface of the solid–fluid interface the bounce-back boundary condition is used for the fluid particles. Here incoming particle densities, denoted by n_i, are simply reversed. Consequently, the force acting on a surface point is equal to the local momentum change ($dp = -2 \times n_i \times c_i$, where, again, c_i is the velocity along link i). The total force acting on a solid obstacle is obtained by summing the local forces over the complete surface of the obstacle. Notice that if the solid obstacle is moving in the fluid, e.g. suspension flows, the force calculation procedure should be modified such that the momentum transfer due to the motion of the obstacle is taken into account (Ladd 1994).

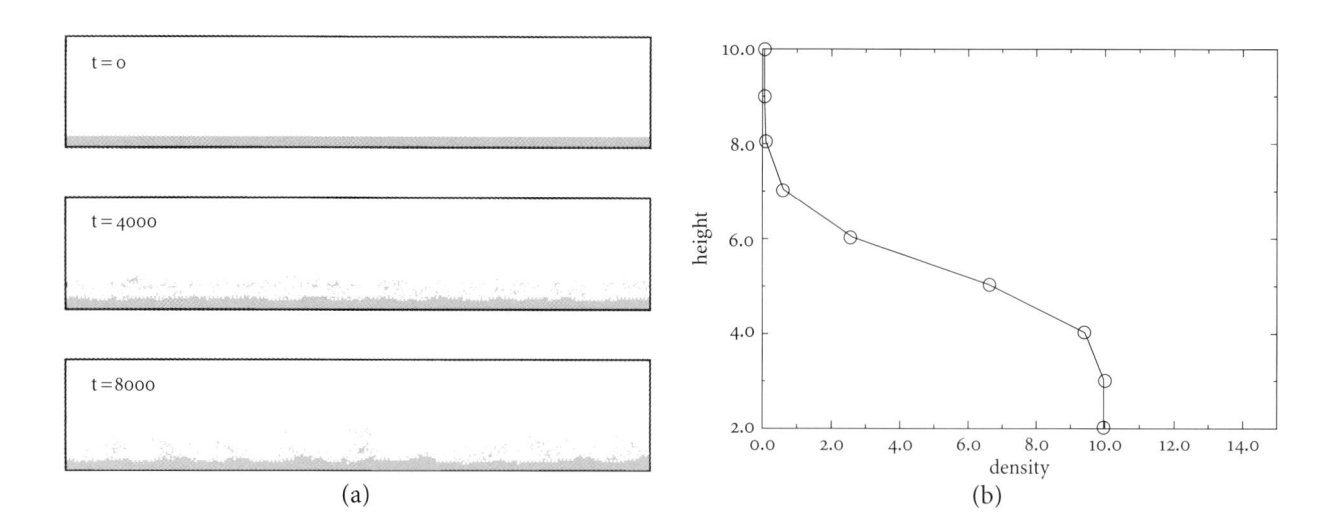

Fig. 4.15a,b. Simulation of erosion of a flat bed of particles. On the left, we show three snapshots of the system at different times and, on the right, the vertical density profile of suspensions in the stationary regime. The initial bed height is 5 sites and N_{thres} is 10.

4.3.3 Transport and Sedimentation

Now we describe in more detail the case of an erosion–deposition process produced by a fluid streaming over a bed of particles. Fig. 4.15 shows several snapshots of the transport process, as well as the vertical distribution of eroded grains.

Under appropriate flow conditions the evolution of the bed results in a surface instability called ripples (Cornish 1914). Ripples are similar to dunes but on a much smaller scale and appear spontaneously when the wind blows fast enough over sand or fresh snow beds, or underwater, as shown in Fig. 4.16.

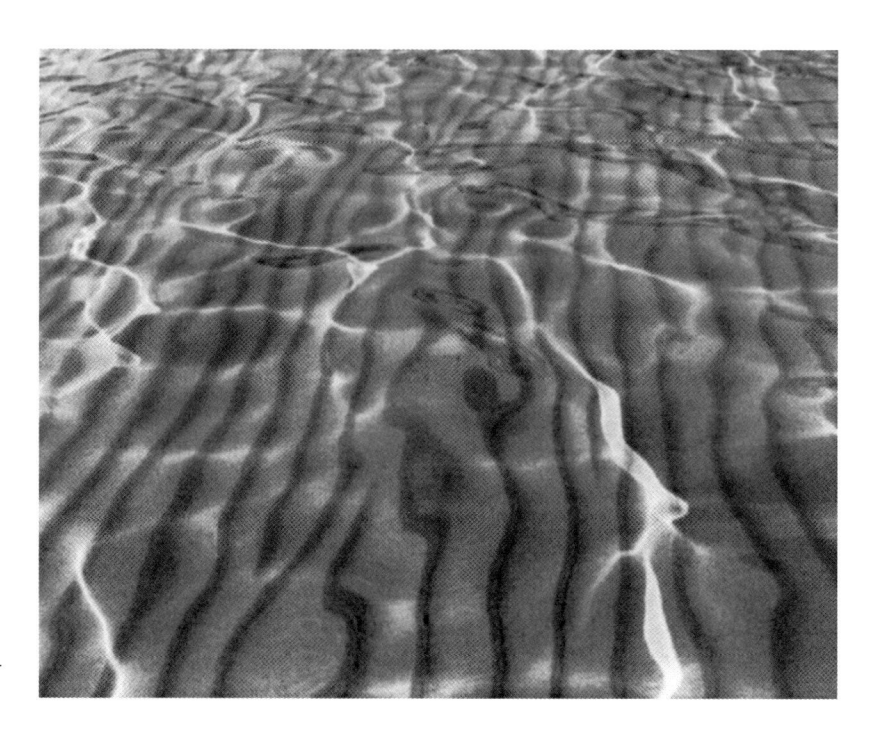

Fig. 4.16. Example of sand ripples forming under water

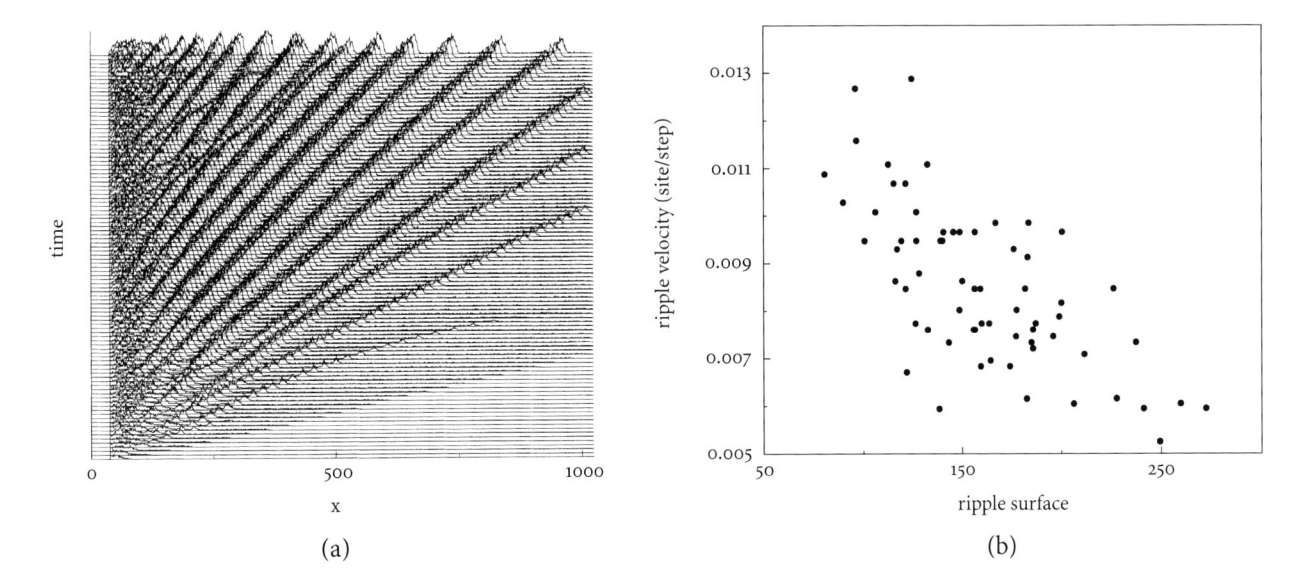

The formation process of ripples is still a controversial topic among the experts of this field. Several models have been proposed to explain how ripples form (see for example Anderson and Bunnas 1993). The interest of the LBM-CA approach compared with many others is that it simulates the complete process (fluid flow plus grains), without including any ad hoc hypotheses. The exact same dynamics produces realistic ripples, as well as realistic large-scale deposits, simply by modifying the boundary conditions and parameters such as p_{erosion} and N_{thres}.

Several properties of ripples are known (Cornish 1914): they form if the wind is fast enough and then move slowly in the wind direction. Small ripples move faster than the larger ones. This fact is confirmed by the present model, as shown in Fig. 4.17 where the ripple profile is plotted for different time steps. Note that in the ripple simulations no toppling rule is used, which has certainly an impact on the ripple shape and possibly on other quantitative features.

Fig. 4.17a,b. *Left*: Simulated evolution of an initially flat bed of particles under the action of a fluid flowing from left to right. The horizontal axis represents the spatial extension of the bed and the vertical axis corresponds to time. From this representation we observe the ripples and their motion which depends on the ripple size. *Right*: measured ripple velocity as a function of ripple size. When a fast ripple collides with a larger but slower one, the two coalesce.

4.4 A Laplacian Model of Branching Growth

4.4.1 Laplacian Growth

Branching marine invertebrates feed off particles suspended in the water. Provided the currents are not too strong, it is advantageous for a filter-feeding organism to grow away from boundaries into the open flow, as the higher water movement means they are able to catch more particles. It is possible that they are able to use the flow itself to organize their growth form, in much the same way as a terrestrial plant is able to use light to control the growth of its shoots. In order to illustrate how this may work, a model is presented of branching growth in response to the gradient in a diffused quantity. The model presented here is a proof of concept, demonstrating that a realistic branched form may develop through the iteration of locally applied rules, in response to an environmental gradient. The diffused quantity can be thought of as representing the time-averaged kinetic energy of the flow, which increases away from the boundaries. In order to further simplify the representation of the physical environment the model is restricted to a two-dimensional surface, with coordinates x and z, where z represents height above the substrate. For convenience we wrap the x coordinate around on itself, so the computational domain is a vertical tube. A diffusing substance ϕ is supplied by setting $\phi = 1$ along the upper edge of the tube and $\phi = 0$ along the lower edge, which represents the substrate. Into this environment a model organism is introduced. It is attached to the lower boundary and allowed to grow following rules which are discussed below. The diffusion of ϕ is governed by the differential equation

$$\frac{\partial \phi}{\partial t} = D\nabla^2 \phi \qquad (4.16)$$

where D is the diffusion coefficient and $\nabla^2 = \partial^2/\partial x^2 + \partial^2/\partial z^2$ is the Laplacian operator. The diffusion is much faster than the growth process, and so to a close approximation the concentration at any time satisfies the Laplace equation

$$\nabla^2 \phi = 0 \qquad (4.17)$$

There are many physical realizations of Laplacian growth processes. A classic example is the discharge of a spark from a high voltage conductor. The electric potential around the spark satisfies the Laplace equation. The growth of the spark is proportional to the potential gradient. This is highest at the tips of the spark, and so the tips propagate fastest. The electric potential guides tips away from one another, leading to sparks which have a self-avoiding branched form. The analogy between sparks and the growth of marine organisms was also made by Pettigrew (1908), as was shown in Fig. 1.6. In this section the analogy is clarified, showing how branched structures, similar to those of marine organisms, may develop in response to a Laplacian field.

4.4.2 The Numerical Model

A concentration Phi(i,j) is defined on a grid of points, with indices i,j=0,...,n (see Fig. 4.18). The periodic boundary conditions are set by requiring that Phi(0,j)=Phi(n,j), for all j, and the top and bottom boundaries of the domain are fixed to be Phi(i,n)=1, Phi(i,0)=0, for all i. At each iteration all the points which are within a distance BranchRadius of the model organism are identified as internal to the organism and Phi is held to zero there. The Laplace equation is then solved using a suitable numerical scheme such as successive overrelaxation (see for example Press et al. 1988). Initially Phi is just a linear gradient between the top and bottom, but as the organism grows gradients develop around it. These gradients are used to guide the growth, in the way described below.

It is assumed that the organism has a defined structure consisting of linear branch elements and dichotomous branch vertices. These are defined by a list of x, y points which take values between 0 and n, but are not restricted to be integers. This architecture is applicable to a wide variety of organisms, an example being the sponge *Raspailia inaequalis* whose growth was discussed in Sect. 2.2.2. In the model, growth is assumed to occur only through the addition of branch elements to the end of the existing branches, the length

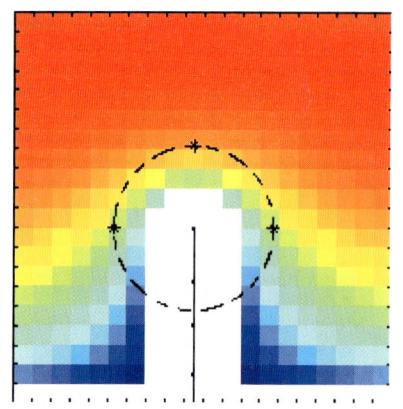

Fig. 4.18. The Laplacian model after 4 iterations. The central line and dots mark the skeletal elements of the organism. The colored pixels show the values of Phi(i,j), warmer colors being higher values, with Phi = 0 along the substrate and within a distance of BranchRadius from the skeleton. The region used to calculate the gradient near the tip is shown by the dashed circle, of radius GradRadius. The stars on the circle mark the points used to determine the values TipPhi, LeftSidePhi, and RightSidePhi. A ratio of these values is used to decide whether the tip should branch, as described in the text.

and orientation of the elements being determined by the gradients of `Phi`. For many organisms, thickening growth occurs all along the branches. This type of response is not modeled here as it is intended to represent only the branching pattern.

At each iteration the concentration gradient is calculated in the vicinity of each tip. The gradient of `Phi` at a point `i,j` is `GradPhi= (Phi(i+1,j)-Phi(i-1,j), Phi(i,j+1)-Phi(i,j-1))/2`. The gradient in the vicinity of a branch tip is then calculated by taking the vector average of `GradPhi` over all the non-zero points which are within a distance `GradRadius` of the tip.

Before new skeletal elements are added a decision is made whether to branch each tip. This is decided by calculating three concentrations at a distance `GradRadius` from the tip, using a linear interpolation of the gridded data. The first value `TipPhi` is in the direction of the gradient vector; the two others `LeftSidePhi` and `RightSidePhi` are at right-angles to this direction. The ratio `Branch = min(LeftSidePhi,RightSidePhi)/TipPhi` is used to decide whether or not branching should proceed. As a single tip grows further away from the substrate the gradients at the tip increase and the contours of `Phi` become pushed down around the sides of the branch, increasing the ratio `Branch`. If `Branch` becomes larger than a value `BranchThreshold` then the tip splits into two. Because `Branch` depends on the ratio of two `Phi` values it is independent of the magnitude of `Phi`. Instead, it gives geometric information on how far a given tip is poking away from its neighbors.

If the tip has not branched then the new skeletal element lies in the direction defined by `GradPhi`. If the tip has just split into two then the direction of growth of the two tips are 60 degrees to either side of the direction of increasing gradient. This is needed to separate the two tips so that they grow away from one another.

Two versions of the model are considered. In the first version of the model, the growth rate is set to be the same across all tips and constant with time. In the second, the growth rate of each tip is set to be proportional to the magnitude of the gradient vector `GradPhi`. The gradient is highest near tips which poke clear of their neighbors and they then grow faster, leading to an escalation of the growth rate. In order to control this, the growth rates are scaled by a factor which is the same for all tips. This factor is chosen to keep the maximum growth rate below an upper bound.

The length of each new skeletal element is set to be proportional to the growth rate multiplied by a random factor. This factor is chosen for each tip from a distribution which has a mean of 1 and, in the model runs presented here, a standard deviation of 10%. The inclusion of the small random factor is necessary to break the determinism in the model and so to allow a branching form to develop which varies, in detail, from simulation to simulation.

The two simulations presented have been run on a grid with $n = 600$. The radius of the branches is `BranchRadius` = 2.4 pixels; the gradient near each tip is calculated within a distance `GradRadius` = 4.2 pixels of each tip; the maximum growth rate is 2.5 pixels per iteration; and the organism begins as a single tip at $x = 300$, $y = 0$. With these values the choice `BranchThreshold` = 0.5 gives a suitable branch density, allowing resolution of `Phi` in between the branches.

Fig. 4.19a–d. A sequence of snapshots from the Laplacian growth model, with the tip-growth rate proportional to the concentration gradient. The colors indicate the concentration, `Phi`, which increases towards the top. The white pixels were within a distance of `BranchRadius` pixel units from the organism's skeleton, and have been set to zero. This is a 400 × 400 pixel subscene from the full simulation.

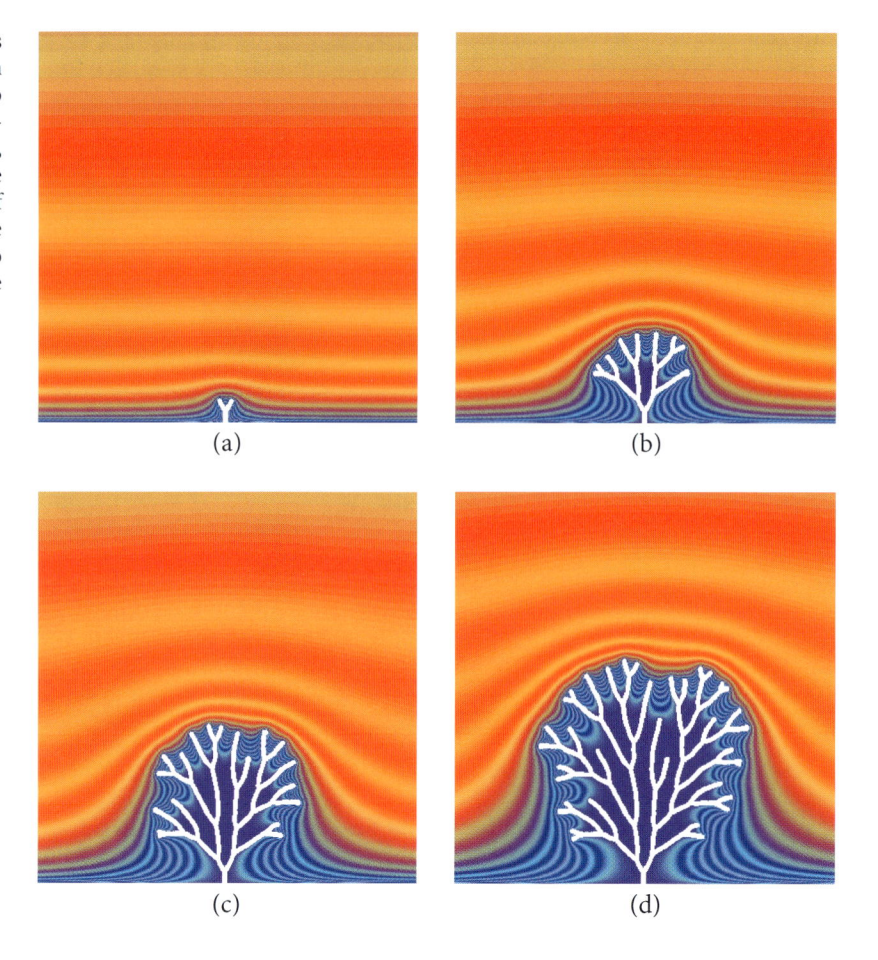

(a) (b)

(c) (d)

4.4.3 Model Results

The Laplacian model produces forms which have an indeterminate dichotomous branching pattern, with branches that are evenly spaced (Figs. 4.19 and 4.20). If the growth rate of each tip is proportional to the concentration gradient then the longer tips grow faster, suppressing the growth of the tips which are in the interior. This results in a branching pattern with a shrubby appearance. If instead the tip growth rate is uniform across the organism then the resulting branching pattern is very different, with a well-organized fan-like arrangement. These two model runs show how strikingly different branching patterns may be achieved simply by changing the growth response of the tips. Horton analysis shows that the sponge *Raspailia inaequalis* has a growth form which is more fan-like than shrubby, with branching ratios close to those seen in Fig. 4.21b. This does not imply that the model correctly represents the growth process of this species, but more likely is an effect of the constraints on branching pattern which are imposed by radiative fan-like growth.

The model presented in this section is in some ways intermediate between approaches such as L-systems, which have a highly structured biological growth form, and models of growth as an aggregative process, which have a much more complex representation of the fluid environment. While

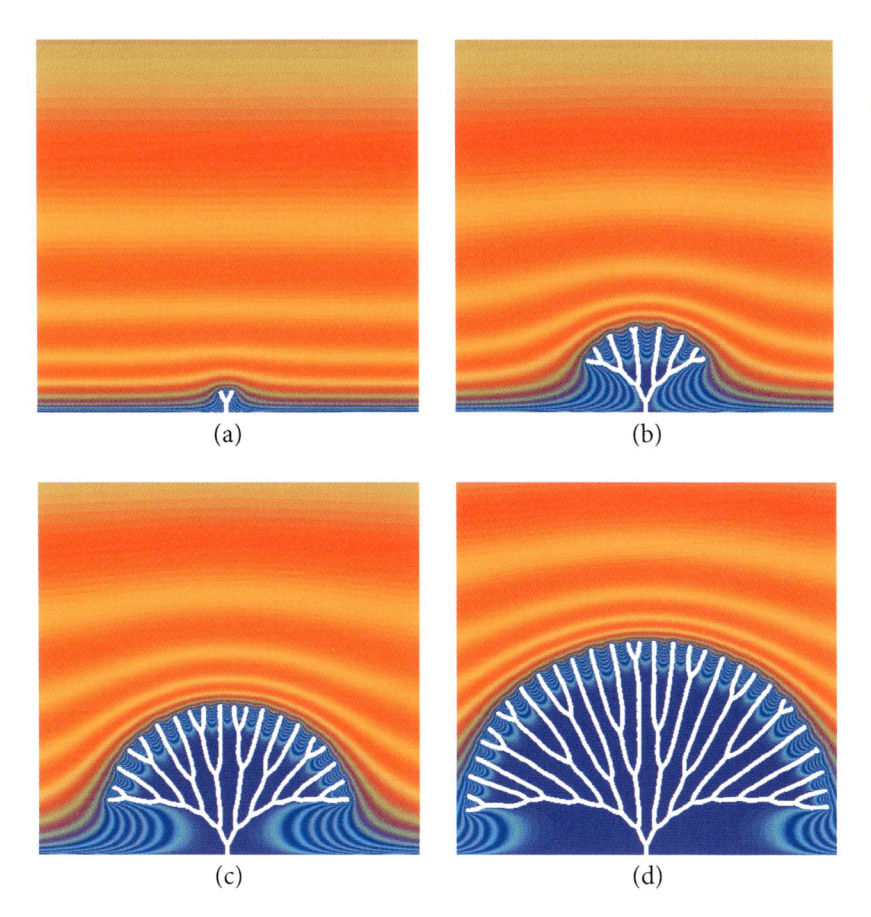

Fig. 4.20a–d. The same model as in Fig. 4.19, but with the tip-growth rate being constant. Changing this rule produces a growth form which has a very different branching pattern.

providing a useful method for generating artificial indeterminate growth forms against which to compare the branching patterns of real organisms, a more detailed understanding of the growth process and a more realistic representation of the interaction between the flow and the organism would be required to make models of this kind able to satisfactorily represent actual growth.

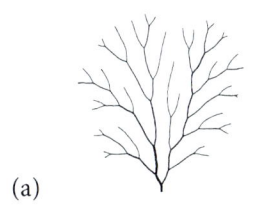

(a)

Order	Branch number	Branch length
1	39	178
2	14	265
3	4	306
4	1	90

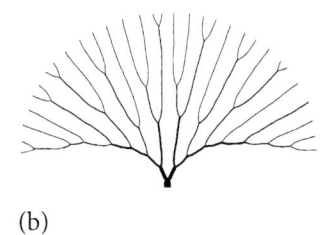

(b)

Order	Branch number	Branch length
1	32	228
2	12	267
3	4	190
4	2	27
5	1	7

Fig. 4.21a,b. Horton analysis of branching patterns of the modeled organisms, after 100 time steps. The width of the branches is drawn proportional to their Horton order and the tables summarize the Horton data: (a) If the tip-growth rate is proportional to the gradient (Fig. 4.19) then the branching has a shrubby appearance. This is reflected in the Horton ratios, which are $R_n = 3.4$ and $R_l = 1.2$. (b) If the tip-growth rate is independent of the concentration (Fig. 4.20) then the branching is fan-like, with Horton ratios $R_n = 2.4$ and $R_l = 0.3$.

4.5 Growth by Aggregation

4.5.1 Morphological Plasticity and the Influence of Hydrodynamics

In Sect. 3.3 the morphological plasticity of some marine sessile organisms, for example the sponge *Haliclona oculata*, the hydrozoan *Millepora alcicornis*, and the stony coral *Pocillopora damicornis*, were related to the impact of hydrodynamics. There is a strong impact of hydrodynamics on the growth process. In a number of cases it is possible to arrange growth forms of sponges, hydrocorals, and stony corals along a gradient of the amount of water movement, as shown in Figs. 2.16 and 1.1. In the case of the stony coral, the growth form gradually transforms from a compact shape under exposed conditions, to a thin-branching one under sheltered conditions. In growth forms of the sponge *Haliclona oculata*, plate-like, more compact shapes emerge at exposed sites. This shape gradually transforms into a thin-branching form when the exposure to water movement decreases. A similar trend can be observed in growth forms of the hydrozoan *Millepora alcicornis* (de Weerdt 1981). In this species the shape changes from plate-like forms at shallow and exposed sites, to thin-branching forms at deeper and sheltered locations.

Water flow has a strong influence on the local supply of food particles in suspension feeders. In Sect. 2.1.1 the effects of hydrodynamics on particle capture by suspension-feeding invertebrates was discussed in detail. In several studies (Frechette et al. 1989, Buss and Jackson 1981, Pile et al. 1997) it has been demonstrated that locally around a sessile suspension feeder, exposed to various flow velocities, areas may occur which are depleted from food particles. It has been found by several authors that there exists an asymmetric type of food capture. Patterson (1984) demonstrated that the largest amount of food capture is found at the upstream side of the octocoral colony for low flow velocities, while the situation reverses for higher flow velocities where the largest amount of food is captured at the downstream side of the colony. A similar phenomenon was observed by Sebens et al. (1997) for the stony coral *Madracis mirabilis* (see Fig. 2.5), where the largest amount of food particle capture was found at the upstream side of the colony for low flow velocities (below $15 \, cm \cdot s^{-1}$), while the situation again reverses for higher flow velocities (see Fig. 2.6). In studies on branching stony corals (Sebens et al. 1997, Chamberlain and Graus 1977) and the influence of hydrodynamics it is demonstrated that branch spacing is a crucial morphological property which determines the microflow pattern inside the colony. The flow direction will basically reverse twice a day due to the tidal movements (see also Sect. 2.1.1) and in branching sessile suspension feeders growth forms emerge which have a roughly radial symmetry or show a tendency to develop a flattened growth form (see also Sect. 2.2.3 on octocorals, Johnson and Sebens 1993, Stearn and Riding 1973, Wainwright and Dillon 1969). In the last case most branches will develop in a plane perpendicular to the governing flow direction. Local food particle absorption patterns which are determined by the direction and the velocity of the flow, local areas which surround the organism and which are depleted from food particles, as well as microflow patterns in branching organisms have important consequences for the growth process and the resulting morphology.

To obtain insight into the influence of hydrodynamics on the growth process of sessile suspension feeders, a morphological simulation model was developed. In the next sections (Sections 4.5.3 and 4.5.4) we will first discuss a very simple model of the growth process of accretive organisms. The growth process is modeled by an aggregation process, where growth is represented by the addition of particles in a cubic lattice. In this model we use an extension of the diffusion limited aggregation (DLA) model introduced by Witten and Sander (1981); see also Fig. 1.10 for an example of the DLA model. In Sect. 4.6 on accretive growth we will extend these ideas further in a model of surface normal deposition processes, in which a model of the influence of hydrodynamics is included.

In the absence of flow, the distribution of nutrients around the growth form can be modeled as a diffusion process in a steady state: there is a source of suspended material and the organism continuously consumes nutrients from its environment. In general in a marine environment, there will be a significant contribution of the hydrodynamics to the dispersion pattern of the suspended material around the growth form. In this case the distribution of nutrients around the organism will be determined by a combination of flow and diffusion. The contribution of flow to the nutrient distribution can be quantified by the Péclet number shown in (2.4). In the diffusion-limited case *Pe* is very small (nearly zero), while in the flow-dominated case *Pe* is large. The effect of hydrodynamics on the morphology of the aggregates is studied by varying the Péclet number in the simulations.

4.5.2 Modeling the Nutrient Distribution

The method which we have applied to model the nutrient distributions is the lattice Boltzmann method, which is combined with a tracer step. The underlying method is discussed in Sect. 4.3.1, more detailed accounts of this method can be found elsewhere (Frisch et al. 1987, Chen et al. 1992, Ladd 1994, Chopard and Droz 1998).

In the simulations a cubic lattice is used consisting of 144^3 nodes. Each node is connected with 18 other nodes. There are 6 links with the length of 1 lattice unit and 12 links with the length of $\sqrt{2}$ units. All parameters and variables in the lattice Boltzmann method are expressed in lattice units. The mean populations of particles move simultaneously from one node to one of the 18 neighbors. The evolution of the lattice is described by the dynamical rule shown in (4.7) in which the N_is represent the particle densities at the node. The momentums of the particles are changed by adding an external force, the driving force F, to the system. One update step of the lattice now involves a propagation step where N_i particles travel from node x to node $x + c_i$ and a collision step in which the post-collision distribution is computed.

Two types of boundary conditions are used: at the borders of the lattice periodic boundary conditions are applied, while in the nodes adjacent to the nodes representing the obstacle, solid boundary conditions are used. Periodic boundary conditions can be implemented by exchanging the N_is of the links at the borders of the lattice. Solid boundary conditions can be represented by exchanging the N_is between the adjacent node and a neighboring fluid node.

After the lattice Boltzmann iteration, a tracer step is applied where populations of tracer particles are released from source nodes and are absorbed by the sink nodes: the nodes adjacent to the growth form. The tracer particles can travel from a node at site x in the lattice to one of the 18 adjacent nodes $x + c_i$, where the Pe number in (2.4) determines if flow or diffusion dominates. When flow dominates, most particles will move in the direction of the governing flow, while under diffusion dominated conditions the number of particles which travel in each of the 18 directions will be approximately equal. In the simulations the diffusion coefficient D varies, and $\bar{u}$ is kept constant by adjusting the driving force F of the system. Due to the growth of the aggregate the velocity in the free fluid would gradually decrease if the driving force were not adjusted.

4.5.3 Growth by Aggregation in a Monodirectional Flow

In the aggregation process the growth process is modeled as in the Diffusion Limited Aggregation model (Witten and Sander 1981). In Fig. 4.22 the basic construction of the aggregate is shown. The cluster is initialized with a "seed" positioned at the bottom plane of the lattice. In both the cluster and substrate sites solid boundary conditions are applied. The flow in the lattice is directed from the yz-plane at $x = 0$ to the yz-plane at $x = xmax$. The flow pattern about the obstacle (substrate and cluster) is determined in the lattice Boltzmann iteration, using 10 iteration steps followed by a tracer step. In each growth step 10 iteration steps are used to ensure that an equilibrium is obtained in the flow pattern. In the last step tracer particles are released from the source plane, the lattice sites located at the xz-plane at $y = ymax$. The tracer particles are absorbed by the fluid nodes adjacent to obstacle nodes, which can be nodes in the substrate plane (the xz-plane at $y = 1$) and the aggregate nodes. In this growth model it is assumed that both the tracer distribution and flow velocities are in equilibrium and the growth velocity of the aggregate is much slower than the dispersion of the tracer. In the sink nodes the number of absorbed tracer particles is determined and a new node is added to the

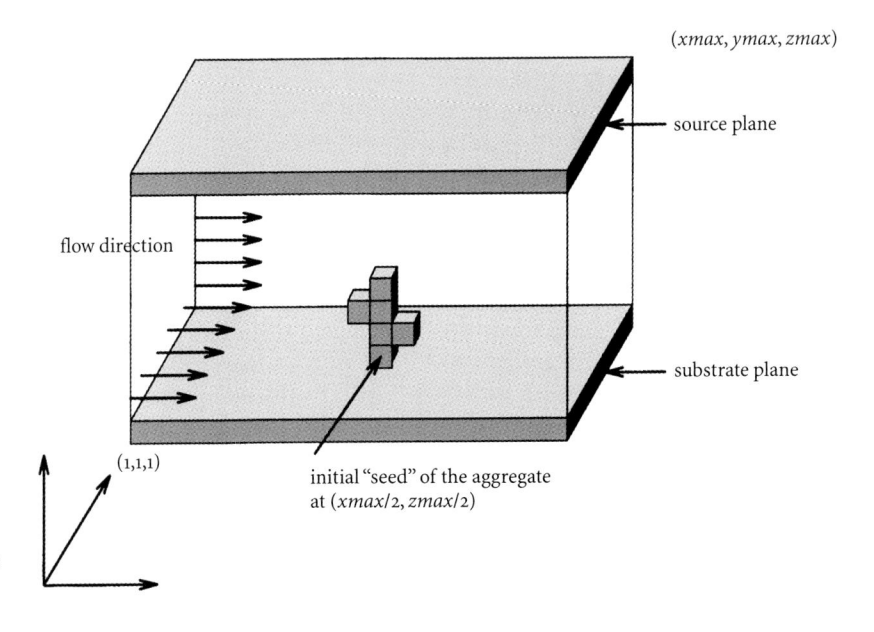

Fig. 4.22. Basic construction of the aggregate

(a)

(b)

(c)

(d)

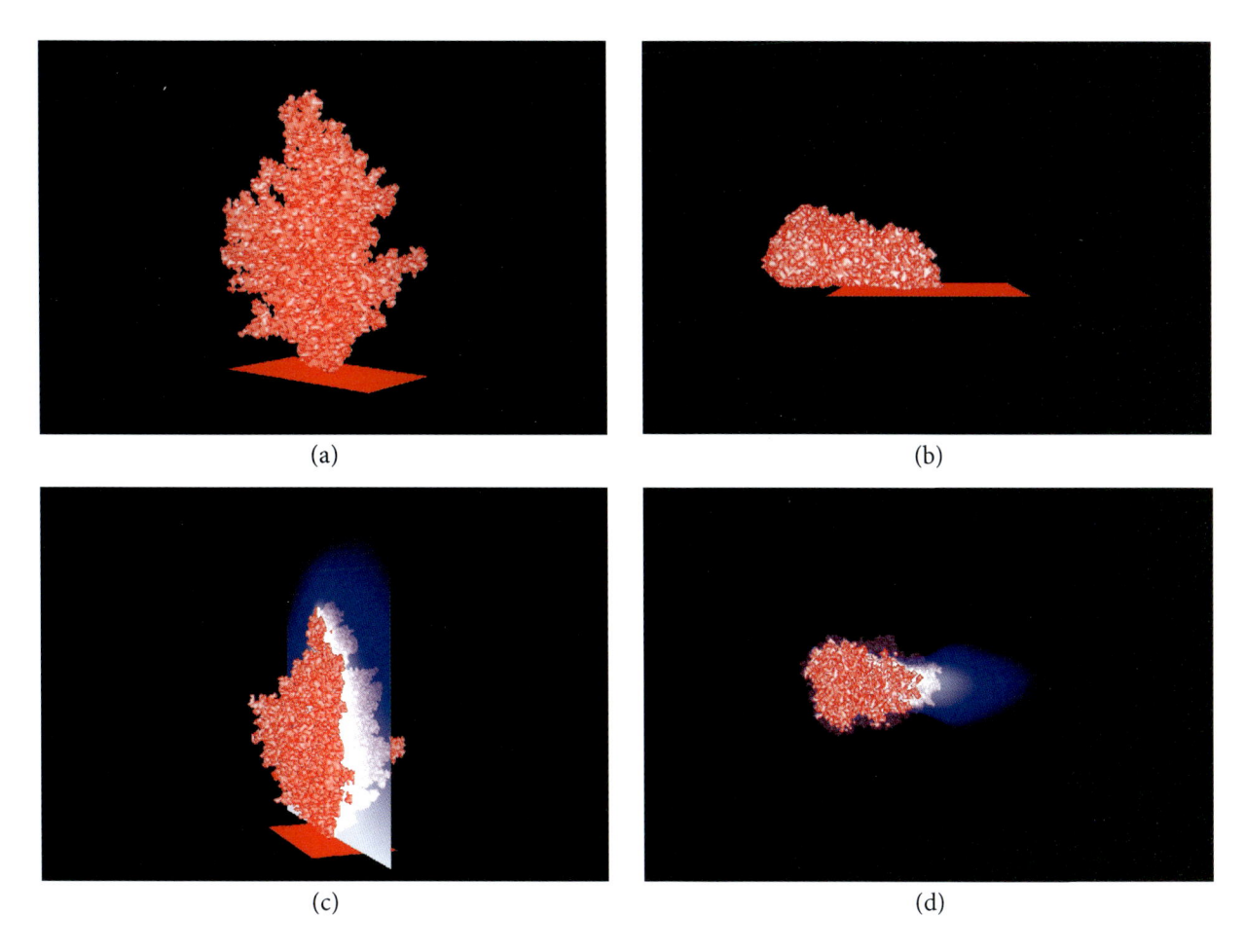

aggregate. The probability p that k, an element from the set of open circles ○ (the adjacent sink nodes), will be added to the set of black circles (the aggregate nodes) is given by:

$$p(k \in \circ \to k \in \bullet) = \frac{(a_k)}{\sum_{j \in \circ}(a_j)} \quad (4.18)$$

where a_k is the absorbed number of tracer particles at position k.

In Fig. 4.23a a 3D aggregate is shown resulting from a simulation in which Pe in (2.4) was set to the value 0.015. The aggregate in the figure is visualized by constructing a surface over the aggregate sites using the marching cubes method. In Fig. 4.24 a series of slices is shown through the middle of the aggregates for a range of increasing Pe numbers. In this picture it can be observed that the aggregates gradually change from a thin-branching form in Fig. 4.24a to a more compact shape in Fig. 4.24h, for increasing influence of hydrodynamics. Furthermore it can be seen that the aggregates tend to grow in the upstream direction. This trend becomes stronger for higher Pe numbers. In Fig. 4.23c the nutrient distribution in a section made in the xy-plane, through the middle of the lattice, is shown. The color shift blue to white indicates a decrease in tracer particle concentration. The shape of the basins of equal nutrient ranges is displayed by coloring the adjacent basins black. In Fig. 4.23d a similar section in the xz-plane is shown through the

Fig. 4.23a–d. Aggregates resulting from simulations where Pe is set, respectively, to the values 0.015 (a) and 3.000 (b) in the monodirectional flow experiment. (c) The nutrient distribution in the xy-plane around the aggregate shown in (a). (d) The nutrient distribution in the xz-plane around the aggregate shown in (b)

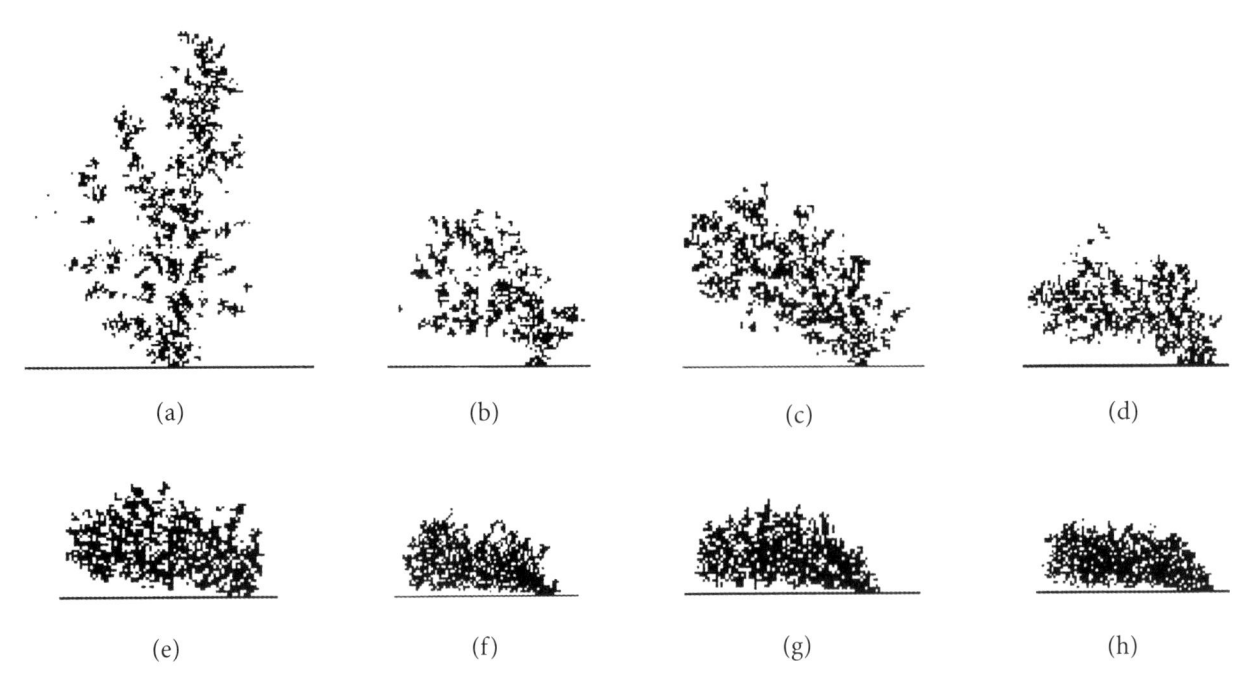

Fig. 4.24a–h. Slices through the middle of the aggregates in the xy-plane at $zmax/2$, monodirectional flow experiment: (a–h), Péclet number increases from approximately 0.0 to 3.0.

nutrient distribution and aggregate; in this experiment Pe is set to the value 3.000. In this picture the aggregate is seen from above.

The results of the simulation experiments are summarized in Table 4.2. We have estimated error bounds on the various measurements by repeating the simulations at $Pe = 0.0150$ and $Pe = 3.000$ four times. The ratio R of the total sink nodes to the total cluster size indicates the compactness of the aggregates. The typical cluster size is about 6.4×10^4. The fractal dimension D_{box} of the surface of the aggregate was determined using a three-dimensional version of the (cube) box-counting method described by Feder (1988) (compare the two-dimensional version in (3.10)). D_{box} measures the space-filling properties of the surface of the aggregate. In three dimensions its value varies from a minimum of 2, for a solid object with a perfectly smooth surface, to a maximum of 3 for a solid with a space-filling surface. In the simulations we have determined the number of sink nodes that have a nutrient absorption between a and $a + \Delta a$ for the aggregates obtained at various Péclet numbers. The number of sink nodes with absorption rate a, $N(a)$, can be related to a by a power law:

$$N(a) \sim a^{-D_{\text{abs}}} \tag{4.19}$$

Table 4.2. Ratio R of the total sink nodes to the total cluster size, the 3D fractal dimension D_{box}, the average absorption $\bar{a}$, the measure D_{abs} of the uniformity of the nutrient distribution, and the average x, y, z coordinate of the aggregate in the mono directional flow experiment for a series of increasing Pe numbers

Pe	0.0150	0.1322	0.2521	0.4918	1.0000	2.0000	2.5000	3.0000
R	2.35 ± 0.10	2.21	1.88	1.64	1.29	1.20	1.31	1.27 ± 0.06
D_{box}	2.27 ± 0.02	2.20	2.27	2.20	2.16	2.10	2.09	2.05 ± 0.04
$\bar{a}$	0.04 ± 0.01	0.85	1.75	3.2	5.5	11.0	13	15 ± 5
D_{abs}	1.7 ± 0.2	1.0	1.0	0.9	0.8	0.7	0.8	0.7 ± 0.1
$\bar{x}$	0.47 ± 0.03	0.38	0.33	0.32	0.32	0.24	0.24	0.24 ± 0.01
$\bar{y}$	0.51 ± 0.04	0.39	0.34	0.33	0.30	0.19	0.18	0.16 ± 0.01
$\bar{z}$	0.48 ± 0.02	0.53	0.50	0.49	0.50	0.49	0.5	0.49 ± 0.02

The value of D_{abs} can be estimated from a log–log plot. The exponent D_{abs} can be interpreted as a measure of the uniformity of the nutrient distribution. In Table 4.2 the average absorption $\bar{a}$ in the boundary nodes, and the values of D_{abs}, are listed for the various Pe numbers. Furthermore, for each cluster the average x, y, z coordinate (the center of gravity) was measured.

4.5.4 Growth by Aggregation in a Bidirectional (Alternating) Flow

In the simulation model discussed in the previous section some important simplifications have been made. One simplification is that it is assumed that the growth form develops under mono-directional flow conditions. As a consequence an asymmetric form develops, as shown in Fig. 4.24, where the aggregates tend to grow in the upstream direction. This trend becomes stronger for higher Pe numbers. In reality, the flow direction will basically reverse twice a day due to the tidal movements. In this section a growth process in a bidirectional flow is studied by using an aggregation model in which the flow direction is reversed after each growth step. The nutrient distribution is computed in a similar way to that discussed in the previous section.

The basic construction of the aggregate in (alternating) bidirectional flow is shown in Fig. 4.25. The cluster is initialized with a "seed" positioned at the bottom plane of the lattice. The bottom plane, "the substrate", is positioned in the xz-plane at $y = 1$, while the seed is one lattice site, located at the position ($xmax/2, 2, zmax/2$). In both the cluster and substrate sites solid boundary conditions are applied. The flow in the lattice is directed, alternating between two phases, where the flow is directed from the left to the right and the right to the left in Fig. 4.25. Initially the flow is directed from the left to the right. The flow velocity in the free fluid nodes is kept to a constant value $\bar{u}$ by adjusting the driving force F of the system. The flow velocity in the system is kept at a low value; all simulations are done in the laminar flow regime. As in Sect. 4.5.3, the flow pattern about the obstacle (substrate and cluster) is determined by a number of lattice Boltzmann steps

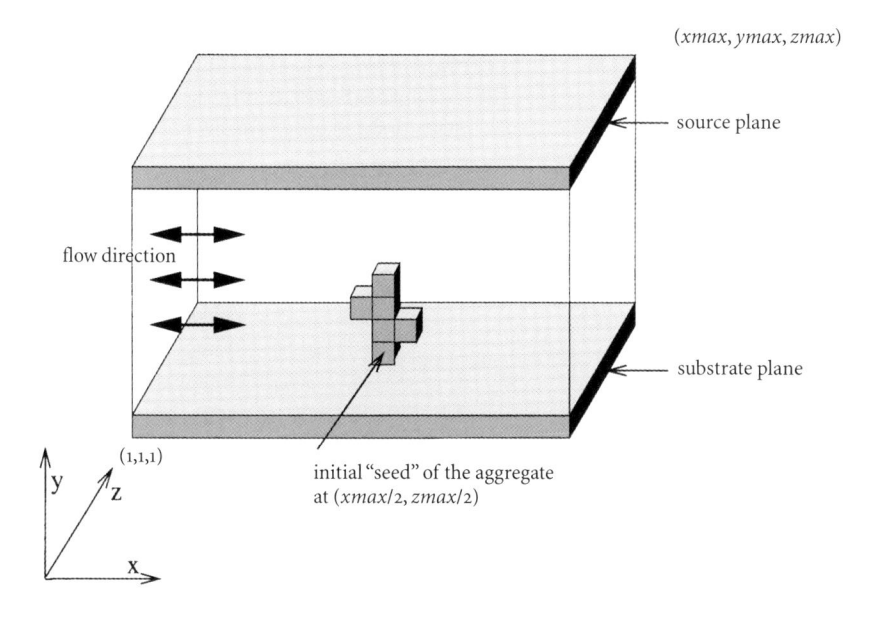

Fig. 4.25. Basic construction of the aggregate in a bidirectional flow

for each phase until equilibrium, followed by a tracer step. To reduce the time for the tracer distribution to attain an equilibrium, the tracer computation uses two separate tracer distributions: a distribution for the case where the flow is directed from left to right (*tracer distribution A*) and a distribution for the case where the flow is directed from right to left (*tracer distribution B*). If the flow at a certain time step is for example directed from left to right, the results of a previous time step where the flow was directed from left to right can be used to speed up the computation time. Tracer particles are again released from the source plane, the lattice sites located at the xz-plane at $y = ymax$ for both phases, and tracer particles are absorbed by the fluid nodes adjacent to obstacle nodes. The probability p that a new node will be added to the aggregate is again computed using (4.18), using the corresponding tracer distribution. After each growth step the flow direction is reversed. The aggregation model for a bidirectional flow is summarized in pseudo-code below:

```
initialize aggregate
initialize flow direction
initialize tracer distributions
do {
  -if (flow from left to right) {
    -compute flow velocities until equilibrium;
    -compute tracer distribution A until equilibrium;
    -select randomly with probability p (4.18), using tracer
     distribution A, one of the growth candidates and
     add it to the aggregate;}
  -else(flow from right to left) {
    -compute flow velocities until equilibrium;
    -compute tracer distribution B until equilibrium;
    -select randomly with probability p (4.18), using tracer
     distribution B, one of the growth candidates and
     add it to the aggregate;}
    -reverse flow direction;
} until ready
```

The results of the simulation experiments in a bidirectional flow are summarized in Table 4.3. As was done in Sect. 4.5.3 the fractal dimension D_{box} of the surface of the aggregate, the average absorption $\bar{a}$, and the uniformity of the nutrient distribution D_{abs} (4.19) were computed. The morphology of the aggregates for various values of Pe is demonstrated in Fig. 4.26 by showing slices through the middle of the lattice (in the xy-plane). In Table 4.3 the average absorption $\bar{a}$ in the boundary nodes, and the values of D_{abs}, are listed for the various Pe numbers. Furthermore, for each cluster the average x, y, z coordinate (the center of gravity) was measured.

Table 4.3. Ratio R of the total sink nodes to the total cluster size, the 3D fractal dimension D_{box}, the average absorption $\bar{a}$, the measure D_{abs} of the uniformity of the nutrient distribution, and the average x, y, z coordinate of the aggregate in the bidirectional flow experiment for a series of increasing Pe numbers

Pe	0.0150	0.1322	0.2521	0.4918	1.0000	2.0000	2.5000	3.0000
R	2.41 ± 0.02	2.29	2.16	1.86	1.49	1.15	1.05	1.01 ± 0.01
D_{box}	2.25 ± 0.01	2.24	2.26	2.29	2.24	2.19	2.17	2.15 ± 0.02
$\bar{a}$	0.03 ± 0.01	0.17	0.17	0.69	1.0	6.3	8.4	10.0 ± 0.1
D_{abs}	2.2 ± 0.2	1.3	1.3	0.9	1.0	0.7	0.8	0.8 ± 0.1
$\bar{x}$	0.50 ± 0.01	0.52	0.50	0.52	0.51	0.51	0.51	0.51 ± 0.01
$\bar{y}$	0.35 ± 0.03	0.23	0.21	0.17	0.14	0.12	0.11	0.10 ± 0.01
$\bar{z}$	0.57 ± 0.04	0.49	0.50	0.49	0.50	0.49	0.49	0.51 ± 0.03

(a)   (b)   (c)   (d)

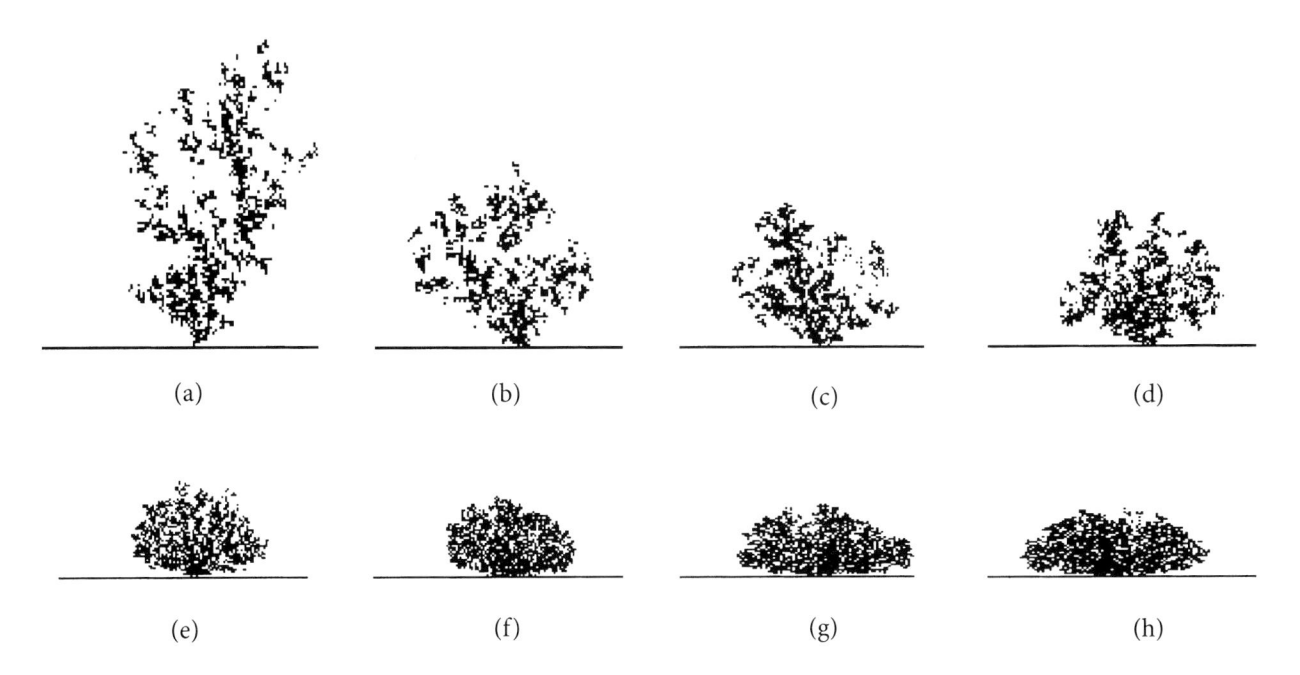

(e)   (f)   (g)   (h)

In Figs. 4.27 and 4.28 slices through the simulation box in the *xz*-plane are shown in which the nutrient distribution is visualized, for respectively the *Pe* numbers 3.000 (flow dominates) and 0.015 (diffusion dominates). The color shift from black to white in these pictures indicates a depletion of nutrients: black indicates the maximum concentration of nutrients, while the regions with nearly zero concentration are shown in white. In both pictures the nutrient distributions *A* and *B* of two successive phases in the bidirectional flow model are shown. In Figs. 4.27a and 4.28a the flow is directed from the top to the bottom, while in Figs. 4.27b and 4.28b the flow direction is reversed. In Fig. 4.29 slices through the simulation box in the *xz*-plane are shown in which the absorption in the boundary nodes around the aggregates is visualized for respectively the *Pe* numbers 0.015 and 3.000, where black indicates a high absorption rate and white an absorption of approximately zero.

Fig. 4.26a–h. Slices through the middle of the aggregates in the *xy*-plane at $zmax/2$, bidirectional flow experiment: (*a–h*), Péclet number increases from approximately 0.0 to 3.0.

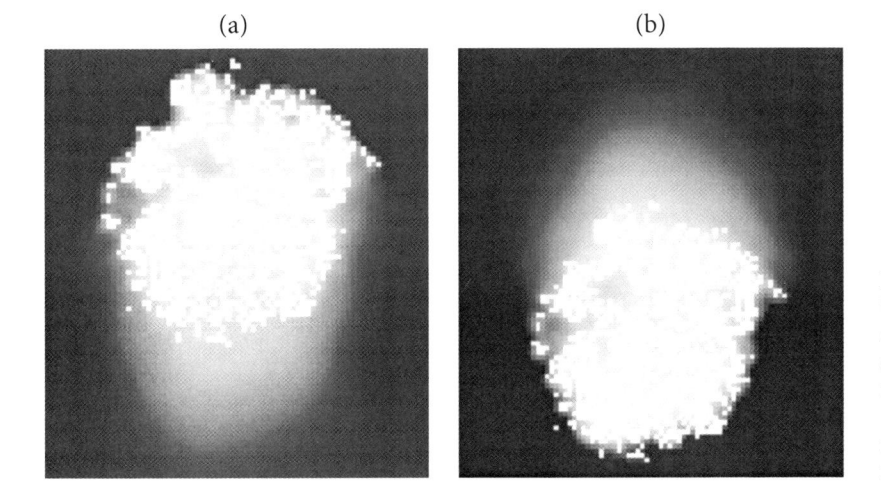

(a)   (b)

Fig. 4.27a,b. Slice through the simulation box in the *xz*-plane showing the nutrient distribution in two successive growth stages in the bidirectional flow experiment in which *Pe* is set to the value 3.000 (flow dominates). The flow is directed from top to bottom in (*a*) and directed from bottom to top in (*b*).

Fig. 4.28a,b. Slice through the simulation box in the xz-plane showing the nutrient distribution in two successive growth stages in the bidirectional flow experiment in which *Pe* is set to the value 0.015 (diffusion dominates). The flow is directed from top to bottom in (*a*) and directed from bottom to top in (*b*).

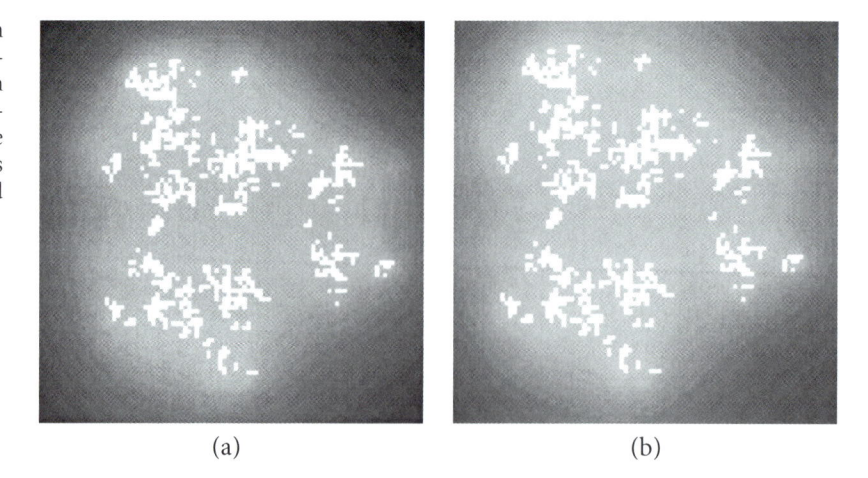

(a) (b)

Fig. 4.29a,b. Slices through the simulation box in the xz-plane showing the absorption in the boundary nodes for respectively the *Pe* numbers 0.015 ((*a*), diffusion dominates) and 3.000 ((*b*), flow dominates)

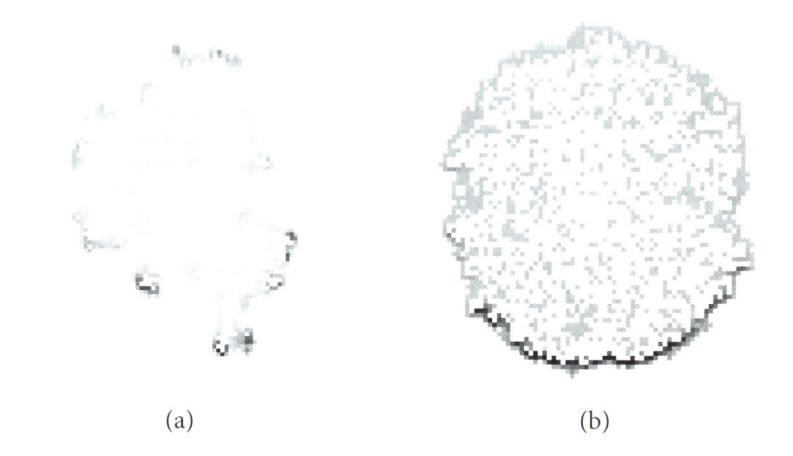

(a) (b)

Fig. 4.30. Aggregate and the corresponding nutrient distribution in a slice made in the xy-plane in the bidirectional flow experiment. *Pe* is set to the value 0.015 (diffusion dominates), and flow is directed from the left to the right.

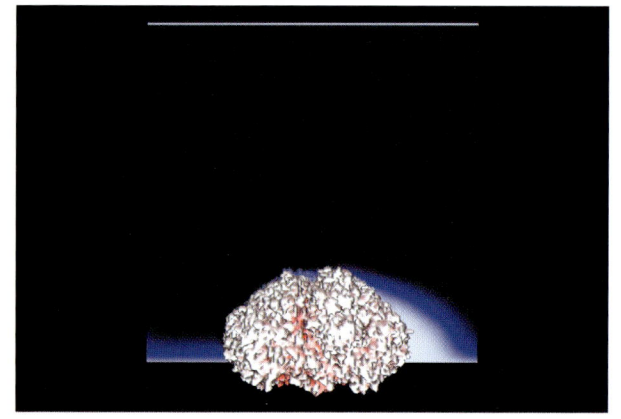

Fig. 4.31. Aggregate and the corresponding nutrient distribution in a slice made in the xy-plane in the bidirectional flow experiment. *Pe* is set to the value 3.000 (flow dominates), and flow is directed from the left to the right.

In Figs. 4.30, 4.31, and 4.32, the three-dimensional morphology of the aggregates is visualized in combination with the corresponding nutrient distribution. The color shift from white to red in the aggregates indicates which parts of the aggregate are added most recently on top of the previous growth stage: red indicates the "oldest" part of the growth form in Fig. 4.30, while (for visualization reasons) the youngest parts of the aggregate are shown in white in Fig. 4.31. The nutrient distribution is depicted using a color shift from blue to white, where blue indicates the highest concentration and white the lowest. In Figs. 4.30 and 4.31 the nutrient distribution is shown in a slice in the xy-plane. In Fig. 4.32 the shape of the depletion zone is depicted by the construction of an isosurface at a level of very low (nearly zero) nutrient concentration.

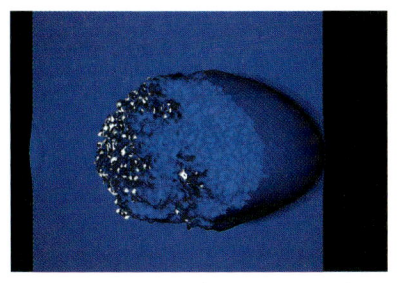

Fig. 4.32. Top view of the aggregate from the bidirectional flow experiment in which *Pe* is set to the value 3.000 (flow dominates), and flow is directed from the left to the right. The depletion zone around the aggregate is visualized by constructing an isosurface at a nutrient concentration of nearly zero.

4.5.5 Comparison Between the Range of Aggregates and the Growth Forms

In the results of the monodirectional flow experiments shown in Fig. 4.23 it is demonstrated that the influence of hydrodynamics on the nutrient distribution occurs at higher *Pe* numbers. In the case of a low *Pe* number the diffusion-limited situation is obtained: an irregular object that is branching towards the nutrient source. At higher *Pe* numbers the influence of the flow becomes visible: a more compact-shaped object emerges where branches tend to develop in the opposite direction to the flow, and in the stream shadow of the aggregate an area depleted of nutrients develops.

The impact of the bidirectional flow model presented in Sect. 4.5.4 can be demonstrated by comparing the measurements carried out for the monodirectional flow model (see Table 4.2) to measurements done in the bidirectional flow model (see Table 4.3). The center of gravity (the average x, y, z coordinate of the aggregate) tends to move in the negative x direction (the upstream direction) in the monodirectional flow experiment, while the center of gravity remains more or less in the center of the xz-plane in the bidirectional flow experiment for increasing *Pe* numbers. When visually comparing the slices through the aggregates shown in Figs. 4.24 and 4.26 it can also be observed that the increasing degree of asymmetry in the aggregate in the monodirectional flow experiment, for increasing *Pe* numbers, has disappeared in the bidirectional flow experiment. In the last experiments, aggregates have developed with a roughly radial symmetry, which corresponds qualitatively to the shape of branching sessile organisms such as *Pocillopora damicornis*. These experiments seem to indicate that a bidirectional flow, a reversal of the flow direction basically twice a day, leads to radial symmetrical growth forms.

The nutrient distributions shown in Figs. 4.27, 4.28, 4.30, and 4.31 demonstrate the main differences between diffusion- and flow-dominated regimes. For low Péclet numbers the distribution of nutrient is roughly symmetric about the center of the aggregate, where the highest concentration resides at the tips of the aggregate and where between the branches an area depleted of nutrients is found with a very low growth probability. In Fig. 4.28 it can be observed that there is hardly any difference between the two successive phases: there is very little influence of the governing flow direction on the nutrient distribution. At higher Péclet numbers a clear asymmetry develops in the distribution with a depleted region developing downstream of the object (see Fig. 4.27). In a top view of the aggregate and nutrient distribution (see Fig. 4.32) it can seen that this depletion zone has more or less a parabolic shape. As a consequence, there exists little probability of growth in the depleted region.

The absorption patterns shown in Fig. 4.29 show an asymmetry in the absorption of nutrients at the boundary nodes of the aggregates for increasing *Pe* numbers. For the diffusion-limited case (Fig. 4.29a) the highest degree of absorption is found at the tips of the object, while near the center hardly any nutrient is being absorbed. In Fig. 4.29b the highest amount of absorption is found at the upstream side of the object. This corresponds to the observations done in experimental work on the absorption of food particles by the stony coral *Madracis mirabilis* (see Fig. 2.5), where the highest amount of food particle capture was found at the upstream side of the colony (see Fig. 2.6). In Table 4.3 it can be seen that the average absorption $\bar{a}$ increases for increasing *Pe* numbers. The absorption pattern characteristic of the diffusive regime changes into a more slowly decaying distribution. In other words, the flow results in a more even distribution of nutrient in boundary nodes. This is illustrated by the decrease in the measure D_{abs} of the uniformity of the nutrient distribution. As a result the probability that boundary sites are added to the aggregate becomes more equal, which accounts for the increasing degree of compactness of the aggregate.

When comparing Figs. 4.30 and 4.31, it can be observed that the added part in Fig. 4.31 (visualized in white) "hides" the previous growth stages, so all boundary sites seem to have a more or less equal probability of being added to the aggregate, while in Fig. 4.30 the "trunk" of the object remains visible and growth occurs only at the tips of the object. A gradual increase of compactness is demonstrated in Fig. 4.26 and by a decrease of both D_{box} and the ratio R of the total sink nodes to the total cluster size in Table 4.3 for an increasing influence of hydrodynamics. This gradual increase of compactness corresponds qualitatively to the observations made in stony corals, hydrocorals, and sponges, where growth forms gradually transform from a compact shape, under conditions exposed to water movement, into a thin-branching one under sheltered conditions. Qualitatively this corresponds to the D_{box} measurements, estimated for the plane-filling properties of the projections of the growth forms, analyzed in Sect. 3.3.2. The other morphological properties, based on the skeleton of the object, cannot easily be compared. In the aggregation model, growth is represented by the addition of discrete sites in a three-dimensional lattice, which is a substantial simplification of the actual growth process. The actual growth process in many sponges, stony corals, and hydrocorals (see also Fig. 1.4) consists of adding layers of material (varying in thickness) on top of the preceding growth stage, and not the addition of particles. In the section on accretive growth (Sect. 4.6) we will try to develop a growth model, based on surface normal deposition, which comes closer to the actual growth process. An accretive growth model, in which layers of material are constructed on top of the previous layers and where the local thickness is determined by the local amount of absorbed nutrients or light intensity, also offers possibilities for a quantitative morphological comparison of simulated and actual growth forms as was discussed in Sect. 3.3.2.

In the bidirectional flow model several other simplification are made. In many stony corals, for example the species *Pocillopora damicornis*, photosynthesis represents a major energy input. Furthermore, there is no mechanism present in the aggregation model for nutrient accumulation at some distance from the object and there are no effects of erosion included in the model.

In our bidirectional flow simulations objects are found with a roughly radial symmetry and where the degree of compactness increases with an increasing influence of hydrodynamics, and the model may be used as a simple explanation of a similar phenomenon which is observed in the morphology of various marine sessile organisms.

4.6 Accretive Growth

4.6.1 Surface Normal Deposition in Marine Sessile Organisms

In many marine sessile organisms a skeleton is formed by a surface normal deposition process. In this growth process skeleton material is deposited on top of previous layers, which remain unchanged, during the growth process as shown in Fig. 1.4. In many cases the skeleton formed in this growth process is characterized by a radiate accretive architecture (see for example the sponge *Haliclona oculata* in Fig. 2.17); a similar architecture is found in many stony corals (Graus and Macintyre 1982) and stromatolites (Grotzinger and Rothman 1996). A diagram of the layered structure formed by the accretive growth process is shown in Fig. 4.33. The growth process in these organisms may be driven almost exclusively by the local availability of light required for the photosynthesis. In predominant autotrophic stony corals there is a direct relation between local deposition velocities and local light intensities on the surface of the colony.

In many stony corals, light represents the major energy source. The availability of light may have a strong influence on the morphology of the stony coral, as was discussed in Sect. 2.2.3. An example of this are the colony shapes of the coral *Montastrea annularis* (see Fig. 2.31), where the colony gradually transforms from hemispherical form into a plate-like colony at deep locations. The branching stony coral *Porites sillimaniani* displays a variation of the whole colony morphology with respect to light availability. In Fig. 2.32a a typical branching morphotype is shown from a shallow site, which gradually changes into a plate-like growth form (Fig. 2.32c) at deeper sites.

In many other marine sessile organisms, for example in many sponges and stony corals, where suspension feeding may represent a significant part of the energy intake, the suspended material from the direct environment is filtered away and locally absorbed. In sponges the absorbed suspended material may be transported through the tissue by the aquiferous system over relatively large distances. The amount of transport depends on both the degree of development of the aquiferous system and the local amount of contact with the environment. The degree of development of the aquiferous system differs greatly between species (Brien et al. 1973). Within stony corals nutrients may be translocated from one polyp to its neighbors, through the living tissue covering the colony, over relatively short distances and is again species-specific. Experiments on the translocation of nutrients in stony corals have been done by Taylor (1977) and Rinkevich and Loya (1983b) (see also Sect. 2.2.4). In sponges and stony corals with a relatively weakly developed transport system the amount of nutrients arriving at a certain site in the tissue, and the local deposition velocity of skeleton material, is limited by both the locally available suspended material and the local amount of contact with the environment. At the protruding parts of the growth form the amount of contact with the environment gives these parts a relatively higher access

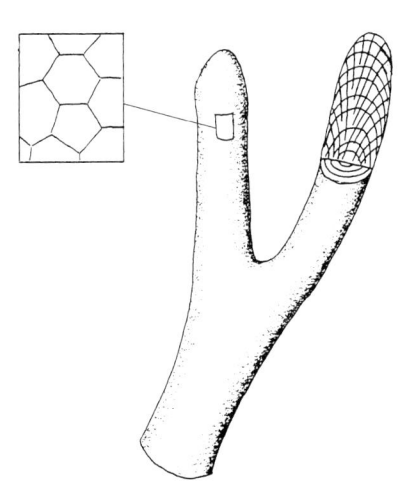

Fig. 4.33. Diagram of a branching tip of an organism with radiate accretive growth

to the suspended material in the immediate environment, and is relatively higher compared with the flattened sites of the growth form.

As was discussed in Sect. 4.5.1 water flow has a strong influence on the distribution of food particles in the immediate environment of suspension feeders and on the resulting morphology of the sessile suspension feeder. In branching sessile organisms growth forms emerge which have a roughly radial symmetry (see also Sect. 2.2.4 and Fig. 2.39a and b) or show a tendency to develop a flattened growth form. In the latter case most branches will develop in a plane perpendicular to the governing flow direction. Furthermore there is a strong influence on the overall degree of compactness of the organism, which can be observed in sessile organisms from various taxonomical groups: in general thin-branching forms tend to be formed in the absence of water movement, while more compact shapes develop when the influence of water movement increases. Local food particle absorption patterns related to the direction and the velocity of the flow, depletion zones surrounding the organism, and micro flow patterns in branching organisms have important consequences for the growth process and the resulting morphology in organisms where suspension feeding represents a significant part of the energy intake.

In the section on the regulation of the growth process of the stony coral *Stylophora pistillata* (Sect. 2.2.4), it was stated that this coral forms colonies with a roughly radial symmetry and that the colonies are approximately spherical. A remarkable property of the growth in *Stylophora pistillata* is that branches never fuse. There is a very regular spacing between the branches; a similar phenomenon can be observed in Figs. 2.5, 3.17, and 3.18 showing the stony coral *Madracis mirabilis*. More or less spherical colonies are formed, where branches do not fuse and where the branch spacing (*br_spacing*, see also Sect. 3.3.2) is remarkably constant. In Rinkevich and Loya (1985a) this phenomenon was explained by proposing a chemical signal which regulates the growth pattern. In their observations, done in field experiments with *Stylophora pistillata*, it was demonstrated that as soon as branches grow towards the other branches, a buffer zone seems to be formed in the immediate vicinity of each branch. Growth of branches in this region is suppressed or the growth direction is changed (see also Fig. 2.39). In the paper by Rinkevich and Loya (1985a) it is proposed that a chemical agent ("isomone") which suppresses the growth process is emitted by the tissue cells and secreted into the water.

In this section several types of models of accretive growth will be discussed, to give an overview of the variety of morphologies which can be produced with this model. The deposition in a surface normal accretive growth process is modeled using a geometrical model. The layers formed in the growth process are represented as layers consisting of triangulated meshes constructed on top of each other. The thickness of a new layer, measured along the surface normal of the previous layer, is determined by a growth function. Several types of models of accretive growth can be constructed by using different types of growth functions.

In the first model, discussed in Sect. 4.6.3, the distribution of growth velocities over a tip of an organism with accretive growth is approximated with a mathematical function. The distribution can be obtained from, for example, figures as shown in Figs. 1.4 or 2.17. In this first model the approximation of the growth velocity distribution is combined with the effect of the local amount of contact with the environment. The local amount of contact with the environment in an object with radiate accretive growth can be related to the local radius of curvature, where a relatively high local radius of curvature indicates a relatively low access to the environment. In Sect. 4.6.3 we will briefly discuss the geometric model and the estimation of the local amount of contact with environment; more details can be found elsewhere (Kaandorp 1994a, 1994b and 1995). Although the growth function, using an approximation of growth velocities, does not provide much insight into the actual growth process, the branching patterns produced with this model demonstrate several important aspects of the branching process. These aspects will be discussed in more detail in Sect. 4.6.7, when the models are compared to the actual growth forms.

In the second model, described in Sect. 4.6.4, the growth function in the accretive growth model consists of a component in which an estimation is made of the local amount of available nutrient in the environment combined with the effect of the local amount of contact with the environment. In the first component the distribution of suspended food particles due to a combination of flow, diffusion, and absorption at the growth form is modeled. The underlying method for modeling nutrient distributions is the lattice Boltzmann method, combined with a tracer step to study the transport process. The nutrient distributions are computed using this method for various Péclet numbers. This method is discussed in Sect. 4.5.2 and Sect. 4.3.1. With this model it is possible to study the morphologies which develop in an accretive growth process, exclusively driven by the availability of simulated food particles and the influence of hydrodynamics on the distribution of food particles. More detailed accounts of the coupling of the accretive growth model and the nutrient model are provided elsewhere (Kaandorp and Sloot 1997 and Kaandorp and Sloot 2001).

In the third model, presented in Sect. 4.6.5, the growth function in the accretive growth model consists of a component making an estimation of the local available (simulated) light intensity, combined with the effect of the local amount of contact with the environment. With this model accretive growth driven by the availability of light can be studied.

Finally a simple and preliminary model will be shown in Sect. 4.6.6, where a nutrient driven growth process, as discussed in Sect. 4.6.4, is regulated by a simulated chemical agent. In this model the growth function consists of three components: in the first component an estimation is made of the local amount of available nutrient in the environment, in the second component an estimation is made of the local amount of growth suppressor (the "isomone"), and in the third component the amount of contact with the environment is approximated.

4.6.2 A Model of Surface Normal Accretive Growth

In the accretive growth model we have used a three-dimensional geometrical model. The different growth layers in this model are represented by layers of triangles. In Fig. 4.34 the basic construction of a new layer of material (the open triangles) on top of the previous layers (the gray triangles) is shown. The edges of the triangles are nearly equally sized, with basic size s. The triangles around one vertex represent one skeleton element (the corallite) in the model of the stony coral. The triangles are arranged in patterns mainly consisting of pentagons and hexagons. A similar arrangement is observed in

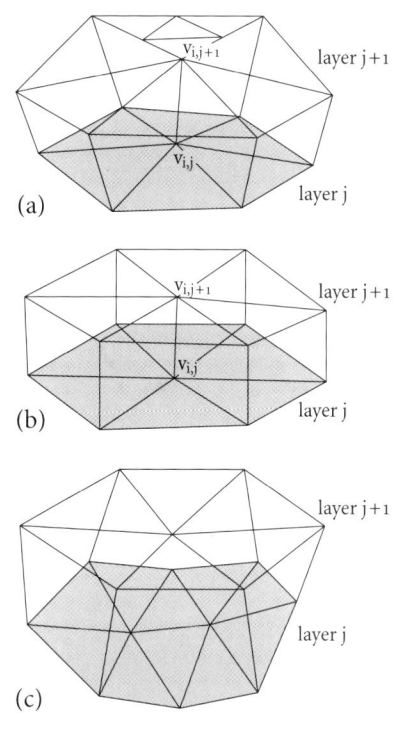

Fig. 4.34a–c. Basic construction applied in the geometrical model of surface normal accretive growth, where a new layer of triangles (the open triangles in layer $j+1$) are constructed on top of the previous layer of triangles (the gray triangles in layer j). In (a) an expanding surface is shown, where one of the triangles is subdivided; in (b) the new surface has more or less the size of the previous surface; while in (c) a shrinking surface is depicted in which 10 smaller triangles are clustered into 6 larger ones.

many sponges and stony corals (see Fig. 4.33). In the basic construction, the thickness l of the new layer varies between a threshold value tr slightly above zero and s the basic size of a skeleton element. In the actual organisms the layers are connected with skeleton elements which may vary in length but cannot become arbitrarily small, since they are made of discrete elements. The longitudinal connection with length l between two vertices $V_{i,j}$ and $V_{i,j+1}$ in two successive growth layers $j, j + 1$ is made along the mean normal vector of the triangles surrounding the vertex $V_{i,j}$ of the previous growth layer.

In the simulations we have used various types of growth functions. The thickness l of a new layer is determined by a growth function $G()$:

$$l = \begin{cases} s \cdot G() & \text{for} \quad G() > tr \\ 0.0 & \text{for} \quad G() \leq tr \end{cases} \tag{4.20}$$

which becomes zero as soon as $G()$ falls below the threshold tr, i.e. no material is being added and growth locally stops.

4.6.3 Accretive Growth Using an Approximation of Actual Deposition Velocities and the Amount of Contact with the Environment

In the first model the distribution of growth velocities over a tip of an organism with accretive growth is approximated with a mathematical function. The distribution was in this case obtained from sections made through growing tips of the sponge *Haliclona oculata* (see for example Fig. 2.17). As discussed in Sect. 2.2.2 this species tends to form flattened branching forms, where most branches develop in a plane which is perpendicular to the governing flow direction. Since the tips of these sponges are flattened, the deposition of material is anisotropic. This deposition process can be approximated with an anisotropic growth function (see Kaandorp 1995):

$$d = w/\cos(\beta) \quad \text{for} \quad 0 \leq \beta < \pi/2 \tag{4.21}$$

$$f(\alpha, \beta) = \begin{cases} 1.0 & \text{for} \quad 0 \leq \alpha \leq (\pi/d) \\ \left(\cos\left(\frac{(\pi/2)}{(\pi/2 - \pi/d)} \cdot (\alpha - \pi/d)\right)\right)^\eta & \text{for} \quad (\pi/2 + \pi/d) < \alpha \leq \pi \end{cases}$$

$$w > 2$$

The growth velocity depends upon the angle α between the surface normal vector and the direction of the growth axis and the angle β. When the flow direction is assumed to be parallel with the yz-plane, β is defined as the angle between the yx-plane and the projection of the surface normal vector on the xz-plane (see Fig. 4.35). The growth function $f(\alpha, \beta)$ has the largest plateau of maximum values when $\beta = 0$, while the widening decreases towards a minimum when $\beta = \pi/2$. The widening effect in $f(\alpha, \beta)$ is controlled with the parameter w and the exponent η. In all experiments w is set to the constant value 8 and only η is varied. The parameter η is used to control the overall shape of the object. Without using this parameter sharp discontinuities may emerge in the objects. In this growth model every vertex is associated with a growth axis with a certain direction.

The anisotropic growth function $f(\alpha, \beta)$ can be combined with a component $h_2(..)$, in which an estimate is made of the amount of contact with the environment, into one growth function $G()$ in (4.20):

$$G() = s \cdot f(\alpha, \beta) \cdot h_2(low\_norm\_curv, av\_norm\_curv) \tag{4.22}$$

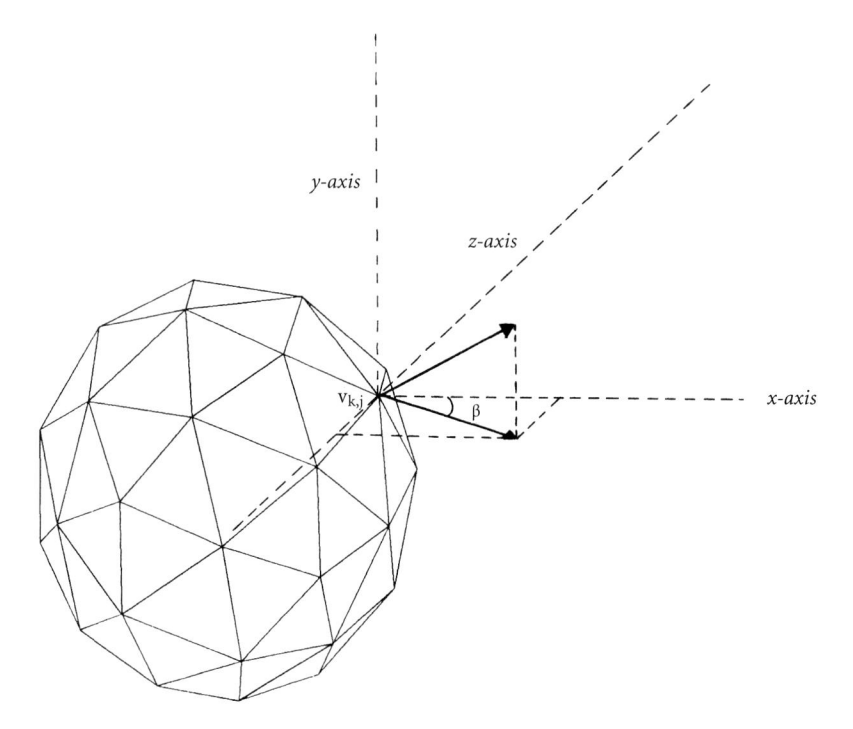

Fig. 4.35. Determination of the angle β between the projection surface normal vector on the xz-plane and the flow direction, which corresponds to the direction of the z-axis

The access to the environment can be described as a quotient of the surface through which nutrient can pass and the volume (with a certain distance from the surface) being supplied with nutrient. The amount of contact with the environment can be estimated by measuring the local radius of curvature *rad_curv* in a vertex in several tangential directions (in the simulated objects 6–8 measurements are done in one vertex) on the triangulated surface of the object. These local radii of curvature can be determined in a set of pairs of vertices, surrounding the considered vertex. At convex sites the value of *rad_curv* is positive and at concave sites *rad_curv* is set to zero. The measurements are summarized in one function $h_2(..)$ (4.24) expressing the amount of contact with the environment. First all radii of curvature are normalized using the function:

$$h_1(rad\_curv) = \begin{cases} 1.0 - (rad\_curv - min\_curv)/(max\_curv - min\_curv) \\ \quad \text{for} \quad min\_curv \leq rad\_curv \leq max\_curv \\ 1.0 \quad \text{for} \quad rad\_curv < min\_curv \\ 0.0 \quad \text{for} \quad rad\_curv > max\_curv \end{cases} \quad (4.23)$$

The final function value $h_2(..)$ is formed by the product of the lowest value (*low_norm_curv*) and the average value (*av_norm_curv*) of the normalized radii of curvature:

$$h_2(low\_norm\_curv, av\_norm\_curv)$$
$$= low\_norm\_curv \cdot av\_norm\_curv \quad (4.24)$$

In all accretive growth simulations shown after this section the parameters *min_curv* and *max_curv* were set to respectively the values 2 s and 20 s.

(a)

axis₁

(b)

(c)

axis₂ *axis₃*

(d)

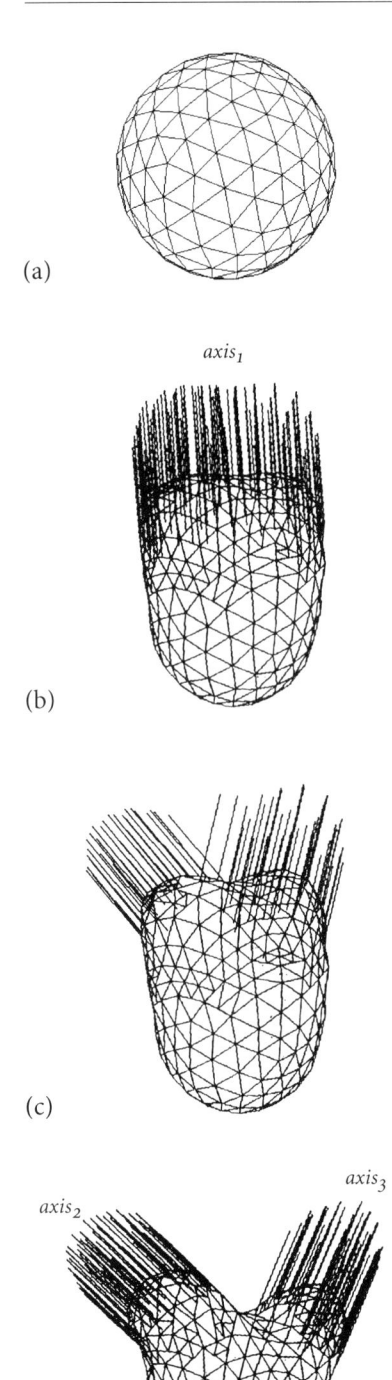

Fig. 4.36a–d. Four successive stages in the iterative geometrical construction; in (*a*) the initial spherical object is shown.

The simulation of the growth process starts with a triangulated sphere (Fig. 4.36a), where the triangles are arranged in a pattern of hexagons and pentagons. By selecting a growth function (for example (4.22)), a new layer is constructed on top of the previous ones and a radiate accretive structure, similar to the one observed in Fig. 4.33, develops. The first stage (see Fig. 4.36b) is a flattened object. In the next stages (c and d), after some iteration steps, a local minimum develops and the patch of active vertices is separated into two new ones, because locally the maximum allowed radius of curvature *max_curv* in $h_1(rad\_curv)$ (4.23) is exceeded and the $h_2(..)$ component in (4.22) becomes zero. For the formation of new growth axes in the model the following rule is applied: the longitudinal element $(V_{i,j}, V_{i,j+1})$ with a length l that is a local maximum in a patch of "active" vertices defines the direction of the growth axis with which all vertices in the patch are associated. In Fig. 4.36b all active vertices are associated with the same growth axis ($axis_1$). In Fig. 4.36d two new patches have developed where the growth process continues independently and two new growth axes $axis_2$ and $axis_3$ are formed.

Due to the growth process the surfaces in Figs. 4.34 and 4.36 will locally expand and shrink at some other points. One of the assumptions in the model is that the surface is covered with triangles with almost equal-sized edges, varying slightly about the basic unit s. To ensure that the surface remains tessellated with almost equal-sized triangles, at some points triangles becoming too large need to be split, while at other points triangles become too small and have to be removed from the system. In Fig. 4.34a at an expanding site a triangle is subdivided into four new ones in the new layer, while in Fig. 4.34c ten triangles are located at a shrinking site and are being clustered into six larges ones in the new layer (see more details on the deletion and insertion of triangles in growing surfaces in Kaandorp 1994b). Finally in each growth step a test has to be made to determine if parts of the object are approaching each other too closely. As soon as these parts are at the point of intersection and a physically impossible situation may occur, growth is inhibited at these sites in the model. More evolved objects, resulting from the accretive growth model using the growth function from (4.22), are shown in Fig. 4.37; in the range a–c the parameters (*max_curv*, η) are set to respectively the values (10 s, 1.0) (a), (30 s, 0.88) (b), and (60 s, 0.7) (c).

The model of radiate accretive growth is very sensitive to small changes in the initial parameter settings. For example, in the simulation experiments small changes in the maximum allowed radius of curvature *max_curv* can be made, leading to different realizations for nearly the same parameter settings (see Kaandorp 1995). We have used this property to generate more realizations for one parameter setting and to make estimations of error bounds in the simulation experiments.

4.6.4 A Model of Accretive Growth Driven by the Local Amount of Available Nutrient and the Influence of Hydrodynamics

In this model it is assumed that the organism is using exclusively suspension feeding as an energy source and that food particles are dispersed by the combined process of diffusion and hydrodynamics. Furthermore it is assumed that transport of nutrients through the tissue is small or negligible and that the local growth velocity is directly related to the local amount of absorbed

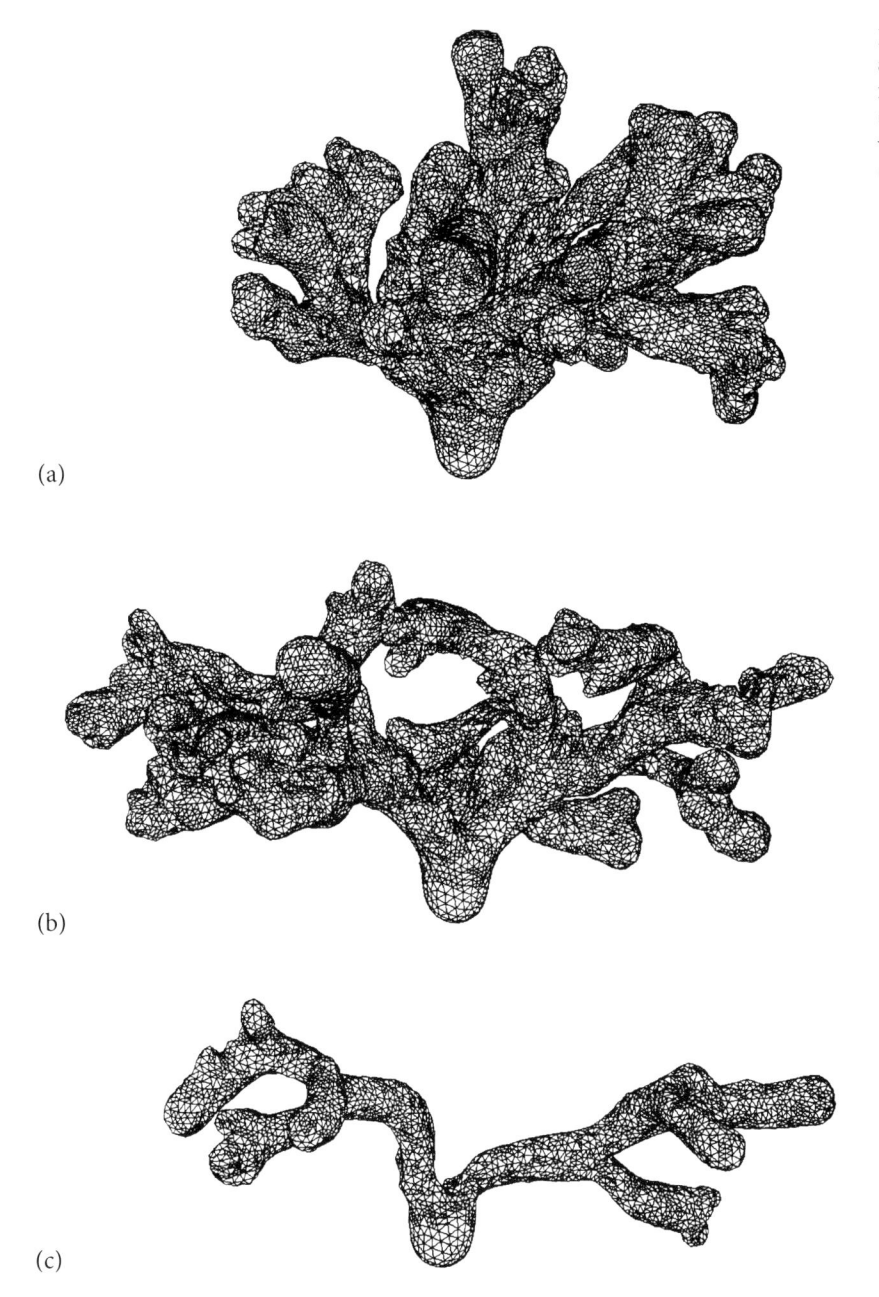

Fig. 4.37a–c. Objects generated with the accretive growth model using the growth function from (4.22). The parameters (max_curv, η) are set to respectively the values (10 s, 1.0) (a), (30 s, 0.88) (b), and (60 s, 0.7) (c).

(a)

(b)

(c)

food particles. In the simulations a bidirectional flow model is applied in order to make the simulated results comparable to the actual situation where the organism is exposed to a bidirectional flow regime due to tidal movements. It is assumed that the time scale of the growth process is very different from the time scale of the hydrodynamic process, i.e. the growth process is much slower than the hydrodynamic process. It is also assumed that both the flow pattern (only laminar flow was studied) and the (simulated) nutrient distribution are in equilibrium.

In the simulations of nutrient-driven accretive growth, we have used two types of growth functions. In the first type the influence of only the local availability of simulated nutrient is included in the growth function $G()$ in (4.20):

$$G() = s \cdot k(c) \qquad (4.25)$$

where $k(c)$ is an estimate of the local nutrient gradient along the mean normal vector on the surface of the object. In the second type of growth function:

$$G() = s \cdot k(c) \cdot h_2(low\_norm\_curv, av\_norm\_curv) \qquad (4.26)$$

where in the component $h_2(..)$ an estimate is made of the amount of contact with the environment.

In the simulations the diffusion coefficient D varies, and $\bar{u}$ is kept constant by adjusting the driving force F of the system. Due to the growth of the object the velocity in the free fluid would gradually decrease if the driving force were not adjusted. In all simulations the corresponding Reynolds number is kept to a very low and constant value, so we are in the "creeping flow" regime.

The simulated growth form is initialized with a spherical object (the object shown in Fig. 4.36a) which was mapped onto the 144^3 lattice and positioned at the bottom plane of the lattice. In Fig. 4.38 the basic idea of the coupled simulation model is shown: the growth form is mapped onto the lattice and the discrete representation is used for computing the simulated nutrient distributions. The bottom plane, the "substrate", is positioned in the xz-plane at $y = 1$, while the center of the sphere is located at the position $(xmax/2, 2, zmax/2)$. In both the growth form and substrate sites solid boundary conditions are applied. The flow in the lattice is directed, alter-

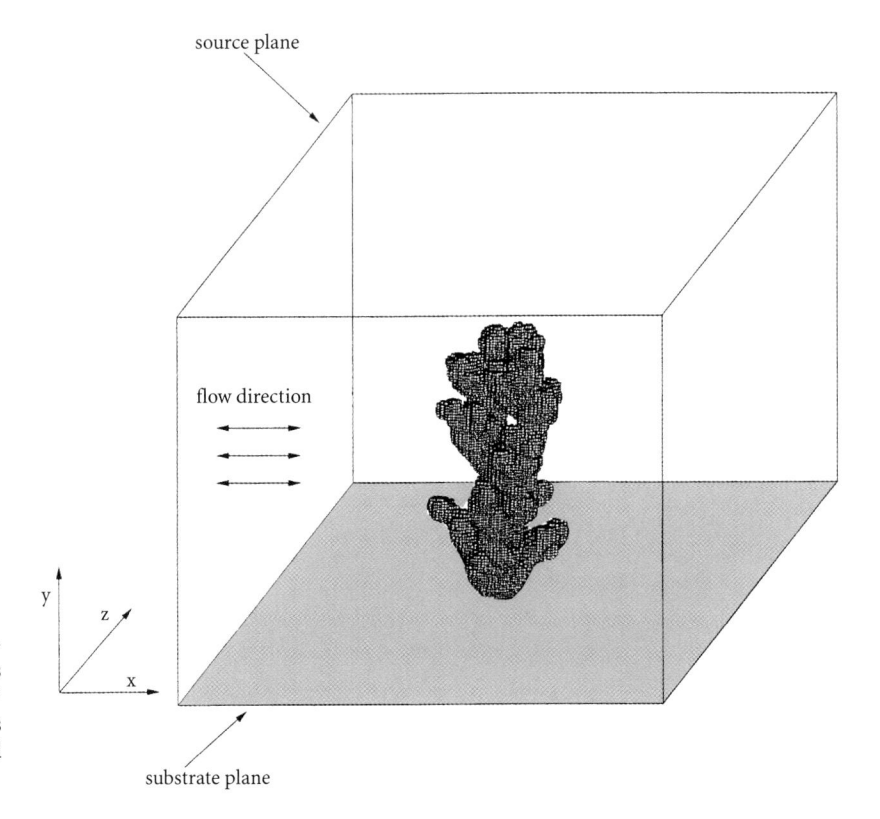

Fig. 4.38. Basic idea of the coupled simulation model. The growth form is represented by discrete sites in the 144^3 lattice; the discrete representation is used for computing the flow pattern and nutrient distribution.

nating from left to right and from right to left in Fig. 4.38. Initially the flow is directed from left to right. The flow pattern about the obstacle (substrate and growth form) is determined in the lattice Boltzmann iteration, using ten iteration steps followed by a tracer step. In each growth step ten iteration steps are used to ensure that an equilibrium is obtained in the flow pattern. In the last step tracer particles are released from the source plane, the lattice sites located in the xz-plane at $y = ymax$. The tracer particles are absorbed by the fluid nodes adjacent to obstacle nodes; these can be either nodes in the substrate plane (the xz-plane at $y = 1$) or nodes representing the surface of the simulated growth form. The nodes representing the object are indicated as object nodes, while the object nodes at the surface of the object are indicated as sink nodes. In this growth model it is assumed that both the tracer distribution and flow velocities are in equilibrium and the growth velocity of the growth form is much slower than the dispersion of the tracer. In the sink nodes the amount of absorbed tracer particles is determined. After each growth step the geometrical object is mapped again onto the three-dimensional lattice with 144^3 sites. In each growth step both the flow computation and the tracer step are done in two phases: left-to-right (*phase1*) and right-to-left (*phase2*). The nutrient gradients $k_1(c)$ and $k_2(c)$, along the mean surface normal vector, are estimated for both phases. The final estimation of the nutrient gradient $k(c)$ is the average of $k_1(c)$ and $k_2(c)$.

The accretive growth model in an alternating flow can be summarized in algorithmic form, using the pseudo-code below:

```
initialize growth form
initialize flow direction
initialize tracer distributions
do {
   -Map the geometrical object onto the lattice model;
   -Phase: flow from left to right {
      -compute flow velocities until equilibrium;
      -compute tracer distribution until equilibrium;
      -determine local nutrient gradient k₁(c) along
       normal vector; }
   -Phase: flow from right to left {
      -compute flow velocities until equilibrium;
      -compute tracer distribution until equilibrium;
      -determine local nutrient gradient k₂(c) along
       normal vector; }
   -compute local mean nutrient gradient
    k(c) = ½(k₁(c) + k₂(c));
   -compute local amount of contact with environment
    h₂(low_norm_curv, av_norm_curv) at the growth form;
   -compute the local thickness l of a new
    layer using either (4.25) or (4.26);
   -Construct a new layer on top of the
    the previous layer, insert new triangles at
    expanding sites, delete triangles at shrinking
    sites;
   -Do a collision detection test, inhibit growth
    in sites which tend to grow through
    sites representing the growth form;
} until ready
```

In Fig. 4.39 the morphologies of objects resulting from the accretive growth model are shown for a deposition process exclusively driven by local nutrient gradients $k(c)$ (4.25) and various Péclet numbers. In Fig. 4.40 slices

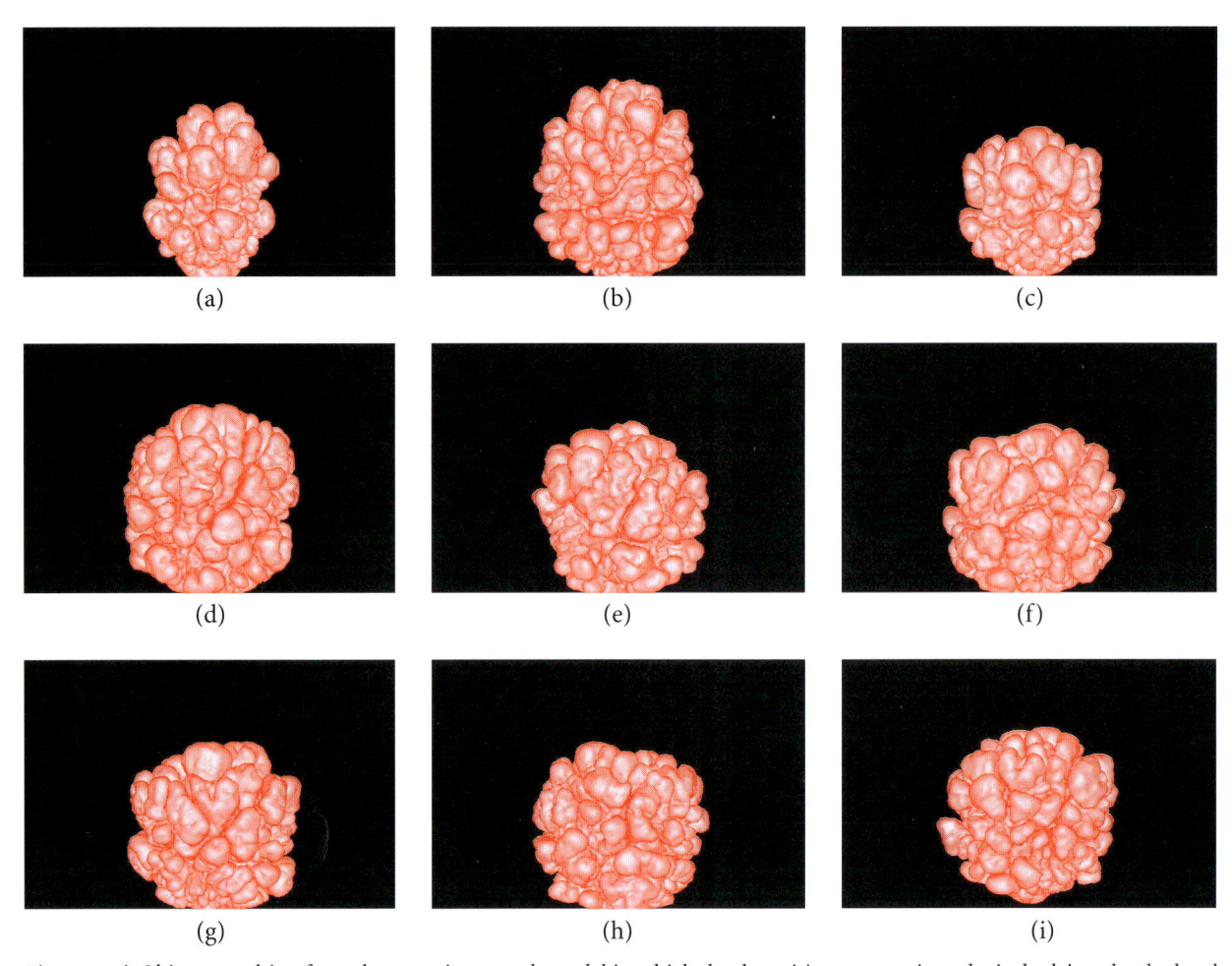

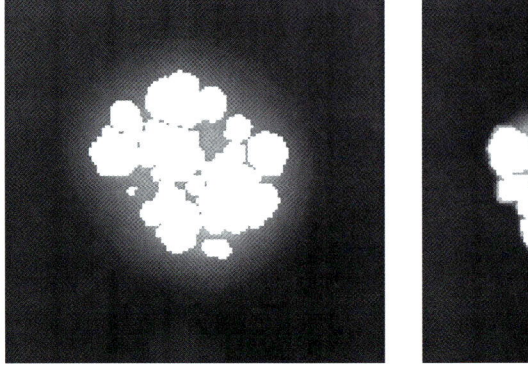

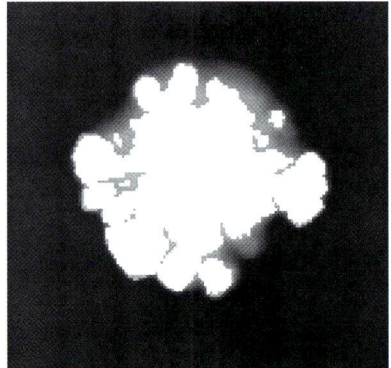

Fig. 4.39a–i. Objects resulting from the accretive growth model in which the deposition process is exclusively driven by the local nutrient gradients $k(c)$ in (4.25), for various Pe numbers

Fig. 4.40. Section through the simulation box made in the xz-plane, showing the nutrient concentration around the growth forms. Black indicates a high concentration, white visualizes a depletion of nutrients, and the section through the object is shown by using white colored lattice sites. In the left picture the nutrient distribution is shown for the diffusion-limited case ($Pe \approx 0$) and in the right picture the flow-dominated case is depicted ($Pe = 3.0$); in both cases the flow was directed from the bottom to the top. The objects were generated with the accretive growth model using only $k(c)$ in (4.25).

through the three-dimensional lattice at the xy-plane are made, showing the concentration of nutrients at the object, for both phases, for the extreme values of the Péclet numbers. Black indicates a high nutrient concentration, while depleted areas are colored white. Fig. 4.41 shows the morphologies of objects generated in the simulations where the deposition process is driven

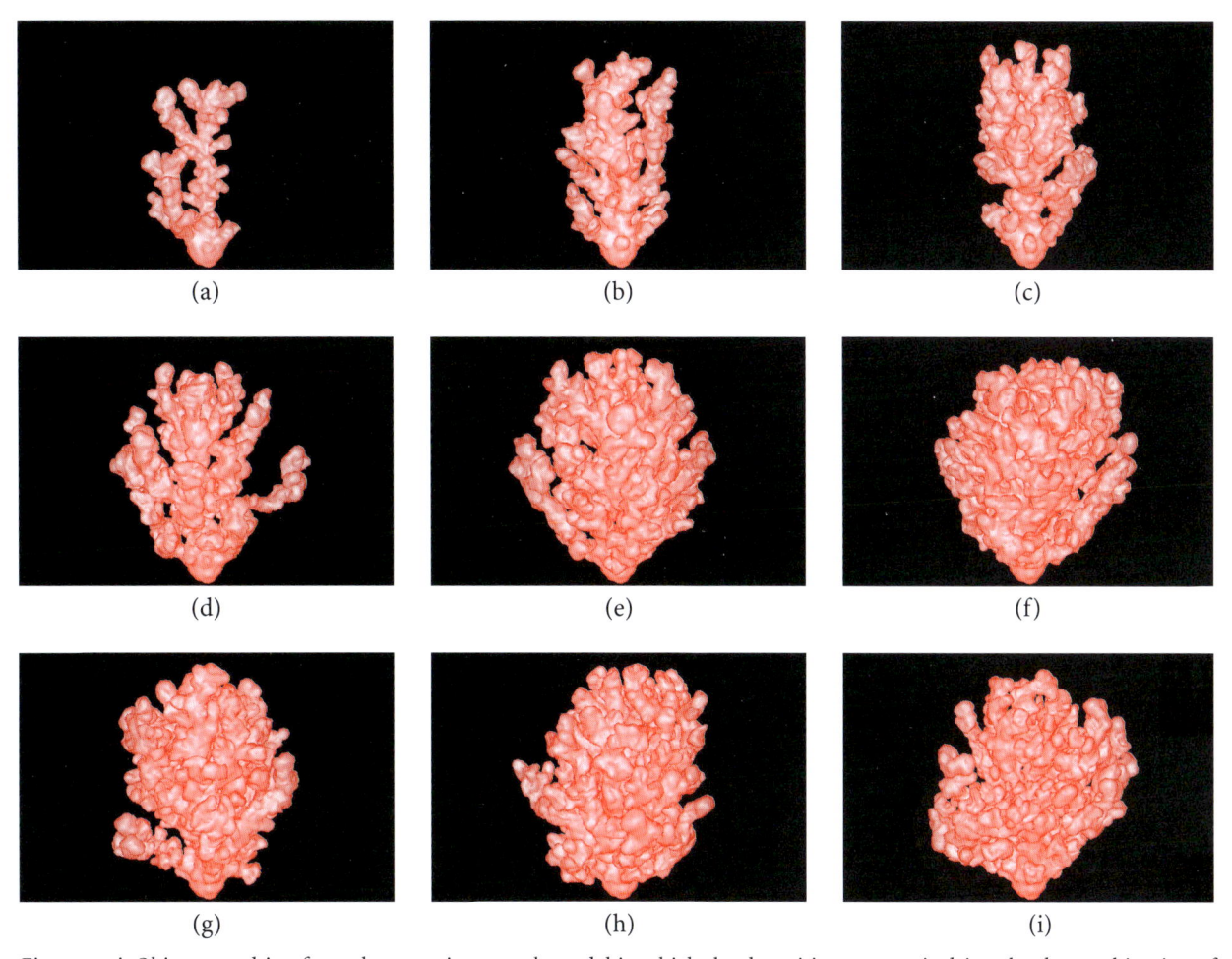

Fig. 4.41a–i. Objects resulting from the accretive growth model in which the deposition process is driven by the combination of local nutrient gradients $k(c)$ and the amount of contact with the environment $h_2(\ .\ .\ )$ in (4.26) for various Pe numbers

by a combination of the local nutrient gradients $k(c)$ and the local amount of contact with the environment $h_2(\ .\ .\ )$ (4.26), for a range of Péclet numbers. In Fig. 4.42 slices through the three-dimensional lattice at the xy-plane are made, showing the concentration of nutrients at the object, for both phases, for the extreme values of the Péclet numbers.

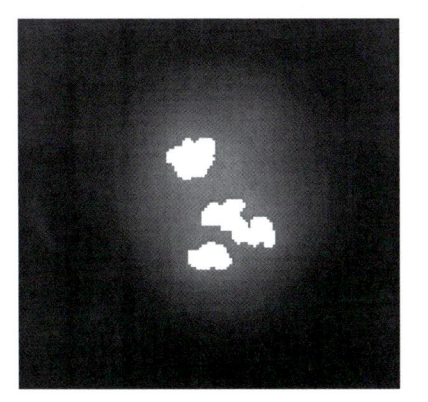

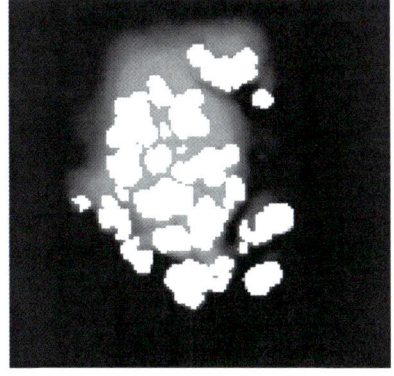

Fig. 4.42. Section through the simulation box made in the xz-plane, showing the nutrient concentration around the growth forms. Black indicates a high concentration, white visualizes a depletion of nutrients, and the section through the object is shown by using white colored lattice sites. In the left picture the nutrient distribution is shown for the diffusion-limited case ($Pe \approx 0$) and in the right picture the flow-dominated case is depicted ($Pe = 3.0$); in both cases the flow was directed from the bottom to the top. The objects were generated with the accretive growth model using $k(c)$ in combination with $h_2(\ .\ .\ )$ in (4.26).

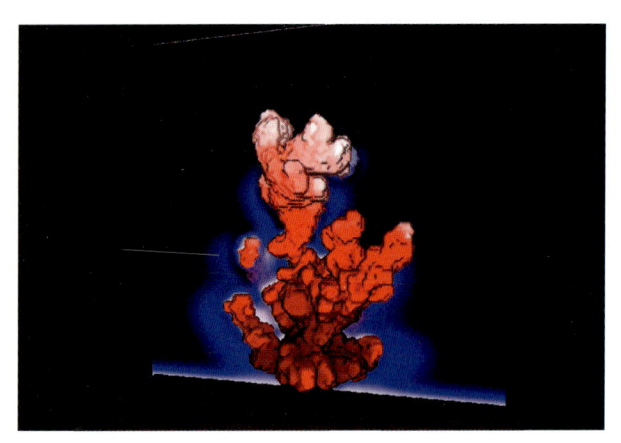

Fig. 4.43. The mean concentration of nutrients, the average of both phases, around an object simulated using the accretive growth model in which the deposition process is driven by the combination of local nutrient gradients $k(c)$ and the amount of contact with the environment $h_2(..)$ in (4.26) for a Pe number set to the value 3.0. The color gradient blue-white indicates a depletion of nutrients, the white-red gradient at the object visualizes the age of the growth layers, and the oldest layers are displayed in red.

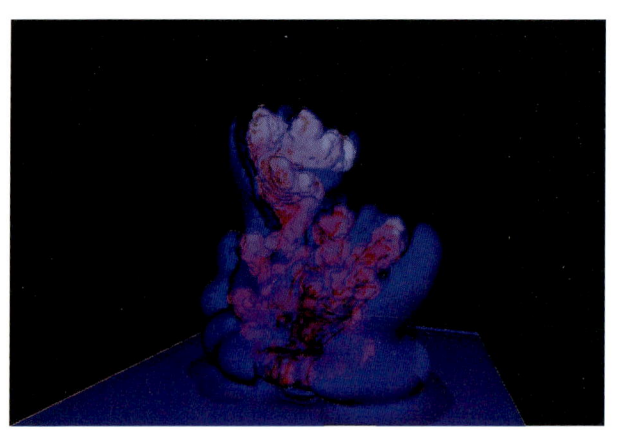

Fig. 4.44. The shape of the depletion zones in the average nutrient distribution around the object shown in Fig. 4.43, visualized by constructing an isosurface at a level where the average nutrient concentration is slightly above zero

The mean concentration of nutrients, the average of both phases, around an object simulated for a Pe number set to the value 3.0 is visualized in Figs. 4.43 and 4.44. The dark blue region in Fig. 4.43 indicates regions with a high nutrient concentration, while depleted regions are displayed in white. The object itself is in this picture depicted in a white and red gradient, where white sites indicate regions where layers were added on top of the object in the most recent growth stages, while the older parts of the object are displayed in red. In Fig. 4.43 at the left and the right of the simulated growth form areas emerge which are depleted from nutrients. The three-dimensional shape of these depletion zones in the average nutrient distribution around the object is visualized in Fig. 4.44 by constructing and visualizing an isosurface at a level where the average nutrient concentration is slightly above zero.

The results of the simulation experiments are summarized in Tables 4.4 and 4.5. Error bounds on the various measurements have been estimated by repeating ten times the simulations at $Pe \approx 0$ and $Pe = 3.0$. The ratio R of the total sink nodes to the total number of object nodes indicates the compactness of the growth form. The fractal dimension D_{box} of the surface of the object was determined using a three-dimensional version of the (cube) box-counting method described by Feder (1988). In the simulations we have determined the number of sink nodes that have a nutrient absorption between a and $a + \Delta a$ for the objects obtained at various Péclet numbers. The number of sink nodes with absorption rate a, $N(a)$, can be related to a by a power law (see (4.19), Kaandorp et al. 1996). The value of D_{abs} can be estimated from a log-log plot. The exponent D_{abs} can be interpreted as a measure of the uniformity of the nutrient distribution. In Tables 4.4 and 4.5 the average absorption $\bar{a}$ in the boundary nodes and the values of D_{abs} are listed for the various Pe numbers. Furthermore, for each object the average x, y, z coordinate (the center of gravity) was measured in lattice coordinates, together with the standard deviations sd_x, sd_y, and sd_z, for respectively the x, y, z coordinates

Table 4.4. The average absorption $\bar{a}$ at the object nodes, the absorption dimension D_{abs}, the box counting dimension D_{box}, the ratio R between the total sink nodes and the total number of object nodes, the center of gravity $\bar{x}, \bar{y}, \bar{z}$, the standard deviations sd_x, sd_y, and sd_z of the x, y, z lattice coordinates of the object nodes, and the corresponding Pe numbers of the simulated growth forms using the accretive growth model driven exclusively by the local nutrient gradient $k(c)$ in (4.25)

Pe	0.0150	0.0718	0.1322	0.2521	0.4918	1.0000	2.0000	2.5000	3.0000
$\bar{a}$	0.42	1.50	3.02	4.34	6.0	9.4	12.6	13.4	14.5
D_{abs}	0.93	0.44	0.38	0.63	0.66	0.45	1.23	0.76	0.32
D_{box}	2.22	2.36	2.22	2.37	2.35	2.35	2.35	2.35	2.36
R	0.30	0.21	0.21	0.25	0.27	0.12	0.12	0.12	0.12
$\bar{x}$	71	72	71	72	72	73	71	73	71
$\bar{y}$	38	42	31	41	39	39	39	38	39
$\bar{z}$	71	71	72	72	71	71	70	71	71
sd_x	20	21	20	21	21	21	21	22	22
sd_y	23	24	19	23	22	22	22	22	22
sd_z	19	21	20	21	21	21	21	22	22

Table 4.5. ▼ The average absorption $\bar{a}$ at the object nodes, the absorption dimension D_{abs}, the box counting dimension D_{box}, the ratio R between the total sink nodes and the total number of object nodes, the center of gravity $\bar{x}, \bar{y}, \bar{z}$, the standard deviations sd_x, sd_y, and sd_z of the x, y, z lattice coordinates of the object nodes, and the corresponding Pe numbers of the simulated growth forms using the accretive growth model driven by the local nutrient gradient $k(c)$ and the amount of contact with the environment $h_2(..)$ in (4.26)

Pe	0.0150	0.0718	0.1322	0.2521	0.4918	1.0000	2.0000	2.5000	3.0000
$\bar{a}$	0.23 ± 0.02	0.55	0.80	1.51	1.9	2.9	4.9	6.1	6.6 ± 0.4
D_{abs}	1.23 ± 0.05	0.69	0.37	0.46	0.59	0.63	0.30	0.02	0.15 ± 0.08
D_{box}	1.99 ± 0.02	2.08	2.13	2.19	2.29	2.31	2.31	2.31	2.29 ± 0.03
R	0.68 ± 0.02	0.61	0.60	0.59	0.56	0.55	0.54	0.53	0.54 ± 0.03
$\bar{x}$	72 ± 2	73	72	71	73	73	71	69	70 ± 2
$\bar{y}$	35 ± 5	42	49	50	55	55	56	56	63 ± 9
$\bar{z}$	71 ± 4	72	71	75	69	70	68	72	70 ± 1
sd_x	25 ± 1	22	23	21	23	22	22	22	21 ± 2
sd_y	32 ± 2	31	31	30	29	30	29	28	28 ± 1
sd_z	5 ± 1	22	21	23	22	21	23	22	23 ± 2

of the object nodes. A nearly equal sd_x and sd_z indicates the formation of an object with roughly radial symmetry, with the y-axis as an axis of symmetry. A relatively lower sd_x compared to sd_z indicates the formation of a flattened form, where the largest plane of the flattened form is perpendicular to the flow direction (the flow is parallel to the x-axis).

4.6.5 A Model of Accretive Growth Driven by Local Light Intensities

A simple light model (Foley et al. 1990) is shown in (4.27) (see also (2.6)). In this model the light intensity I (W/m^2) on a surface is determined by $\cos(\theta)$, where θ is the angle of incidence of the light beam to the surface normal, and the intensity I_S of the light source (see Fig. 4.45). The light beam corresponds to the vertical.

$$I = I_S \cdot \cos(\theta) \qquad (4.27)$$

The light model can be extended by including diffuse reflection from the environment. There is reflection from the bottom as well as the surrounding water, due to scattering (see Roos 1967). A simple growth function $G()$ in

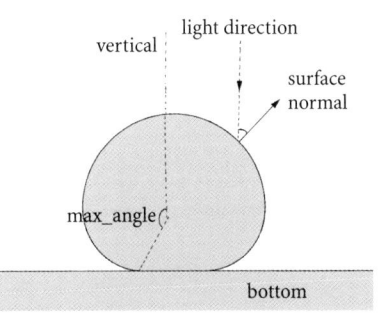

Fig. 4.45. The light intensity I on a surface is determined by the cosine of θ, the angle of incidence of the light beam to the surface normal, and the intensity I_S of the light source.

Fig. 4.46a,b. Objects generated with the accretive growth model using the growth function from (4.28). (*a*) *rest_term* is set to 0.0. (*b*) *rest_term* is set to 0.3.

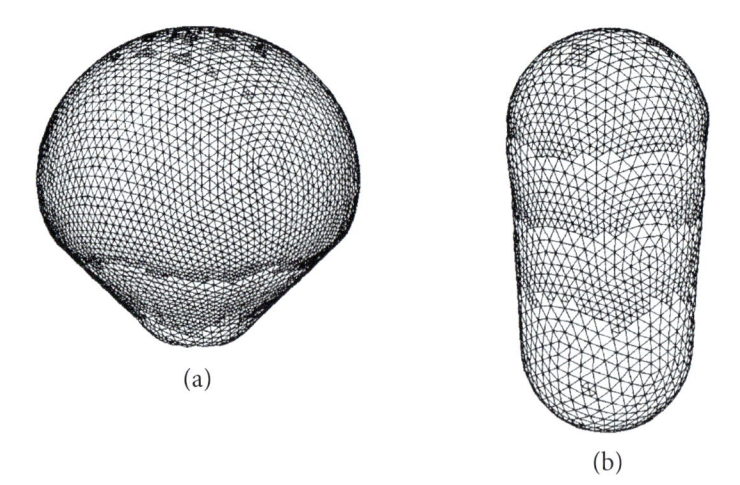

(a)

(b)

(4.20) in which the local light intensity is included is:

$$L(\theta) = (1 - rest\_term) \cdot \cos(\theta) + rest\_term \quad (4.28)$$

$$G() = s \cdot L(\theta)$$

In the function $L(\theta)$ a rest term (*rest_term*) is used which describes the diffuse reflection from the environment. This *rest_term* represents, in most cases, a small percentage of the maximum light intensity (which is set to 1.0).

In the simulations of light-driven accretive growth an estimate of the local light intensity $L(\theta)$ at every vertex on the surface of the object is made, and this can be combined with an estimate of the amount of contact with the environment $h_2(..)$:

$$G() = s \cdot L(\theta) \cdot h_2(low\_norm\_curv, av\_norm\_curv) \quad (4.29)$$

Fig. 4.47a,b. Two views of an object generated with the accretive growth model using the growth function from (4.29) (*rest_term* = 0.3); in this model all local light intensities are corrected for self-shading effects.

A complication in this model is that, due to the formation of branches, there is a self-shading effect of branches. In Kaandorp (1994b) a computational procedure is described which takes this self-shading effect into account in the accretive growth model using the growth function from (4.29). Without the $h_2(..)$ component in (4.28), this model produces column-like objects

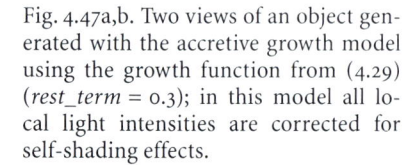

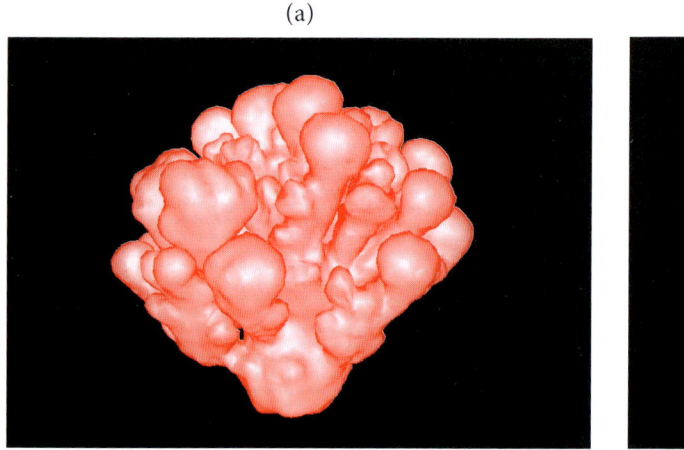

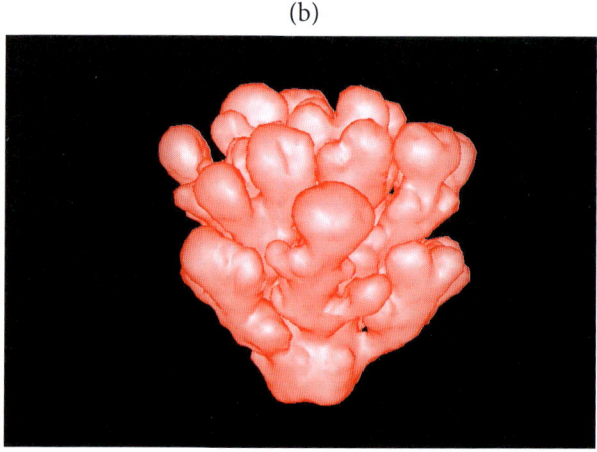

(a)

(b)

(*rest_term* = 0.0) or hemispherical objects (*rest_term* = 0.3), as shown in Fig. 4.46. With the amount of contact with the environment component, this model produces highly complex branching patterns (see Fig. 4.47); in this case it is also necessary to take the effects of self-shading into account.

4.6.6 A Model of Accretive Growth Driven by Local Nutrient Availability and Regulated by a Growth-Suppressing Isomone

In the final model we have combined the model described in Sect. 4.6.4, using the growth function shown in (4.26), with a third component suppressing the growth process. In the lattice model, independently of the nutrient, a chemical agent (the simulated isomone) is being dispersed through both diffusion and flow, using the computed flow velocities in the algorithm described in Sect. 4.6.4. After the (nutrient) tracer step, a second (isomone) tracer step is done for both phases and the local mean isomone gradient $i(c) = \frac{1}{2}(i_1(c) + i_2(c))$ is computed. The sources of isomone are all the voxels in the discrete representation of the object (see Fig. 4.38) neighboring to the fluid nodes, excluding the growing tips of the object. In the simulations only the voxels representing the object and added *delay_isomone* growth steps ago are used as source nodes. Without this restriction all surface nodes would become source nodes and growth would be suppressed everywhere and stop the complete growth process immediately. In all simulations this parameter *delay_isomone* was set to the value 5. In the simulations it is assumed that in every time step in the algorithm in Sect. 4.6.4 there is a constant rate of removal *decay_isomone* of isomone for both phases. The isomone tracer distribution is computed until an equilibrium is reached between the source and decay of isomone. The local mean isomone gradient $i(c)$ is used in a version of the growth function $G()$ in (4.20), shown in (4.30), where the growth velocity is suppressed for values of $i(c)$ above zero.

$$G() = s \cdot k(c) \cdot (1 - i(c)) \cdot h_2(low\_norm\_curv, av\_norm\_curv) \quad (4.30)$$

In Fig. 4.48 the isomone distribution is shown around an object generated with the model using (4.30) for the diffusion-limited case ($Pe \approx 0$) and the flow-dominated case ($Pe = 3.0$); here the color gradient white-black indicates the concentration of isomone where the high concentrations are colored black. Two objects generated with this model are depicted in Fig. 4.49.

4.6.7 A Comparison Between the Accretive Model and the Growth Forms

In the objects generated with the $f(\alpha, \beta) \cdot h_2(..)$ model (4.22) shown in Fig. 4.37 flattened forms are generated. The anisotropy in the object is caused by the $f(\alpha, \beta)$ component in the growth function, where the growth velocity is relatively higher when the angle β between the surface normal and the flow direction is higher. The $f(\alpha, \beta)$ component in the model from (4.22) does not provide much insight into why certain shapes develop in the accretive growth process. For example, the influence of hydrodynamics or light is not included in this model and there is no mechanism in this model preventing branches from self-intersection during the growth process. The branching in these objects is caused by the $h_2(..)$ component in (4.22); the growth axes mechanism in the $f(\alpha, \beta)$ component gives a relatively regular branching pattern. The thickness of the branches, as well as the branching pattern,

Fig. 4.48a–d. Isomone distribution around objects generated with the accretive growth model using (4.30). The gradient white–black indicates the concentration of isomone where the high concentrations are colored black. (a) Section in the xz-plane, the diffusion-limited case ($Pe \approx 0$). (b) Section in the xz-plane, the flow-dominated case ($Pe = 3.0$). (c) Section in the xy-plane, the diffusion-limited case ($Pe \approx 0$). (d) Section in the xy-plane, the flow-dominated case ($Pe = 3.0$)

(a)

(c)

(b)

(d)

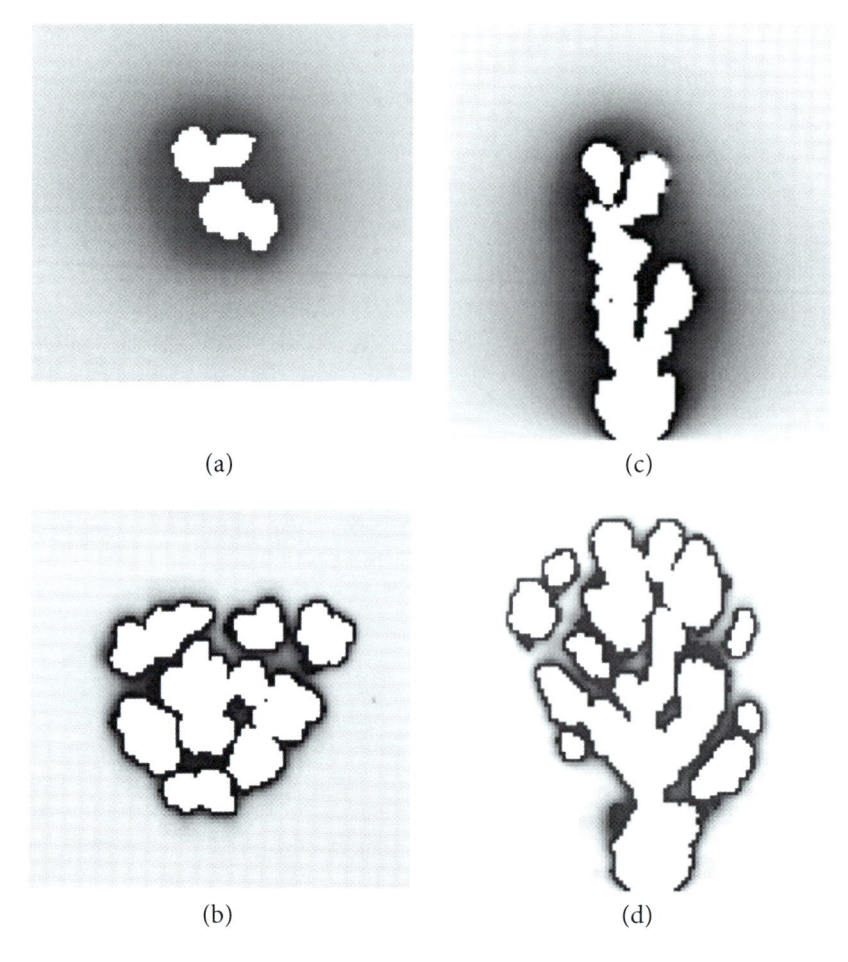

Fig. 4.49a,b. Objects generated with the accretive growth model using (4.30). (a) The diffusion-limited case ($Pe \approx 0$, $decay\_isomone = 0.1$). (b) The flow-dominated case ($Pe = 3.0$, $decay\_isomone = 0.01$)

can be controlled in this model with the max_curv parameter in (4.24). For higher values of max_curv the branching rate and the thickness of branches increases. Objects such as those shown in Fig. 4.37 resemble growth forms of the sponge *Haliclona oculata* (see Fig. 2.16 and Kaandorp 1995). The former can to a certain extent be morphologically compared with the latter using the techniques discussed in Sect. 3.3. By tuning the max_curv it is possible

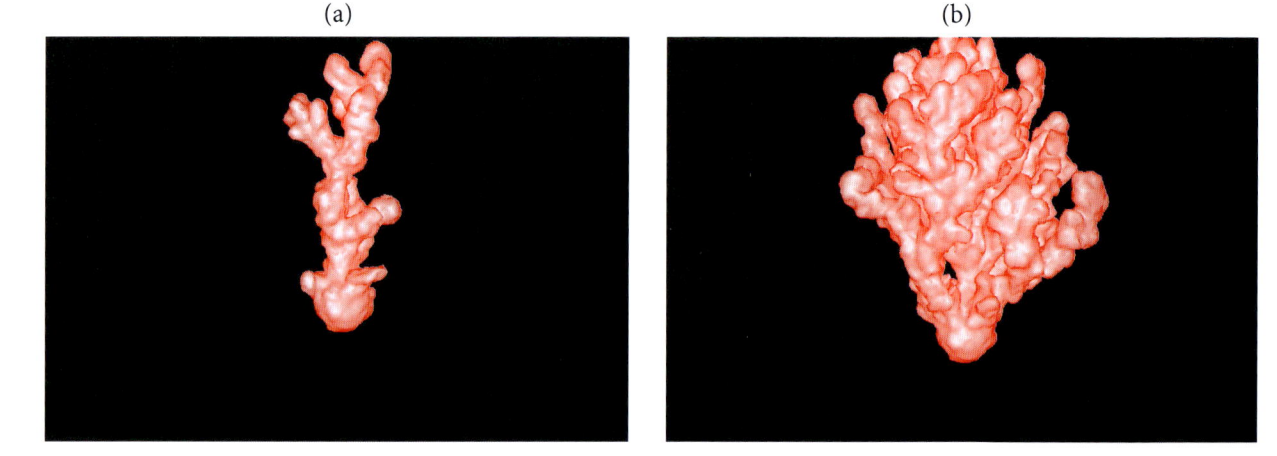

(a)

(b)

to generate a gradient in morphologies similar to that observed in *Haliclona oculata*. The growth axes mechanism applied in the $f(\alpha, \beta)$ component, where the growth direction of previous growth layers partly determines the growth direction of a new growth layer, may be a realistic mechanism. In for example the sponge *Axinella polypoides* (see Fig. 2.19), the dense central axis of skeleton material may be the biological representation of such a growth axis. In organisms with an accretive architecture (Figs. 1.4 and 2.17), where usually the direction of growth of new layers is again determined by the previous directions, the biological representation of growth axes might be the location of cells which are secreting skeleton material.

The sphere- and column-shaped forms generated with the $L(\theta)$ model (4.28) resemble the growth forms of the stony coral *Montastrea annularis* shown in Fig. 2.31a and b. In this species photosynthesis represents the main energy source (see Sect. 2.2.3). The growth forms in the range shown in Fig. 2.31c and d, where the tapered hemisphere gradually transforms into a substrate covering sheet cannot be captured with the current version of the accretive growth model. In this version it is assumed that the object is a manifold, where each triangle in the mesh is connected to a neighbor. A simulation model of light-driven, accretive growth, in which substrate covering sheets can be simulated, requires a modified version of the model allowing for triangular meshes with free boundaries. In such a modified version growth may occur by the addition of new polygons at the free edges of the sheet. The branching form generated with the $L(\theta) \cdot h_2(..)$ model (4.29) resembles growth forms of some types of stony corals, for example the Caribbean species *Acropora palmata*. In this model branching, umbrella-shaped forms emerge, where branches do not self-intersect, since the light-driven growth process stops as soon as neighboring branches are at the point of self-intersection.

When comparing the objects generated with the $k(c)$ model in Fig. 4.39 and with the $k(c) \cdot h_2(..)$ model in Fig. 4.41, it can be observed that in the simulations where growth is exclusively driven by local nutrient gradients (using growth function (4.25)), highly complex surfaces develop with a D_{box} of around 2.35. In contrast the surfaces which develop in the experiment (see Fig. 4.41) where growth is driven by local nutrient gradients and the local amount of contact with the environment (using growth function (4.26)) are relatively smooth and vary between 2.00–2.31. In Fig. 4.39 lobed objects without discernible branches are formed. A comparison between Figs. 4.39 and 4.41 demonstrates the effect of the amount of contact with the environment $h_2(..)$, causing the formation of branches in Fig. 4.41. Both the lobed and the branching morphologies can be found in actual sponges and stony corals. This result suggests that sensitivity to the amount of contact with the environment, for example by a relatively low contribution of translocation of nutrients from the place of absorption to more remote sites, could play a role in the formation of branches.

In a comparison between the ranges shown in Figs. 4.39 and 4.41 using the ratio R of the total sink nodes to the total number of object nodes, listed in Tables 4.4 and 4.5, for both ranges R becomes lower for increasing Pe numbers. The decreasing R ratios indicate an increasing degree in compactness. It is quite remarkable that for the range shown in Fig. 4.41 the values for D_{box} increase with increasing degree of compactness, while based on the observations done on the R ratios alone the opposite would be expected. Probably this is due to the relatively low number of branches which have developed in the objects in Fig. 4.41 close to $Pe \approx 0$, which makes the objects too small for an accurate estimation of D_{box}. Growth of the object in the diffusion-limited regime is relatively much slower than growth in the flow-limited conditions, resulting in smaller objects for the same number of growth steps in the simulations. In the range with branching objects, the object gradually transforms from a thin-branching form in Fig. 4.41 for the value $Pe \approx 0$, into a compact shape for $Pe = 3.0$. In the range with lobed objects in Fig. 4.39 the more or less columnar objects for low Pe numbers transform more abruptly into more spherical forms. The gradual increase in compactness for an increasing influence of hydrodynamics in Fig. 4.41 corresponds to observations of ranges of growth forms of marine sessile organisms collected along a gradient of increasing influence of exposure to water movement (see Sect. 3.3). In the same ranges it is also observed that D_{box} values, estimated from two-dimensional pictures, tend to decrease with an increasing degree of compactness. A similar observation was done in three-dimensional aggregates, where the growth process was controlled by the local availability of simulated nutrients (see Sect. 4.5): for increasing Pe numbers more compact aggregates develop, while the D_{box} value decreases from 2.27 to 2.05.

In both ranges shown in Figs. 4.39 and 4.41 the average absorption $\bar{a}$ (see Tables 4.4 and 4.5) in the sink nodes of the simulated objects increases going from the diffusion-limited ($Pe \approx 0$) to the flow-limited regime ($Pe = 3.0$). The exponent D_{abs}, which can be interpreted as a measure of the uniformity of the nutrient distribution, decreases for the range of branching objects in Fig. 4.41; for increasing hydrodynamics the nutrient is distributed more evenly over the system. The same phenomenon is not observed in the range of lobed objects in Fig. 4.39; the underlying reason could be the complex geometry of these objects where an intricate system of narrow crevices is formed preventing a uniform distribution of nutrients. In Figs. 4.40 and 4.42 slices in the xz-plane through the simulation box are shown where the nutrient distribution is visualized around the object for $Pe \approx 0$ and $Pe = 3.0$. The most obvious difference between the diffusion-limited nutrient distribution and the flow-limited distribution is the asymmetry due to the flow. In the diffusion-limited case for the lobed and the branching objects a roughly radial symmetric nutrient pattern is found; when comparing the left-to-right and the right-to-left flow phases there are hardly any effects of the flow to be seen. In the flow-limited case for both experiments there is a clear difference between the two phases; a depleted region downstream of the object is observed, while upstream the nutrient pattern is not disturbed by the object. In the more open branching objects in Fig. 4.42 more nutrient arrives between the branches compared with the lobed objects in Fig. 4.40, where hardly any nutrient arrives between crevices. This may explain the relatively high D_{abs} for high Pe numbers in Table 4.4. In Figs. 4.43 and 4.44 it can be seen that in the average nutrient distribution, depletion zones develop at both the left and the right side of the simulated object, for the flow-dominated case. Both zones emerge at certain distances from the object; close to the object there is still a region with a relatively high nutrient concentration.

For both experiments there is also a clear difference in the absorption patterns (see Figs. 4.40 and 4.42) for the diffusion-limited and the flow-limited case. In the $Pe \approx 0$ case most nutrients are being absorbed at protruding parts of the object, while in the $Pe = 3.0$ case most nutrients are being absorbed upstream of the object. The absorption patterns show an asymmetry in the absorption of nutrients at the sink nodes of the objects for increasing *Pe* numbers. For the diffusion-limited case, the highest degree of absorption is found at the tips of the object, while near the center hardly any nutrient is being absorbed. The observation that the highest amount of absorption is found at the upstream side of the object corresponds to the observation recorded in experimental work on the absorption of food-particles by the stony coral *Madracis mirabilis* (see Fig. 2.5), where the highest amount of food particle capture was found at the upstream side of the colony.

Within the given error bounds in Tables 4.4 and 4.5 from the standard deviations sd_x and sd_z no particular flattening in the lobed and branching forms can be detected. In both cases due to the effect of bidirectional flow in the range of both the lobed and branching forms, objects with a roughly radial symmetry develop.

The objects shown in Figs. 4.39 and 4.41 differ still in many aspects from the growth forms shown in Fig. 1.1. One important difference between the actual growth forms and the simulated ones, where the range of branching forms in Fig. 4.41 seems to best approximate the actual forms, is the high degree of irregularity in the formation of branches in Fig. 4.41. A simple explanation for this irregularity is that in the model the highest deposition velocities occur at the highest gradients of nutrients and are only slightly limited by the deposition velocities in the previous layers. The main direction of growth is not determined by the direction of growth in the previous layer and influenced only by local nutrient gradients. Consequently the main direction of growth may vary strongly after each deposition step, resulting in a highly irregular form. In Fig. 4.37 more regular forms were generated by applying a growth axes mechanism in the accretive growth model using the growth function (4.22), where the local growth velocity is also determined by the direction of growth in the previous growth layers. As mentioned in the first paragraph of this section, it might be plausible to apply a growth axes mechanism in the accretive growth models.

An important issue, in the comparison with experimental work, is to verify whether the simulated *Pe* numbers approximate the actual conditions. In order to make the *Pe* numbers mentioned in the text above comparable to *Pe* numbers in experimental work, all *Pe* numbers shown for the simulations have to be multiplied by 144 (the size of the simulation box, the characteristic length scale L in (2.4)). This yields an upper limit of $Pe = 432$ in our simulations. When we compare this result to the low flow velocity experiments done by Sebens et al. (1997), with an average flow velocity of 15 cm·s^{-1} and using a characteristic length scale of 10 cm for a *Madracis mirabilis* colony, this gives a value of $D \approx 3.5 \cdot 10^{-5} \text{ m}^2\text{s}^{-1}$ in (2.4) for the diffusion coefficient of the food particles. Assuming that there is a direct relation between the diffusion coefficient and the size of the diffusing particle (Vogel 1988), this would roughly correspond to food particles with a size of about $5 \cdot 10^{-3}$ m. This size corresponds to the order of magnitude of food particles which *Madracis mirabilis* colonies capture. For sponges the typical size of the food particles is within the range $0.2 \cdot 10^{-6}$ m–$50 \cdot 10^{-6}$ m (Brien et al. 1973), showing that the *Pe* numbers in our simulations are below realistic *Pe* values for sponges. In the paper by Ghorai and Hill (2000) an estimate for the value of D is provided for *Chlamydomonas nivalis* cells. Suspensions of *Chlamydomonas* have been used for culturing sponges under laboratory conditions (Brien et al. 1973).

The body diameter of these cells is within the range $10 \cdot 10^{-6}$ m–$20 \cdot 10^{-6}$ m. They estimate the value of D, in which an estimate of the swimming speed of the individual cells is also included, at $2.5 \cdot 10^{-8}$ m^{2}s^{-1}. In our model, simulation of nutrient distributions representing *Chlamydomonas* suspensions would require a *Pe* number of 600000.

The objects generated with the $k(c) \cdot (1 - i(c))h_2(..)$ model (4.30) in which the nutrient-driven growth is regulated by a simple suppression mechanism, shown in Fig. 4.49, differ in several aspects from the objects formed in a process exclusively driven by the supply of nutrients (Fig. 4.41). In Fig. 4.49b, where there is a relatively low decay rate of the suppressing agent (*decay_isomone* = 0.01) compared with Fig. 4.49a (*decay_isomone* = 0.1), this difference is particularly visible. The simulated form, when compared with the object shown in Fig. 4.41 for the same *Pe* number (*Pe* = 3.0), has become somewhat more regular. The branches tend to grow more upwards, away from the substrate, since the older part of the object produces relatively more growth-suppressing agent (see also Fig. 4.48). Although the objects shown in Fig. 4.49 represent only a few realizations and a further, systematic exploration for different parameter settings (various settings for *decay_isomone*, *delay_isomone*, and *Pe* numbers) is required, this demonstrates the important role of possible regulation mechanisms. Compared with the growth forms of actual stony corals (see for example Figs. 2.39 and 3.18) where an isomone regulation mechanism might play a fundamental role in the emergence of colony shapes with a highly regular branch spacing, the objects in Fig. 4.49 do not show an obvious regular branch formation. As mentioned above in the discussion on the $k(c) \cdot h_2(..)$ objects, a possible explanation for this is that growth velocities in a growth layer depend only to a limited extent on the previous growth velocities.

4.7 Gastrovascular Dynamics of Hydractiniid Hydrozoans

In Sect. 2.2.3 the characteristics of fluid transport in the gastrovascular system and redox chemistry were identified as the physiological mechanisms in colonial hydrozoans that control morphological plasticity. As shown in Fig. 2.27 hydractiniid polyps prior to feeding, or long after regurgitation, behave simply. Contractions are infrequent, they lack periodicity, and the amplitude of volume exchange with the stolon is small. The behaviors of polyps between ingestion and regurgitation can be classified into three distinct phases. These phases of behavior reflect differing input-output relationships between the polyp and either the external (via the mouth) or internal (via the polyp–stolon junction) environment. Moreover, these phases carry the signatures of characteristic dynamical behavior that can be expressed in terms of frequencies and amplitudes of polyp and stolon oscillations (Dudgeon et al. 1999). Wagner et al. (1998) used this conceptualization of post-feeding behavior to develop a model of the following ordinary differential equations to characterize the dynamical behavior of a single isolated polyp:

$$\frac{ds}{dt} = f_s(s, u) \qquad \frac{dn}{dt} = f_n(s, n) \qquad (4.31)$$

$$\frac{dx}{dt} = xg(n, x^2, y^2) - y \qquad \frac{dy}{dt} = yg(n, x^2, y^2) + x$$

The first two equations represent inputs to the polyp. It is not necessary to provide the mathematical details of the functions f_s and f_n, except that it is required that the amount of nutrient n, increases from zero after feeding, reaches a plateau when a steady state between production and metabolism of n is reached, and subsequently declines when the food source is depleted. The function f_s represents the amount of solid food substance, s, remaining in the polyp, given a monotonic decrease, after time t, and u represents the number of food items fed to the polyp. In the function f_n, n represents nutrients available to the polyp. The function determines the net rate of nutrient generation: the difference between the rate at which nutrients are generated by the digestion of solid food and the rate at which they are metabolized or exported by the polyp. The time course of n starts at $t = 0$ at the time of feeding, rises as digestion commences, reaches a plateau when a balance between production and utilization is reached, and subsequently declines towards 0 as the food source is depleted. The dynamics of the nutrients, described by the function f_n, trigger the polyp oscillations modeled by the function g. In g the two state variables of the polyp, x and y, represent an underlying nonlinear oscillator whose action is sensitive to n. The oscillations of x and y are reflected in corresponding oscillations in polyp length (or volume). Such a model is in accord with qualitative aspects of polyp behavior, including the rise in oscillation amplitude and increase in polyp length shortly after ingestion and the observation that polyps eventually return to a quiescent state when the nutrient level drops below the threshold level (Wagner et al. 1998). The empirical data from a mensurative experiment on *Podocoryne* in Fig. 4.50a shows the characteristic build-up in polyp length and oscillation amplitude shortly after feeding. Within several minutes polyp oscillations show limit cycle behavior. Likewise Fig. 4.50b, representing data derived from the model shown in (4.31), shows the characteristic build-up of polyp length and limit cycle oscillations of *Podocoryne* polyps.

In hydrozoans, the physiology of fluid transport within a colony is thought by many to be the locus of control of growth and form. Based on this idea, the quantitative model above was developed to characterize and predict the dynamics of that transport. At present, a gap exists between the "conceptual model" that explains the control of colony form by vascular transport and the "quantitative model" that characterizes the dynamics of that transport, making verification of the latter model difficult. The gap revolves around the scales of resolution of each model. The gastrovascular dynamics model has

Fig. 4.50. (*a*) Experimental time series showing oscillations in the length of a *Podocoryne* polyp during the first 2.5 hours following feeding. (*b*) Predicted oscillations in the polyp length during the first 2.5 hours following feeding, using the model shown in (4.31)

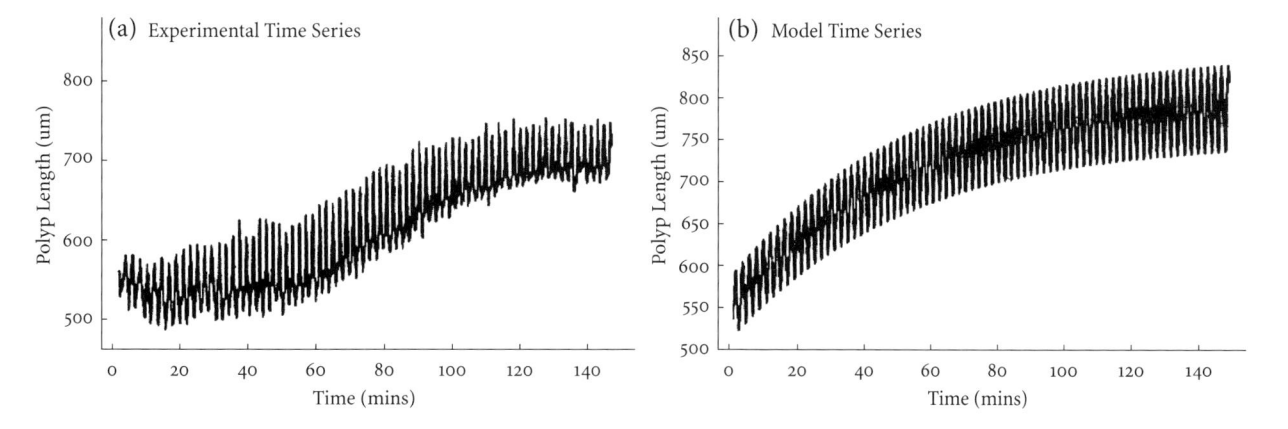

been developed only for single polyps and characterizes their behavior at very high resolution. The output of this model is the time-varying behavior of oscillations in polyp length or volume, not a morphogenetic pattern. In contrast, conceptual models that associate gastrovascular transport or redox chemistry with the regulation of form have emerged from lower resolution studies of whole colonies that focus on only a portion of gastrovascular behavior hypothesized to be most important. Verification of each type of model requires different experimental evidence.

The gastrovascular dynamics model needs to be tested for multi-polyp cases arrayed in different geometries. The test of this model is a test of whether, for a given geometry of colony architecture and input of food, the collective oscillatory behavior of polyps generates predictable patterns of vascular transport and redox variation within a colony. If so, then the second critical test is whether specific patterns of flow and redox variation give rise to specific and predictably spaced morphogenetic events.

There is considerable evidence demonstrating that patterns of vascular transport and redox variation within a colony control morphological development, but the specificity of cause and effect is not yet resolved. In other words, how much of a change is needed, and over what spatial scale, in the characteristics of flow or redox potential to initiate a morphogenetic event is unknown. Nevertheless, the link between gastrovascular physiology and colony development is clear (Blackstone and Buss 1992, Blackstone 1996, 1997, 1998, 1999, Dudgeon and Buss 1996).

This single polyp model can be scaled up to a multiple polyp colony, ideally with polyp behaviors expressed in volume dynamics. Doing so would necessitate equations to represent the behavior of each polyp in the colony whose oscillations are triggered by a threshold concentration of nutrients circulating in the gastrovascular system, and equations that represent volume transport through the stolon, perhaps with parameters that represent the length, diameter, and branching of the tube. Such a model would abstract the gastrovascular system of a hydrozoan colony to a spatially distributed system of coupled nonlinear oscillators.

5. Verifying Models

Experimental evidence is required for validation of simulation models of morphogenesis. Testing the predictive value of the simulation model experiments is crucial and the design of validation experiments requires close collaboration between experimentalists and modelers. An important test is to try to predict the consequences of perturbations of the growth process using the simulation model. Comparison of the predicted results and the actual results of a perturbed growth process can be used to test the degree to which the simulation model approximates the organism. Transplantation and other perturbation experiments, for example, experimentally induced changes in the physical environment and other manipulations in which the growth process is affected, represent an important option for studying the growth process. In this chapter several examples of these experiments done in stony corals (Sect. 5.1), sponges (Sect. 5.2), and hydrozoans (Sect. 5.3) will be presented. In the second section, some examples will be given as to how the experimental manipulations compare to the predictions done using simulation models, discussed in Sect. 4.6.

In most cases detailed information on the growth process of marine sessile organisms is not available in the literature; experiments in which growth and form of the organism is followed for a longer period provide crucial information for the construction of simulation models of growth and form. Through transplantation experiments in which organims are transplanted from an environment with certain physical conditions to another one, it becomes possible to distinguish between morphological variations induced by physical environments and those by genetic differences. On the other hand, simulation models can be used as a guide in experiments to identify missing crucial experimental information and to indicate interesting experimental manipulations and also allow for detailed estimations of, for example, concentration gradients or microflows (see Sections 4.5 and 4.6), which are experimentally not feasible.

5.1 Transplantation Experiments with Stony Corals

Transplantation is an important experimental procedure to examine mechanisms controlling growth forms in stony corals. By comparing the fragments before and after transplantation, we can distinguish between morphological variations induced by physical environments and those by genetic differences, and know which environmental condition causes the plastic changes of morphology and how the mechanism of development varies between the different environments. In stony corals, skeleton architecture shows the history of the colony growth. X-rayed photographs of the slab of the colony section (see for example Fig. 2.34) provide us information on the growth process, such as the growth rate and the angle of corallite growth. The details of the skeleton band will be described in Sect. 6.2. The transplantation experiments can be used to test the predictive value of the simulation models of growth and form discussed in Chap. 4.

A transplantation experiment combined with modeling coral growth was conducted by Graus and Macintyre (1982) using *Montastrea annularis*. *M. annularis* is a species widely distributed in the Caribbean sea, and its growth form varies with the water depth, as shown in Fig. 2.31. This species exhibits a hemispherical morphotype in shallow water, but gradually changes the growth form to column-shaped and then to a tapered form as the light intensity decreases. Eventually it becomes a plate-like morphotype around 30 m in depth. Colonies stained with Alizarin Red S were transplanted reciprocally, i.e. from shallow water to deep water and vice versa, and left for three years. X-ray photographs of the slice of collected colonies showed the surface of the colony at the moment of staining as a colored band. Graus and Macintyre (1982) measured the maximum skeletal growth rate and the maximum angle of corallite growth. Both decreased with increased water depth, and were correlated positively with each other. On the other hand, skeletal density and corallite spacing increased with increased water depth. These four morphological changes resulted in the observed value in the population at the transplanted site. This result strongly suggests that morphological change in growth form in *M. annularis* is a plastic response to light. The model proposed by Graus and Macintyre (1976, 1982) succeeded in simulating the development of various growth forms in *M. annularis* (see also Sect. 4.6.5 on accretive growth driven by photosynthesis). In their model, the calcification rate was assumed as a function of light intensity, and the maximum angle of corallite growth was given by the regression curve obtained from the transplantation experiment.

Reciprocal transplantation has been conducted to reveal the morphological change induced by environmental conditions. Oliver et al. (1983) investigated the morphological variation in *Acropora formosa* along depth gradients. Growth form of *A. formosa* consists of short and close branches at the shallow site, but longer and more widely spaced branches at the deep site. The morphological parameters, the rates of linear growth (branch extension) and the apical growth of branch, were measured. The transplants at the deep site extended linearly twice as fast as those at the shallow site. However branch initiation occurred only at the shallow site. The total growth rate including the extension of new lateral branches was higher at the shallow site. Oliver et al. (1983) suggested that the faster elongation of branches at the deep site was the consequence of translocation from a greater volume of tissue. In contrast, the tips of transplants at the shallow site probably received translocation from a smaller volume of tissue. As a result, the differences in the growth process between the sites, such as frequent branch initiation and slower extension rate at the shallow site and lower branching rate and higher extension rates at the deep site, caused the differences of growth form in *A. formosa*. Light, water movement, or a combination of the two are supposed as probable controlling factors of branch extension and initiation in this experiment. The details of translocation of photosynthetic products between the polyps are described in Rinkevich and Loya (1983b), Oren et al. (1997) and in Sect. 2.2.4.

EXAMPLE OF PORITES SILLIMANIANI. In this section an example is given of transplantation experiments with the stony coral *Porites sillimaniani* which examine the plastic change of growth form in response to light conditions. *Porites sillimaniani* is a common species in a variety of reef environments of the tropical Pacific, and it displays a striking variation of the whole colony morphology with respect to light availability (see Fig. 2.32). The transplantation experiment, in which clonal replicates are exposed to different light conditions, can solve the problem of whether the morphological variation is based on genetic differences or due to phenotypic plasticity. Details of this transplantation experiment can be found in Muko et al. (2000).

Two transplant sites were established side by side within 3 m on the sea bottom 6 m in depth in Okinawa, Japan. Each site had two iron frames in order to manipulate the light intensity. The light intensity was controlled using plexiglass covers. The other environmental factors, such as water velocity, salinity, nutrient level, and so on, were kept unchanged. The irradiance on the horizontal plane at the high-light site was significantly higher than that in the low-light site for various weather conditions (Table 5.1).

Seven plate-like morphotypes were collected from the shaded site at 4 m in depth. The average irradiance of the collected sites was intermediate between the two manipulated sites (Table 5.1). Each colony was considered to represent a different genotype. All colonies were divided into several fragments measuring approximately $3 \times 3\,\text{cm}^2$. The fragments from each colony were evenly allocated to the two sites (total 26 fragments in each site), and each was attached to a substrate with epoxy glue.

Of 26 fragments initially set in the high-light site, only 13 could be used for analysis because half of them were destroyed by a typhoon. All 26 fragments in the low-light site were used for analysis. Of the remaining clonal replicates, some fragments were lost partly by natural death during the eight months from August 1996 to May 1997 but at least one fragment survived in both sites for each of the 7 genotypes. No significant difference in survival rate was detected between the two sites.

From the sequential pictures of each transplanted fragment, we can follow the development of the growth form in *P. sillimaniani*. Fig. 5.1 illustrates the examples at the high-light site and those at the low-light site, respectively. Each row indicates the fragments after one week, and after eight months from the initiation of the experiment. The projected area of each transplanted fragment was measured to the nearest $0.1\,\text{cm}^2$. The growth rate in the projected area was calculated for each fragment as: $100 \times$ (increment in the area during August 1996 to May 1997) / (initial area in August 1996). The number of visi-

site	minimal irradiance	maximal irradiance	median irradiance	n
high-light site	15.4	31.28	22.12	5
low-light site	1.51	14.35	3.22	5
collected site	2.47	18.17	6.18	7

Table 5.1. Irradiance at the two experimental sites and at the collected site. The irradiance differed significantly between the two experimental sites (Mann–Whitney U-test, $z = -2.6$, $P = 0.009$). n indicates the number of observed times for each transplanted site and the number of observed sites for the collected site, respectively.

Fig. 5.1a,b. Close-up pictures of transplanted fragments of *Porites sillimaniani* (after Muko et al. 2000). (*a*) In high-light site. (*b*) In low-light site. Left column indicates after one week (August 1996). Right column indicates after eight months (May 1997). The bar indicates 1 cm.

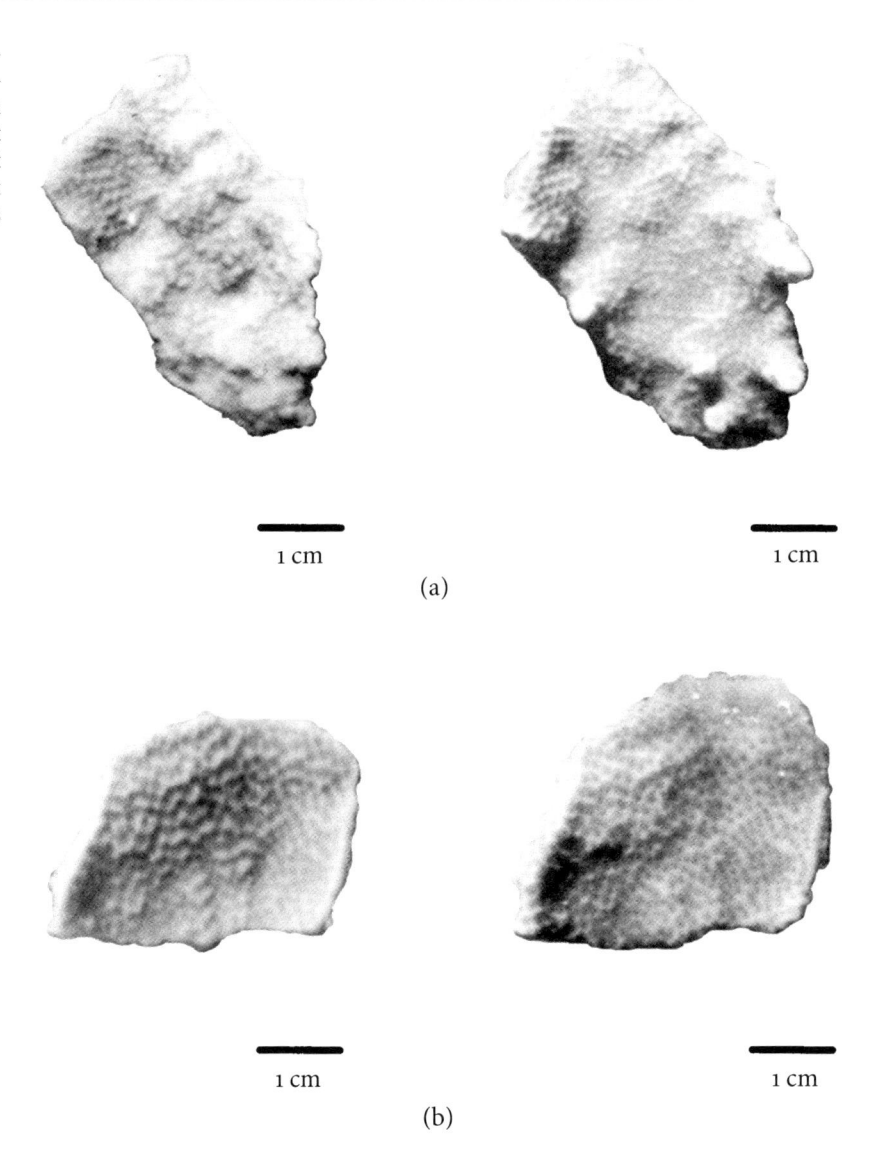

(a)

(b)

ble branches was counted for each fragment in May 1997. The diameter and length of each branch were measured with a calliper and a ruler to the nearest 0.1 mm. The surface area of each branch was calculated approximating to a cylindrical column. The surface area of each fragment was estimated by the sum of the projected area and the surface area of branches, from which the top surface was excluded as it was already included in the projected area. The growth rate in the surface area was calculated in the same way as that in the projected area.

In Fig. 5.1, fragment (a) transplanted at the high-light site showed branch initiation. The growth on the substrate was very slow. In contrast, fragment (b) transplanted at the low-light site grew widely on the substrate. Table 5.2 shows the results of this experiment. To examine the effects of the light conditions on morphology, a nonparametric analysis by the Wilcoxon paired-sample test was performed on each genotype placed at high-light and low-light sites for four parameters. The surface area of transplants before

	min high-light	max high-light	med high-light	min low-light	max low-light	med low-light	n
number of buds	0	8	2	0	0	1	7
initial surface area (cm^2)	3.17	16.96	4.93	5.41	8.87	7.44	7
growth rate of surface area (%)	17.26	86.23	49.98	−4.16	119.22	56.48	7
growth rate of projected area (%)	13.64	76.92	46.25	−4.12	115.75	56.48	7

the experiment did not differ significantly between the two sites. Moreover, significant differences were not detected between the two sites either in the growth rate of the surface area or in that of the projected area. In these three parameters, there was no significant difference among genotypes (S. Muko unpublished data). On the other hand, the number of buds differed significantly between the two light conditions.

Although it remained at an early stage of development after eight months of the transplantation, the formation of the branches was induced by high light intensity independent of genotypes. A morphological change was observed in *P. sillimaniani* in which the transplants developed morphologies similar to the morphotypes found at various light conditions (see Fig. 2.32). This tendency has also been reported for other species (Graus and Macintyre 1982, Foster 1980, Bruno and Edmunds 1997).

Morphological variations in stony corals are induced either by plastic response to the environmental conditions or by genetic constraints. Transplantation experiments can distinguish between the two possibilities and also provide much information on the development of growth form. In a transplantation experiment with two stony coral species Willis (1985) revealed the existence of phenotypic plasticity in *Turbinaria mesenterina* but not in *Pavona cactus* by reciprocal transplantation between the two depths. Later, Willis and his co-worker found that the two extreme growth forms of columnar and convoluted morph in *P. cactus* were genetically determined (Willis and Ayre 1985, Ayre and Willis 1988). Unfortunately, we do not have many studies in which the transplantation experiment is combined with a mathematical model of growth form. Using measured morphological features, models can reveal the growth process and be tested with transplantation experiments (Graus and Macintyre 1976, 1982, and Sect. 4.6.5) or the significance of plastic changes (Muko et al. 2000).

Table 5.2. Four parameters of growth form at the two experimental sites. The Wilcoxon paired-sample test was performed on each genotype ($df = 6$). Minimal (*min*), maximal (*max*), and median (*med*) values are given for low-light and high-light sites. *n* indicates the number of transplants.

5.2 Transplantation and Other Perturbation Experiments with the Sponge *Haliclona oculata* and a Comparison to the Simulation Models

In this section examples will be given of transplantation experiments and other experimental manipulations done with the sponge *Haliclona oculata*. These experiments are described in detail elsewhere (Kaandorp and de Kluijver 1992, Kaandorp 1994b). In this section we will discuss the impact of the perturbation experiments on the morphology briefly and compare, qualitatively, some of the morphologies found in the experiments to morphologies produced by the simulation models discussed in Sect. 4.6.

Fig. 5.2a,b. Contours of sponges before and after the transplantation experiments. Sample (*a*) was transplanted from a sheltered site to an exposed site; with sample (*b*) the reverse experiment was carried out ((*a*) 3.5 and (*b*) 1.5 month experiment). Parts of the sponges are marked with small steel needles (see also Fig. 2.21).

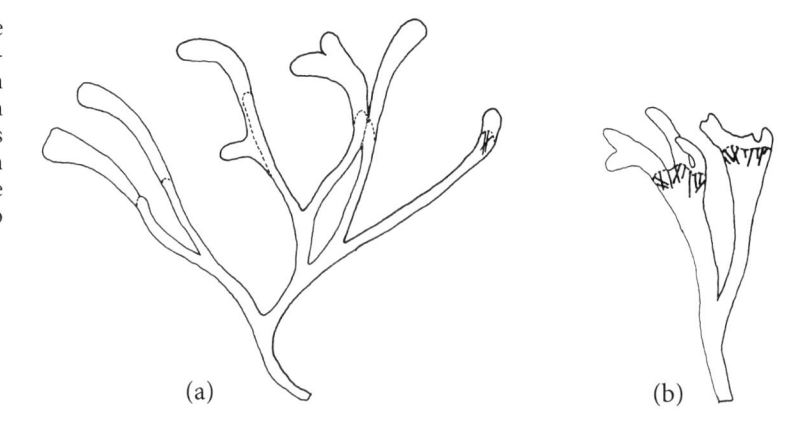

Fig. 5.3a,b. Examples of transplanted *Haliclona oculata* sponges. Sponge (*a*) was transplanted from a sheltered site to an exposed site; with sponge (*b*) the reverse experiment was carried out ((*a*) 3.5 and (*b*) 1.5 month experiment).

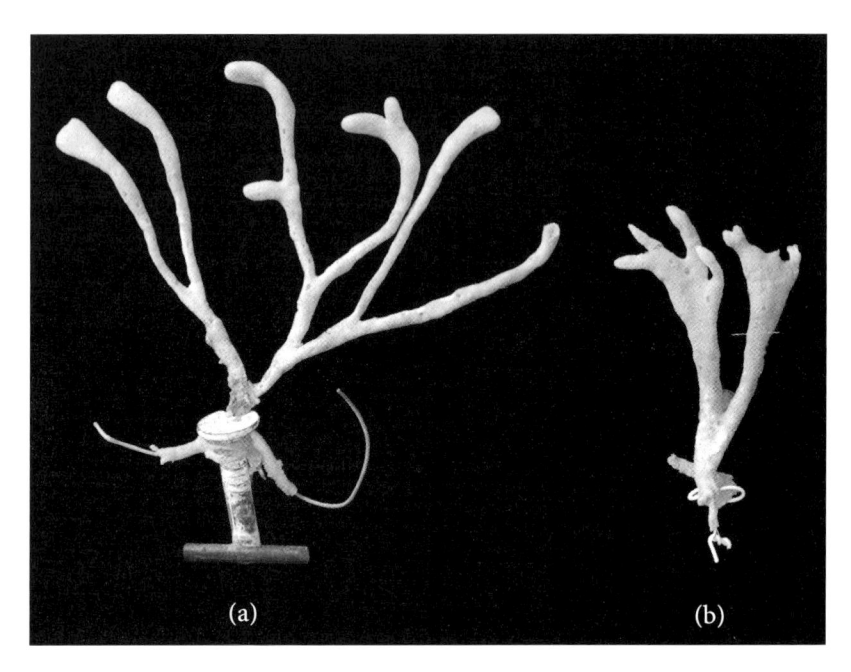

With the sponge *Haliclona oculata* transplantation experiments have been carried out where thin-branching growth forms (for example Fig. 2.16a) were transplanted from a sheltered environment to an environment exposed to water movement, while with plate-like growth forms (for example Fig. 2.16c) the reverse experiment was done. The resulting impact of the transplantation on the morphology is summarized in Figs. 5.2 and 5.3. In both figures it can be observed that the thin-branching sponge has developed plate-like ends, while the plate-like sample from the exposed site has developed thin branches after the transplantation. The experiment shows that the morphological plasticity in *Haliclona oculata* is strongly influenced by differences in exposure to water movement. Furthermore, the experiments as shown in Fig. 5.2 and the marking experiments depicted in Fig. 2.21 provide detailed information on the growth velocities in the sponge. In theory from both figures it is possible to construct a function approximating the distribution of growth velocities over the tips of the sponge.

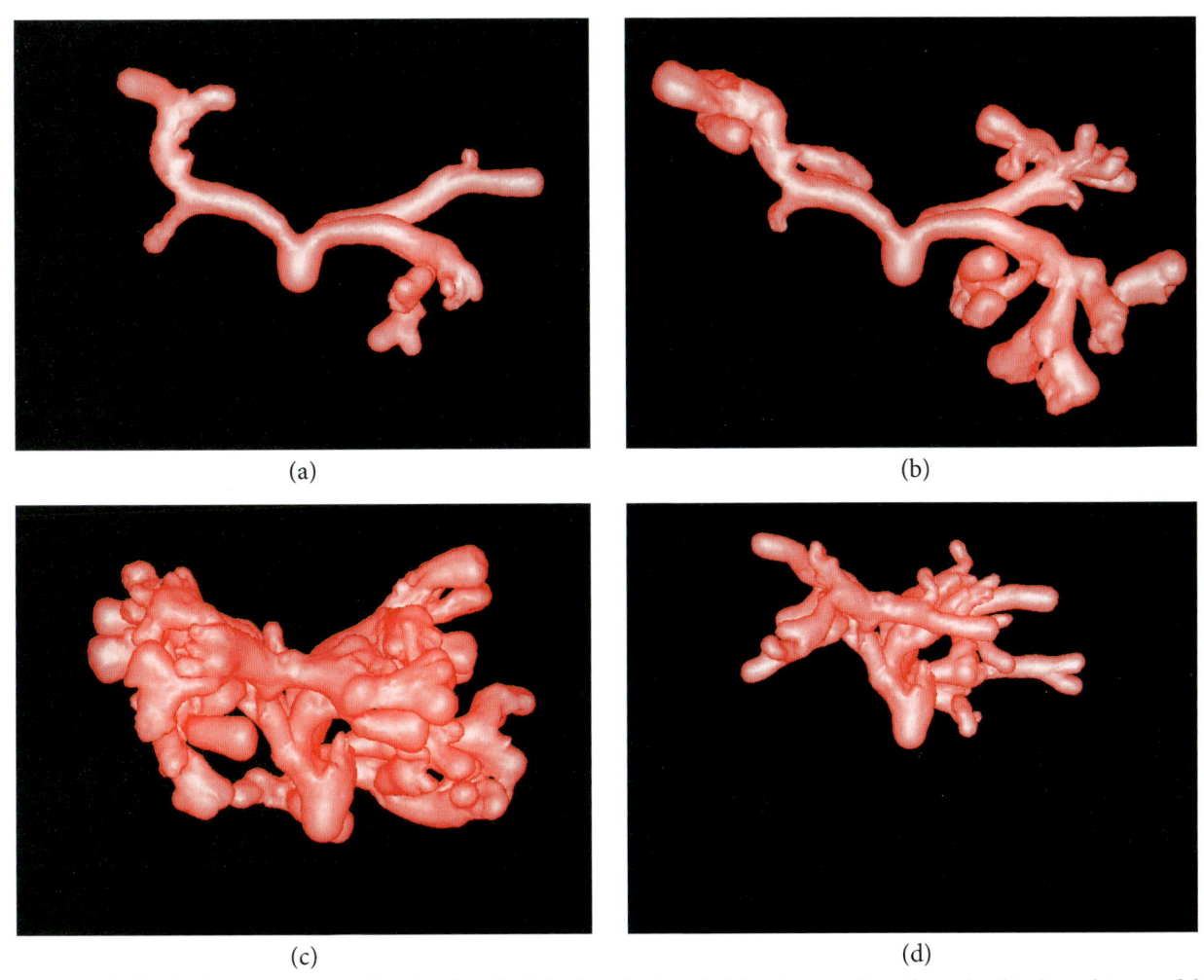

Fig. 5.4a–d. Simulation experiment using the $f(\alpha, \beta) \cdot h_2(..)$ model (4.22). (a) object resulting from the thin-branching model without perturbation, max_curv (4.23) is set to 10 s. (b) A thin-branching model changes into a plate-like one, by changing max_curv from 10 s to 60 s after 60 iteration steps. (c) Object resulting from the plate-like model without perturbation, max_curv is set to 60 s. (d) A plate-like model transforms into a thin-branching one, by changing max_curv from 60 s to 10 s after 60 iteration steps.

Approximated growth velocity distributions can be used to construct the $f(\alpha, \beta) \cdot h_2(..)$ model, discussed in Sect. 4.6.3. The transition in morphologies, shown in Figs. 5.2 and 5.3, can be mimicked in a simulation model based on approximated growth functions. In the simulation experiment shown in Fig. 5.4 a thin-branching model develops plate-like ends after 60 iteration steps by changing the parameter max_curv in (4.23), which determines the maximum allowed radius of curvature at the surface, from 10 s to 60 s. The max_curv represents a threshold above which the "local amount of contact of the object with the environment" becomes sub-optimal, and the biological interpretation of this increase is that in an exposed environment the supply of nutrients is relatively larger and the amount of contact with the environment is less critical. In the simulations using the $f(\alpha, \beta) \cdot h_2(..)$ model, a relatively less critical amount of contact with the environment can be represented in the model by applying a relatively larger value for max_curv. As

Fig. 5.5. Example of a transplanted *Haliclona oculata* which was positioned horizontally and again fixed to the substrate (3.5 month experiment)

Fig. 5.6a,b. Simulation experiment using the $k(c) \cdot h_2(..)$ model (4.22). In (*a*) the *Pe* was set initially to (approximately) 0, and after 80 iteration steps the *Pe* parameter was set to the value 3.0. In (*b*) the reverse experiment was performed and *Pe* was changed from 3.0 to (approximately) 0, after 80 iteration steps.

was discussed in Sect. 4.6.3, the $f(\alpha, \beta) \cdot h_2(..)$ model does not provide much insight into the influence of the environment on the growth process, since there is no model of hydrodynamics present in the model.

In for example Fig. 2.16 it can be observed that the sponges tend to form branches away from the substrate and few or no branches grow towards the substrate. This phenomenon can be demonstrated in a perturbation experiment in which the sponge is rotated 90° with respect to the current growth direction, and again fixed to the substrate (see Kaandorp 1994b for details). The resulting morphology is shown in Fig. 5.5. In this figure it can be observed that the growth direction of the tips of the sponge has changed: after the rotation the tips reorient themselves and bend away from the substrate. This effect cannot be predicted with the $f(\alpha, \beta) \cdot h_2(..)$ model; for simulating this type of phenomena a model of the physical environment is also required.

The transplantation and rotation experiments are repeated in simulation experiments using the $k(c) \cdot h_2(..)$ model (see Sect. 4.6.4), which uses a model of the distribution of food particles in the environment due to the influence of hydrodynamics. In Fig. 5.6a an object is shown which was initially formed under conditions where diffusion dominates and the influence of hydrodynamics is low. In this case the *Pe* parameter in the $k(c) \cdot h_2(..)$ model is set initially to (approximately) the value 0, and after 80 iteration steps the *Pe* parameter was set to the value 3.0 (the flow-dominated case). In Fig. 5.6b the reverse simulation experiment was done. In Fig. 5.6 it can be observed that the thin-branching object in (a) forms club-like branches after the perturbation, while in the reverse experiment on the compact-shaped object thin branches are formed after the *Pe* change. In these figures basically the same phenomena can be seen as demonstrated in the actual experiments: under exposed (to water movement) conditions, the thin-branching transplant starts to form club- or plate-like branches, while in the reverse experiment thin branches are formed after the perturbation. In Fig. 5.7 the rotation experiment is carried out in a simulation experiment: an object is generated with the $k(c) \cdot h_2(..)$ model, and after 80 iteration steps (with the parameter $Pe \approx 0$), the object was rotated 90° with respect to the *y*-axis in the simulation box (see Fig. 4.38). Before the geometric object was mapped onto the lattice in Fig. 4.38, the simulated nutrient distribution in the lattice was completely reinitialized, after the rotation. In Fig. 5.7 the influence of the simulated nutri-

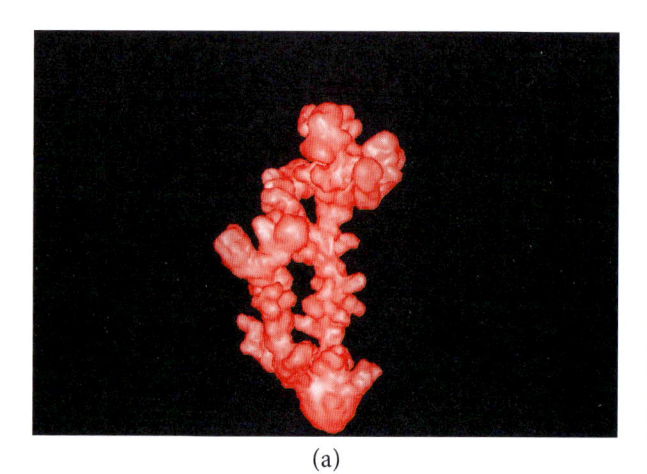

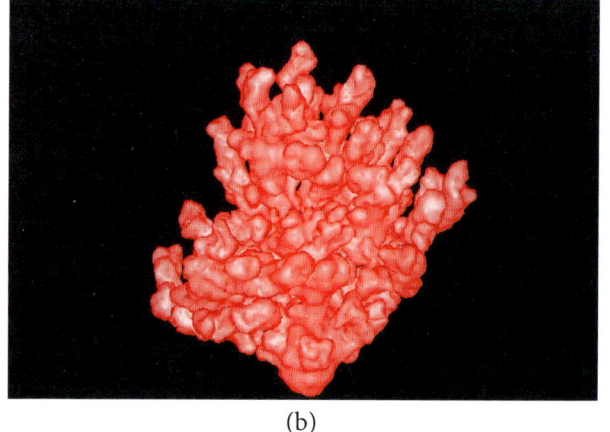

(a)

(b)

ent distribution becomes visible; the branches of the object after the rotation bend away from the substrate and grow towards the source of nutrient in the simulation box (the top plane). With the $k(c) \cdot h_2(..)$ model it is basically possible to predict the effects of manipulations where experimental changes are made in the nutrient distribution around filter feeding organisms with accretive growth.

It should be noted that the comparison between the $k(c) \cdot h_2(..)$ simulations is only valid to a limited extent. In Sect. 4.6.4 it was noted that one of the properties of the $k(c) \cdot h_2(..)$ model is that it produces branching patterns with a relatively high irregularity (the branching pattern is often more irregular than the growth forms shown in Fig. 2.16, which complicates (or may even prevent) a morphological comparison using the techniques discussed in Chap. 3. An example of a quantitative comparison between a version of the $f(\alpha, \beta) \cdot h_2(..)$ model, using a two-dimensional morphological comparison, can be found elsewhere (Kaandorp 1995). Furthermore the $k(c) \cdot h_2(..)$ model generates objects which display a roughly radial symmetry, such as that found in many stony corals, while the growth forms of *Haliclona oculata* exhibit a more or less flattened morphology. In addition it should be noted that the size of the food particles captured by filter-feeding sponges is much lower compared with stony corals (see Sect. 4.6.7), and as a consequence in the simulations much higher *Pe* values are required; this is currently beyond the computational capabilities.

In this section examples are given of some manipulations which can be predicted to a certain extent using simulation models. In the models discussed in Sect. 4.6 the effects of various other manipulations could be predicted, at least in theory. Examples are manipulations in which branches of the stony coral *Stylophora pistillata* are brought in close contact with neighboring branches (Fig. 2.39), using the "isomone" simulation model discussed in Sect. 4.6.6. Another example could be manipulations with light intensities in photosynthetic organisms (for example the shading experiments discussed in Sect. 5.1) in branching and massive stony corals, using the $L(\theta) \cdot h_2(..)$ or $L(\theta)$ models discussed in Sect. 4.6.5.

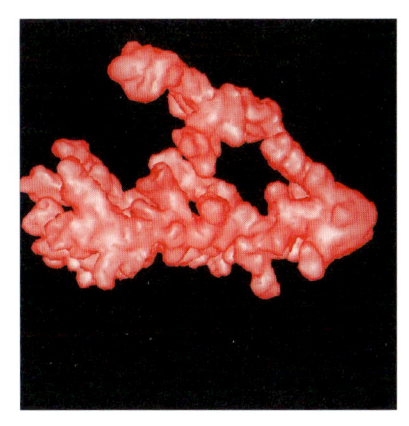

Fig. 5.7. Simulation experiment using the $k(c) \cdot h_2(..)$ model (4.26). The *Pe* parameter was set initially to (approximately) 0, and after 80 iteration steps the object was rotated 90° with respect to the *y*-axis.

5.3 Colonial Hydrozoans: Perturbation Experiments of Gastrovascular Physiology and Effects on Colony Development

In Sect. 2.2.3 the characteristics of fluid transport in the gastrovascular system and redox chemistry were identified as the physiological mechanisms in colonial hydrozoans that control morphological plasticity. As shown in Fig. 2.27 polyps of the hydractiniid hydrozoan *Podocoryne carnea* prior to feeding, or long after regurgitation, behave simply. The behaviors of polyps between ingestion and regurgitation can be classified into three distinct phases. These phases of behavior reflect differing input-output relationships between the polyp and either the external (via the mouth) or internal (via the polyp-stolon junction) environment. Moreover, these phases carry the signatures of characteristic dynamical behavior that can be expressed in terms of frequencies and amplitudes of polyp and stolon oscillations (Dudgeon et al. 1999). Based on these observations a quantitative model was developed to characterize and predict the dynamics of the transport within the hydrozoan colony discussed in Sect. 4.7.

Several lines of research indicate the dependence of morphological development on characteristics of gastrovascular physiology during phase 3 polyp behavior. As discussed in Sect. 2.2.3, phase 3 behavior is that characterized as the period of maximum volumetric transport of gastrovascular fluid through the stolonal network. The importance of this phase of behavior is not surprising. It represents the period in which physiological integration of the colony is strongest due to the interaction between all (or nearly all) of the polyps in a colony. Moreover, morphogenetic signals based on maximal metabolic activities presumably would convey less ambiguous information to cells compared to signals during other phases of behavior where their magnitudes are less and their variance is greater.

Experiments perturbing characteristics of gastrovascular flow and/or redox state during phase 3 strongly influence the timing of polyp and stolon tip morphogenesis. Blackstone and Buss (1992, 1993) manipulated flow and redox state by treatment with loose couplers of oxidative phosphorylation, such as 2,4-dinitrophenol (DNP), and assayed the effects on colony development. Such compounds diminish ATP production thereby diminishing the energy available to transport fluid and, consequently, the volumetric rate of transport declines. The effects on morphology are dramatic. In *Podocoryne carnea*, diminished flow accelerates polyp and stolon tip morphogenesis so that treated colonies appear much more sheet-like. Manipulation of gastrovascular physiology with loose couplers of oxidative phosphorylation or excessive overfeeding (which also slows fluid transport) generates a physiological state of relative oxidation, and in all experiments using *P. carnea* similarly accelerates the production of polyps and stolon tips. These perturbations also show a dose–response effect: an increase in the level of treatment increases the effect on morphology (Blackstone 1997).

Treatment with inhibitors of electron transport, such as azide, also diminish flow rate, but have the opposite effect on redox state; colonies become relatively reduced (Blackstone 1999). In this case, colonies of *P. carnea* produce fewer, larger polyps and fewer stolon tips and become more runner-like. This suggests that in *P. carnea*, variation in redox chemistry associated with gastrovascular transport regulates morphological development.

The results from perturbation experiments in *Podocoryne carnea* parallel the patterns of morphology and gastrovascular physiology in phase 3 observed in other studies. Colonies that exhibit higher volumetric rates of flow also exhibit greater redox variation and are associated with more rapid elongation of stolons and lower rates of stolon branch and polyp bud formation (Blackstone 1996, 1998). For instance, colonies of *Hydractinia symbiolongicarpus* show a more sheet-like form than does *P. carnea*. They also show lower rates of flow than *P. carnea* following the formation of stolonal mat tissue early in development and thereafter (Blackstone 1996). Finally, colonies of *P. carnea* derived from inbred lines and selected on the basis of morphology show the predicted patterns of flow and redox variation (Blackstone 1998). Runners have both greater volumetric rates of flow and greater variation in redox state than sheet colonies.

In neither case of reduced flow rate or redox state change is it known whether both polyp buds and stolon tips are regulated directly. It may be that they are regulated separately by the different putative causes, or one may be regulated directly and the other may be a consequence of the change in the regulated event. Consider, for example, the spacing of polyps along stolons. It has long been established that the number of polyps per unit stolon length is roughly constant for a given clone (Braverman 1963). If redox state, for instance, regulates stolon tip morphogenesis, then the acceleration of polyp morphogenesis may be a consequence of the greater total length of stolon (that is, there is no change in the relative investment in polyps and stolons). A simple explanation for the covariation between polyp number and stolon length or tip number is that polyp spacing may be related to flow. At some distance from a polyp another polyp is needed to continue transport of fluid to the growing margin. In other words, fluid exported into the stolon will travel only so far before being stopped by shear forces along the wall of the stolon. Thus, one might expect that greater forces exerted by polyps when pumping (or less resistance to flow) should result in greater spacing between them. Evidence exists that is consistent with this hypothesis (Buss, in press). As predicted by the equation balancing pressure and shear forces, Buss detected a linear relationship between polyp size (assumed to be linearly related to the pressure polyps can exert) and twice the stolon length between polyps divided by stolon radius (measured during phase 3).

The complex relationship between redox state, flow, and morphology is further demonstrated by considering *Hydractinia spp*. In *Hydractinia*, treatment with DNP shifts the redox state towards oxidation without diminishing flow rates (Blackstone and Buss 1993). In this case, runners maintained a low rate of polyp bud and stolon tip formation (i.e. continued as runners), whereas sheet colonies (in which flow to stolon tips increased) showed a decline in rates of bud and tip formation (i.e. subsequently developed as runners). What appears particularly important in the case of *Hydractinia* is how the vascular architecture (the sizes and arrangement of stolons) constrains the distribution and hydrodynamic characteristics of flow within a colony.

Evidence from experiments manipulating the sizes and arrangement of stolons supports the hypothesis that vascular architecture determines the characteristics of flow which regulate further morphological development (Dudgeon and Buss 1996). Runner colonies are characterized by highly variable stolon diameters, whereas sheet colonies exhibit little variation in diameters of stolons. Thus, runner colonies have an unequal distribution of flow to the growing periphery, whereas sheet colonies have equivalent volumes of flow delivered to the growing tips. These flow patterns reinforce their respective growth trajectories, as in a self-maintaining system. Surgical manipulation of stolon connections that change flow patterns in the gastrovascular system changes the subsequent morphological trajectory. Diminishing variation in flow rate among stolons within runner colonies results in subsequent sheet growth. Conversely, increased variation in flow among stolons within sheet colonies results in subsequent runner growth. The DNP experiment of Blackstone and Buss (1993) using *Hydractinia* can be explained in the context of this model: the effect of DNP on the transport of fluid was to diminish the energy available to pump fluid and to make the delivery of fluid to stolon tips inconsistent and highly variable, thereby initiating a runner growth trajectory.

In summary, the behavior of polyps, the redox state of cells, and the architecture of the vascular system provide information to cells lining the gastrovascular system about the physiological state of the colony. Polyps behave simply: they are relatively inactive when not feeding and they oscillate vigorously following ingestion for many hours thereafter. This generates a dynamic pattern of fluid transport throughout a colony and metabolic activity within cells. These simple rules of polyp behavior therefore provide local information that cells can use to make morphogenetic decisions. In this way, information about the physiological state of the colony emerges from the communication between polyps that are each sensing the local environment and this integrated translation of information generates an adaptive morphogenetic response.

6. Applications

Simulation models of growth and form and the analysis of growth forms provide insight into one of the most fundamental questions in biology: how is genetic information, in combination with environmental influences, physically translated into the actual form. These models also have are several other important applications. In this section we will briefly discuss one potential application, the design of aquacultures, and give two more detailed examples of another (potential) application, the analysis of bioarchives.

In marine organisms a large diversity of metabolites of great pharmacological and toxicological importance has been identified. The percentage of anticancer leads with significant cytotoxic activity in preclinical drug discovery is much higher in marine organisms than in terrestrial organisms (Braekman and van Soest 1993). Among the marine animals, the sponges represent an especially rich source of novel biologically active compounds. For example, compounds with cytotoxic, antibiotic, antifungal, antitumor, antiviral, antifouling, and enzyme-inhibitory activities have been found in sponges. Compared to the cnidarians and the algae, which are another important source of bioactive compounds, twice as many of these compounds have been described for sponges (van Soest and Braekman 1999). One of the underlying reasons might be that both sponge cells and sponge microsymbionts tend to produce bioactive compounds.

In most cases harvesting of natural populations is not an option and would very soon deplete the natural source. Furthermore species which tend to produce the most bioactive compounds are usually slow-growing species, which depend on these chemical agents for defending themselves against both predators and spatial competition. The alternative is to set up aquacultures of organisms producing these compounds, with the advantage that potentially high amounts of bioactive compounds can be produced under controlled conditions. An attractive option in sponges would be for example to use suspensions of cells in bioreactors. Unfortunately setting up aquacultures of marine organisms is extremely difficult (Osinga et al. 1999); it has been said that "trying to do invertebrate aquaculture is a nightmare" (see Pain 1996). Important issues in aquacultures are: the growth process of the organism is frequently not well understood; it is difficult to create the right hydrodynamic conditions and food supply in filter feeders; biological regulation mechanisms are not well known, for example the factors which keep the sponge cells in suspension and those which prevent them from forming aggregates. One potential application of the simulation models that we have discussed in the previous chapters is in engineering aquacultures, just as simulation models are being used for designing chemical reactors.

The second application of simulation models of growth and form is in the analysis of bioarchives. Separation of unnatural environmental change from natural change and understanding the causes of natural change both require records longer than those available from instruments, which cover roughly the last 100 years (see Eddy and Oeschger 1993). Longer records are recovered from substitute or proxy environmental records. For example, certain documentary records – such as diaries, chronicles, and government reports – provide information about the environment in the past. Biological and geological processes also create proxy environmental records. Seasonal growth of stalagmites and stalactites has provided information about variations in rainfall. Accumulation of seasonal snow in distinct layers over many thousands of years in the Arctic, the Antarctic and at high elevations has provided data for precipitation. Air bubbles trapped in the snow have provided information about atmospheric composition, and the oxygen isotopic composition of the snow has provided information about temperature (e.g. Delmas 1994, Thompson et al. 1995). Annual rings in trees are perhaps the most familiar and certainly the most successful source of high-resolution proxy environmental records. Annual rings form because, in sub-polar and temperate regions, trees stop growing in winter. The distance between consecutive rings is then a measure of the excellence of the summer growing season. Since many factors may affect tree growth, environmental information is usually recovered from trees at sites predominantly affected by one environmental factor. Trees growing in semi-arid regions provide information about rainfall or soil moisture. Trees growing close to the upper tree line record temperature over the short summer growing season (Fritts 1976, Schweingruber 1988, Cook and Kairiukstis 1990). By matching rings in living and dead trees, such records have been extended back for thousands of years.

There is a close relation between the growth process in marine sessile organisms and the influence of the physical environment. There are many examples of marine sessile organisms which are very suitable as proxy environmental recorders: stromatolites (Walter 1976), coralline algae (Bosence 1976), and coral skeletons. In Chap. 1 the example of the deep sea coral *Desmophyllum cristagalli* was mentioned as an example of an environmental recorder. In the second section of this chapter the analysis of the deposition of growth layers in the stony coral *Porites* will be discussed in more detail. In the first section sponges from the Antarctic will be given as another example of potential bioarchives. Although there are no clearly distinguishable growth layers present in these organisms, because of their (estimated) enormous age and their location, these sponges might be a very important source of climatological information.

6.1 Antarctic Sponges

Antarctica has been called "the sponge kingdom" (Koltun 1968). More than 300 different species have been found in Antarctica and many benthic communities on the Antarctic shelf are clearly dominated by sponges (> 90% of biomass) (Voß 1988, Dayton 1974). Like trees in a tropical rainforest, sponges play a major role in structuring the habitat. Sponges provide living quarters "on the second floor" (Arntz et al. 1994) on the otherwise essentially flat

Fig. 6.1. Large hexactinellid sponges (depth 283 m) with crinoids and holothurians living "on the second floor". (Copyright: Julian Gutt)

and soft sediment. Underwater photographs show that many motile invertebrates (e.g. crinoids, ophiurids, and holothurians) are regularly sitting on top of sponges (Fig. 6.1). Possibly the current regime slightly above the bottom enhances food availability for suspension feeding organisms. Benthic life thrives in, on, around, and under sponges. Even after death of a sponge individual, the remaining spicule mat has substantial structuring influence in stabilizing the soft sediment. By accumulation over longer periods of time these spicule mats can reach a thickness of up to 1.5 m and provide a biogenic substrate that resembles hard substrata (Koltun 1968). Areas can be colonized by subsequently arriving taxa, which would usually not be able to settle on soft sediment.

Commonly slower growth and a lower productivity can be observed in polar benthic invertebrates when compared with boreal species (e.g. Bluhm et al. (1998) for sea urchins). These have been attributed to low temperatures on one hand and – more important – to a common scarcity of food on the other hand. The benthos experiences a distinct pulse of fresh food in summer (6–8 weeks of intense primary production) with a relative scarcity of fresh input during the rest of the year. Sponges and other suspension feeders may either feed on fresh food particles each summer while starving in autumn and winter and/or utilize pico- and nanoplankton and detritus from resuspension events as a more constant food source throughout the year, albeit of lower nutritional value.

The remarkable stability of water temperature as well as rather low sedimentation rates have for some time led to the erroneous assumption that the Antarctic environment was very stable with a next to complete absence of disturbance (Arntz et al. 1994). The Antarctic benthos, however, is subjected to significant disturbance events. Drifting or overturning table icebergs are touching and ploughing the ground of the shelf and eradicating any fauna in their path. Video transects show various stages of recolonization of these iceberg scour marks (Gutt et al. 1996).

One of the early colonizers after iceberg scouring is the demosponge *Stylocordyla borealis* (Gutt et al. 1996). Its shape and appearance closely

Fig. 6.2. *Stylocordyla borealis* (depth 171 m): often large patches are exclusively populated with this sponge species. Many individuals growing closely together are sometimes called a "lolly meadow". (Copyright: Julian Gutt)

resembles a lollipop: a spherical body (max. diameter 42 mm) is lifted into the water column by a stem (maximum length 435 mm) (see Fig. 6.2). About 80% of the total organic carbon content of an individual is found in the body (Gatti and Brey, in prep.). Occasionally a stem may lose its body, either by predation or by physical disturbance. From body/stem size relationships it seems possible for a stem to grow a new body. So far nothing is known about the mechanisms initiating and governing such a regrowth process.

For most species of Antarctic sponges it is very difficult or impossible to measure the extremely slow growth rates directly (Dayton 1974). Mass-specific oxygen consumption rates were used to estimate metabolic rates (Gatti and Brey, in prep.). Mass-specific proportions of total respiration going into somatic and gonadic production, inferred from measurements in other invertebrate species, were used to model growth rates for *Stylocordyla borealis*. *S. borealis* can apparently grow rapidly when young and newly settled. But after it reaches a minimum size, growth slows down considerably.

In the light of possible global warming, the impact of iceberg disturbance on the Antarctic system needs to be quantified. A scenario with global warming, with a larger number of drifting icebergs and thus a larger number of iceberg scour marks, where all fauna is eradicated, could reach a point when the benthos would be unable to recover. Iceberg scouring might disturb the fauna so often that recolonization processes might not be completed before the next iceberg affects the same area. Furthermore it would be possible to use iceberg scouring as a model mechanism for disturbance. Results could then be transferred to other mechanisms of disturbance (e.g. anthropogenic disturbance). Growth rates (or age) of abundant species would enable us to estimate the minimum time needed for recovery of the system after a disturbance event. If Antarctic sponges were several hundred years old, as we assume right now for some large hexactinellid species, they could act as valuable bioarchives for past environmental conditions in Antarctica.

6.2 Coral Records

An extensive literature indicates that coral skeletons are excellent archives of environmental information and contain a more diverse range of information than any other proxy environmental recorder. The discovery by Knutson et al. (1972) that massive (= rounded) coral skeletons contain annual density bands (Fig. 6.3b) provided a means to date recovered information and promised environmental information equivalent to that obtained from tree rings. Features of massive corals giving them potential as proxy environmental indicators include:

1. skeletons are formed continuously and large colonies represent growth over several hundred years (Fig. 6.4),
2. growth rate is rapid, typically 5–20 mm y^{-1}, allowing possible recovery of sub-annual information,
3. records other than density banding are provided by the isotopic composition of the skeleton and by a wide variety of trace materials included in the skeleton during growth (inclusive records),
4. the skeleton is dead, consequently, records are locked away for several millennia,
5. dead and fossil corals provide windows to the more distant past.

Although the annual nature of coral density bands was quickly established, the intra-annual timing of the bands gave problems. Reports of seasonality associated with sub-annual bands (e.g. Fig. 6.3) largely divided into those that reported dense regions to be formed in summer and those that reported them formed in winter. There were also conflicting reports about the appearance of density bands. Eventually it emerged that appearance could range from a narrow to a broad high density band in the same species at similar locations. To further complicate an already complicated picture, it was found that the annual density banding pattern included, or was made up from, a pattern of narrow, probably lunar, density bands (see Barnes and Lough 1989).

Problems with density banding caused a shift in emphasis towards inclusive records; first to inclusive records of annual cycles and then, as problems

Fig. 6.3a,b. X-raying coral skeletal slices. *Porites* colonies 30–50 years old are collected for investigation of density banding and to provide a data base of coral growth characteristics. (*a*) Removal of skeletal slice from the center of a colony. (*b*) X-ray positive showing a series of density bands – higher density is darker; lower density is lighter. A pair of bands, dark plus light (dense plus less dense), represents one year's growth.

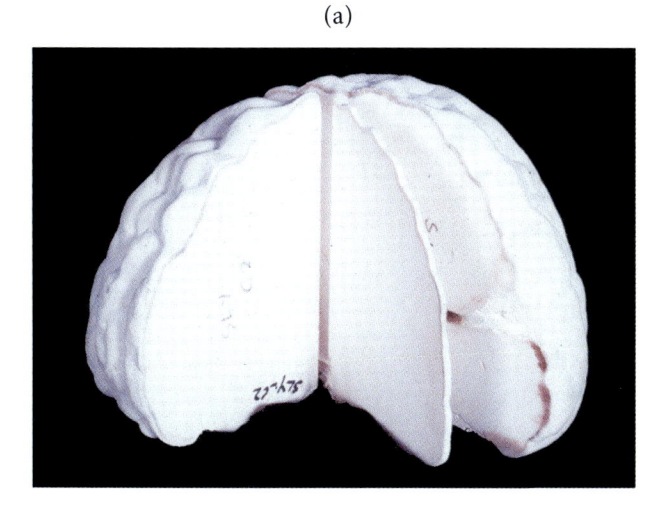

(a)

(b)

⊢ 100 mm ⊣

Fig. 6.4. A very large colony of *Porites*. Most work on coral skeletal records has been carried out using rounded colonies of Porites from the Indo-Pacific. Growth to very large sizes is one important characteristic of *Porites* for recovery of environmental records. A colony of this size probably represents 700–800 years of continual growth. Records are recovered by drilling a core down from the summit of the colony.

emerged with these, towards inclusive records of short-term events. The central, common problem was a lack of understanding of how information becomes stored in coral skeletons. This required that research be directed towards links between the environment and mechanisms of skeletal growth. The science of dendrochronology was founded upon mechanistic understanding of links between treering formation and controlling environmental factors (see Fritts 1976). Such understanding of coral density bands began to emerge in the early 1990s from work on *Porites*, the coral most commonly used for skeletal records.

It was shown that growth of *Porites* skeletons involves three mechanisms (Barnes and Lough 1992a, 1992b): (1) The colonial skeleton extends by growth at its outer surface. (2) The scaffolding created is thickened below the surface because the tissue occupies the skeleton to a depth of 2–10 mm. (3) The lower layer of the tissue is raised every few weeks, maintaining the tissue as a narrow band around the outer surface of the colony. Regions of skeleton no longer occupied by tissue are cut off by very thin, horizontal, skeletal bulkheads known as dissepiments.

Of these three mechanisms, only skeletal thickening through the depth of the tissue layer was a new observation. Indeed, the depth of the tissue layer was first used as a coral parameter by Lough and Barnes (1992) and variations in this depth (tissue thickness) with colony size and position on the GBR were first described by Barnes and Lough (1992a). The three processes explain problems with intra-annual timing of density bands and with their appearance. The model based on these three processes also explains the formation of fine, possibly lunar, density bands in certain corals, including *Porites*. Annual density bands result from variations in the thickness of skeletal elements (Barnes and Devereux 1988). Thickening occurs throughout the depth of the tissue layer and, consequently, density bands form below the surface of a colony. Since all techniques used to date density bands depend upon the position of the outer surface of the colony (Lough and Barnes 1992, Barnes and Lough 1992b), density bands formed below the surface will anticipate their apparent dating. Assuming constant thickening through the depth of the tissue layer, this anticipation could vary between < 1 and > 8 months (Barnes and Lough 1992a). Consequently, failure to link density and environmental variations by correlations is not surprising.

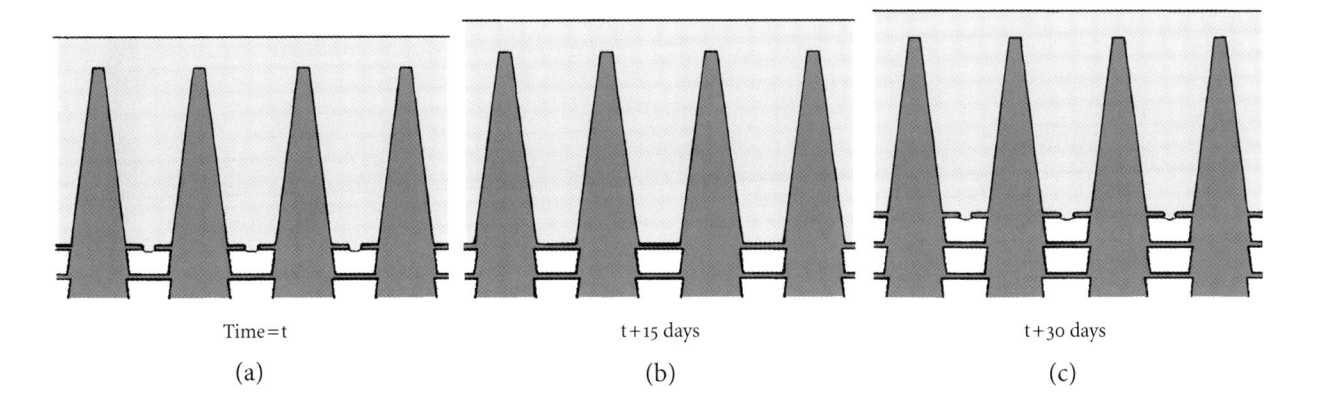

This model also explains the fine density bands in certain corals. Uplift of the base of the tissue layer is accompanied by isolation of newly vacated skeleton by thin horizontal skeletal bulkheads called dissepiments. Fine, probably lunar, monthly density bands and dissepiments have similar frequencies (Barnes and Lough 1989) suggesting that fine band formation and dissepiment formation are linked. Monthly uplift of the base of the tissue layer means that, just above a dissepiment, skeleton will have thickened for one month longer than just beneath a dissepiment. Uplift involves altering the thickness of the tissue layer by 15–25% (Barnes and Lough 1992a). Thus spines just above a dissepiment are likely to be 15–25% thicker than spines just below a dissepiment (Fig. 6.5) – producing a fine density banding pattern coinciding with dissepimental spacing.

Modeling was used as a means to understand the complex interactions amongst annual cycles in extension and thickening, and variations in tissue thickness (Taylor et al. 1993). In the model, calcification rate was assumed to be in the form (Fig. 6.6):

$$AA = AA_0\{1 + AA_1 \sin(2\pi t + \delta)\} \quad (6.1)$$

where AA_0 is the average calcification rate over a year, AA_1 is half the difference between maximum and minimum calcification rates, t is time in years, and δ is a phase factor ($\delta = 0$ if $t = 0$ represents mid-spring). The actual density will have the form:

$$TA = IA + AA_0(\Delta t + t' - t) \quad (6.2)$$
$$+ (AA_0 AA_1/2\pi) \sin(\pi(\{\Delta t + t' - t\}) \sin(2\pi\{\Delta t + ti' + t\} + \delta)$$

where TA is skeletal density, IA is the initial deposition, Δt is the time it would take the skeleton to extend a distance equal to the tissue thickness, $t' = j \cdot t_d$ (j = integer value of (t/t_d)), and t_d = the average time taken for the skeleton to extend the distance between consecutive dissepiments. If the extension rate is not constant, actual distance extended (EXT) is assumed to be:

$$EXT = E_0 t + (E_0 E_1/2\pi)\{\cos(\delta) - \cos(2\pi t + \delta)\} \quad (6.3)$$

Fig. 6.5a–c. Schematic representation of skeletal growth within the tissue layer of *Porites*. The tissue layer is shown as light shading and the skeleton as dark shading. Horizontal dissepiments are shown linking the vertical skeletal elements. (*a*) The most recent dissepiments are within hours of meeting. (*b*) Vertical elements are being extended and thickened and this continues until, after a lunar month, the next dissepiment is formed. (*c*) This growth results in a greater thickening of vertical elements just above a dissepiment than just below a dissepiment producing a pattern of fine, monthly density bands.

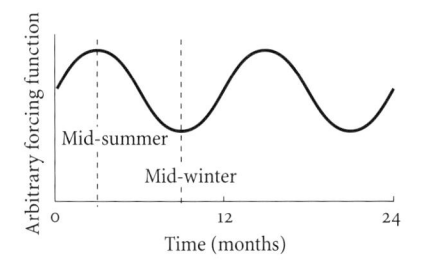

Fig. 6.6. A sine curve representing annual variations in the calcification rate of a coral. This was used as the forcing function in numerical models of skeletal density band formation.

where E_0 is the average extension rate over a year, and E_1 is half the difference between maximum extension rate and minimum extension rate (Fig. 6.7). These simulations showed how interactions between tissue thickness and intra-annual variations in extension compounded difficulties in dating (c.f., Figs. 6.6 and 6.7e; Fig. 6.7e was constructed using average growth parameters for *Porites* from the Great Barrier Reef). They also reproduced the wide variation in the appearance of annual bands and in the shape of density profiles, similar to those reported in the literature. In the model, deposition of skeletal calcium carbonate as extension and deposition as thickening are both set as 50% of the total thickening. Originally, this was an arbitrary choice. However, it was not possible to simulate density variations similar to those reported in the literature with an initial deposition value outside the range 40–60% of the total deposition.

The model also provided information about inclusive records. It was used to simulate the presence in seawater for one month of a material that becomes incorporated in the skeleton (Fig. 6.8). Such "pulse" events include river flows, upwelling and shifts in currents, and trade winds that affect the chemistry of seawater for weeks to months.

Figure 6.7 shows four of the vertical spines that make up the bulk of the skeleton in *Porites*. The spine is drawn with an average tissue thickness of 6 mm and an average growth rate of 12 mm per year. For simplicity, we have made the spine grow the same amount each month. In fact, our results indicate that *Porites* from the central Great Barrier Reef extend 2–3 times

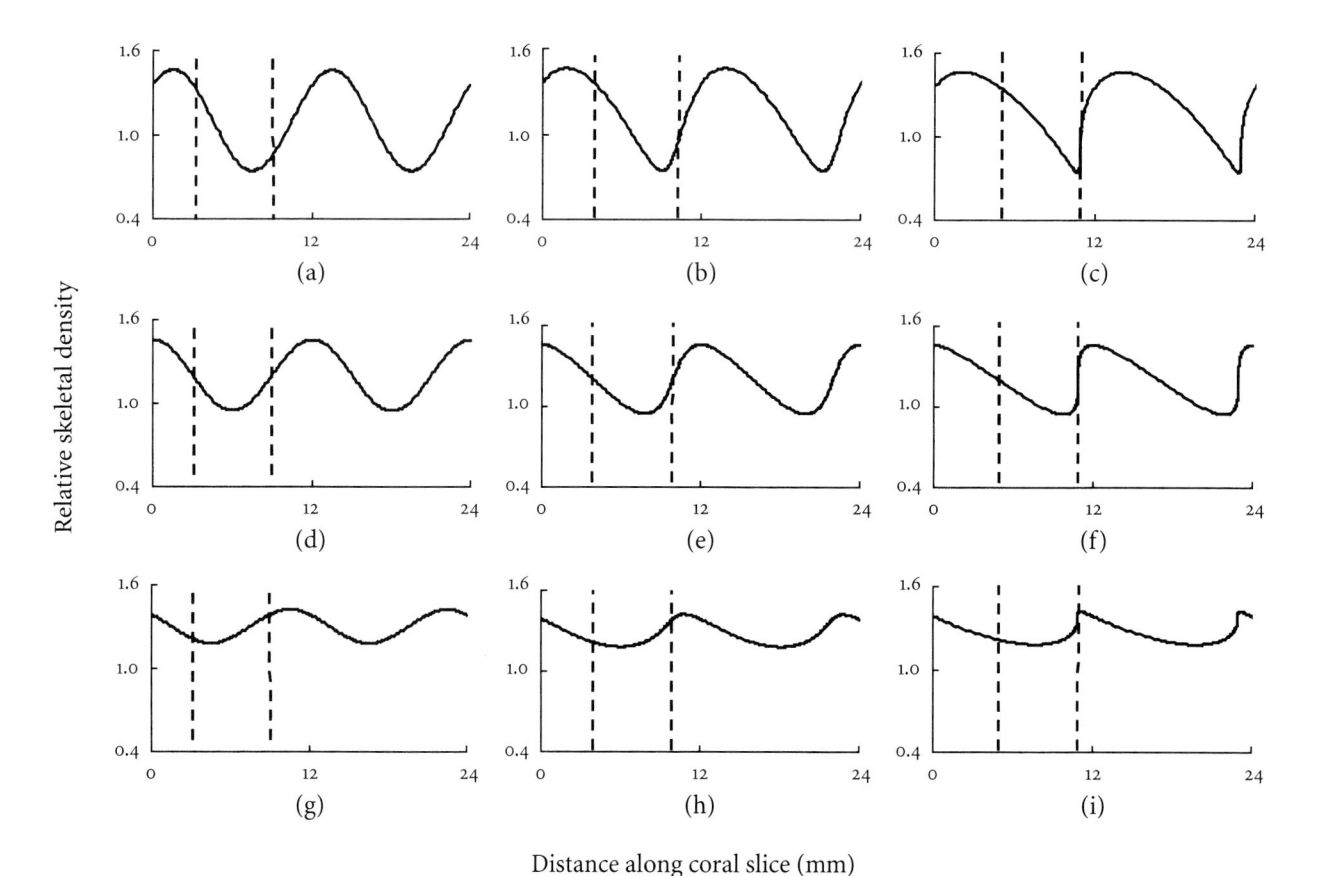

Fig. 6.7a–i. Density profiles generated by numerical models of coral density band formation in which extension and thickening are driven by the forcing function shown in Fig. 6.5. Mid-summer and mid-winter are marked by dashed vertical lines (c.f., Fig. 6.5). Tissue thickness increases down the columns such that it is 25%, 50%, and 75% of annual growth. Intra-annual variation in extension rate increases across the rows such that it changes from no variation to three times faster in summer than in winter to extension just stopping in mid-winter. Note that the lines marking mid-summer and mid-winter seldom align with density peaks and troughs.

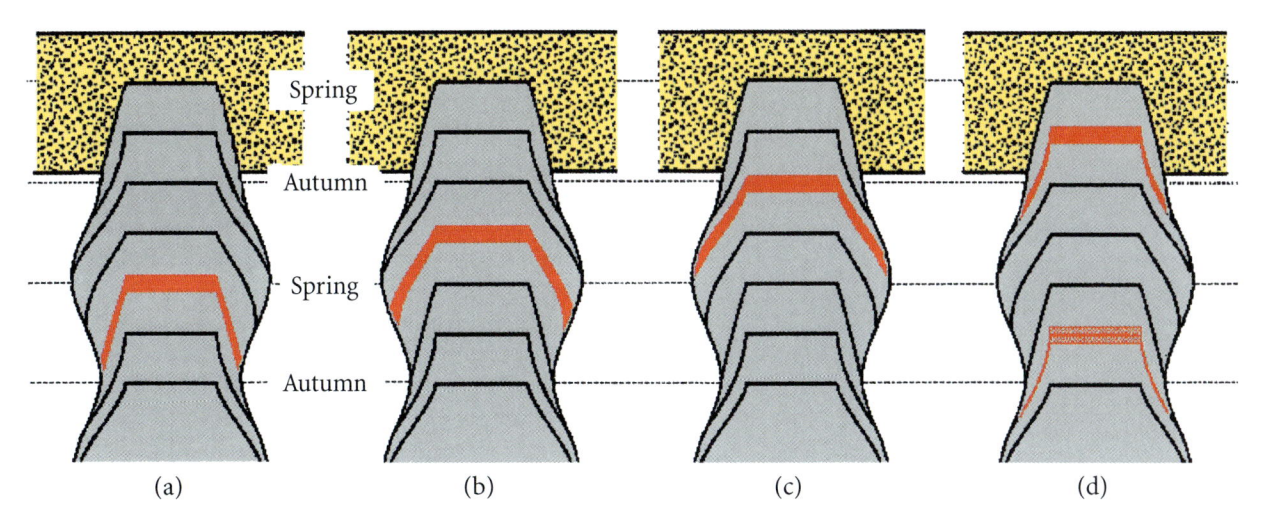

faster in summer than in winter. Changes in the diameter of the spine are caused by variations through the year in the amount of thickening; that is, they represent density variations. The red line indicates the substance included in the skeleton as a result of its presence in the surrounding seawater for one month. There are two components to the inclusion. The flat, horizontal, central bar represents addition to the top of the spine during that one month. The tails back from either side of the bar represent thickening deposits made during that one month (Fig. 6.8). Suppose you could just section the spine horizontally to remove the inclusion marking addition at the top of the spine. Such an inclusion made during spring will be far more diluted by thickening deposits than exactly the same inclusion made during autumn. In fact, the skeletal concentration of a substance deposited in autumn will be twice that of the same substance deposited during spring, even though the causative events were identical. Overall, modeling indicated that recovery of information about the intensity and even the timing of such "pulse" events would be difficult without better information about coral growth (Taylor et al. 1995).

Figure 6.7 can also be used to illustrate problems that will arise in reconstructing annual cycles from trace deposits that follow an annual cycle. For example, changes in the skeletal $O^{16}:O^{18}$ ratio and the Sr:Ca ratio have been used to reconstruct annual temperature cycles. Deposits made in mid-summer are diluted by subsequent deposits in autumn and winter (Fig. 6.7). Similarly, deposits made in mid-winter are diluted by subsequent deposits in spring and summer. Thus, when skeleton is sampled for the annual temperature cycle, peak summer skeletal temperatures will appear to be less than the peak seawater temperatures, and winter skeletal temperatures will appear to be higher than minimum seawater temperatures. Examination of the literature shows that temperature calibrations made by relating annual temperature cycles to inclusions across a year's growth have lower slopes than calibrations made from relating a range of average annual temperatures to average inclusions in a year's skeletal growth. This modeling showed that, even with the present poor understanding of skeletal growth parameters, it is possible to reconstruct reasonable records of annual cycles from skeletal inclusions (Barnes et al. 1995).

Realization of useful records from tree rings depended upon understanding the causes of tree rings and the associated mechanisms (Fritts 1976). This

Fig. 6.8a–d. Output of numerical models showing incorporation of a tracer (*red bands*) from seawater into coral skeleton. Month-long rectangular pulses of the tracer were simulated in mid-spring (*a*), mid-summer (*b*), mid-autumn (*c*) and mid-winter (*d*). The horizontal, red block at the center of each spine represents incorporation of tracer into skeleton deposited in extending the spine. The downward tails from either side of this block mark skeleton deposited in thickening the spine. The stippled, colored area represents the tissue. The outline of the spine indicates annual density variations. Tissue thickness was set to 6 mm and the spines had a constant growth of $12 \text{ mm} \cdot y^{-1}$. Lines within the spines indicate its past outline in mid-spring (Sp), mid-autumn (Au), mid-summer, and mid-winter. Spine (*d*) has not fully thickened over the tracer because the tracer is still within the tissue layer. The eventual ratio of extension to thickening is indicated by shading lower in the spine.

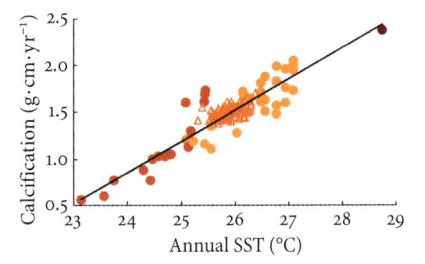

Fig. 6.9. Scatter plot of annual average calcification in *Porites* versus annual average seawater temperature. Data involve 415 colonies from 44 reefs spread over a latitudinal range of 21°. Red points are for data from the Hawaiian Archipelago (Grigg 1981, 1997). Orange points are for data from the Great Barrier Reef (Lough and Barnes 2000) and the dark orange point represents colonies collected at Phuket Island, Thailand (Scoffin et al. 1992). The regression line goes through these points. Also shown (open triangles) are data from cores drilled from 10 very large *Porites* colonies on the Great Barrier Reef covering the period 1903–1982 (Lough and Barnes 1997).

led to two fundamental procedures in recovering information from tree rings: site selection and averaging of information across many trees. Site selection recognized, for example, that the growth of trees growing close to the tree line on a mountain is overwhelmingly affected by temperature. Similarly the growth of trees in semi-arid regions is overwhelmingly affected by availability of water. Averaging of information enhanced any environmental signal common to all trees and lessened "noise" associated with the growth of single trees. Better understanding of coral skeletal growth has similarly led to site selection and averaging of data recovered from coral skeletons. Light acting through symbiotic algae living in the tissues of reef-building corals enhances calcification and the principal factor affecting coral calcification is light. The effect of light can largely be eliminated by collecting corals from shallow water, where they are light-saturated for most of the day (Barnes and Chalker 1990). The principal factor then affecting coral calcification (and increase in colony size) is temperature (Fig. 6.9). There is considerable noise associated with such data. The excellent relationship shown here was obtained by averaging growth characteristics across 415 colonies from 44 reefs (Lough and Barnes 2000). We can then examine the effects of increases in seawater temperature over the past century. Analysis of past records allows predictions that reef development might already be pushing south beyond the present confines of the Great Barrier Reef due to global warming (Fig. 6.10).

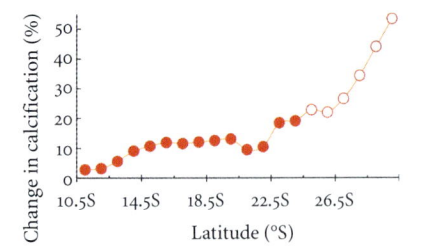

Fig. 6.10. Percentage increase in calcification along the Great Barrier Reef. Calcification estimated from observed differences in sea surface temperature between 1903–22 and 1979–88 and the relationship given in Fig. 6.9. Open points are for regions to the south of the Great Barrier Reef. Brisbane is at latitude 27.5° S and the Solitary Islands, which have extensive coral communities but no reefs, are around 30° S.

7. Epilogue

The models of growth and form discussed in Chap. 4 are defined at various levels of abstraction. The highest level of abstraction occurs in the two-dimensional L-system models and iterative geometric constructions of the seaweeds. The lowest level of abstraction occurs in the three-dimensional models of accretive growth and the influence of the physical environment and the model of fluid transport in the gastrovascular system of a hydrozoan. The Laplacian model represents an intermediate between the two extremes. Which type of modeling is selected depends strongly upon the type of questions to be answered. Growth models of individual organisms might be applied in studies on individual-based population dynamics (see Uchmanski et al. 1999), where it is not necessary to include all biological details. In simulation models applied to study bioarchives (see the examples of the Antarctic sponges in Sect. 6.1, coral records in Sect. 6.2, and biomonitoring studies, for example shown in Sect. 5.2, where changes in the growth forms due to perturbation experiments are compared to simulated results), a model of the influence of the physical environment is required. In simulation models where the genetic regulation of the growth is included, probably even lower levels of abstraction are required and it will be necessary to descend further in the range of possible biological organizational levels (ecosystem, population, organism, module, cell, organelles, complexes, molecules) and to include gradients of morphogens, internal transport mechanisms, models of the physiology of the organisms and biomechanical details.

From the overview on the state of the art of modeling growth and form of marine sessile organisms which we have presented in this book, it might become clear that many parts of this field of research are still in a state of development. Only recently have computational techniques become available which are capable of simulating the influence of the physical environment on the growth process of sessile organisms. Knowledge of developmental biology has increased greatly during the last decade, although it might become clear that one of the final aims of the simulation models, to bridge the gap between the genetic information and the final shape of the organism, is still a highly ambitious goal. The list of items which are still needed to achieve this goal, or items which are still in a preliminary state is very large, but also demonstrates that this field is a rich and promising field of research. A list of items which currently seem to be urgent, and not arranged in order of importance since such a ranking is in itself is controversial, is presented below.

TRANSLOCATION OF NUTRIENTS. In the available literature (see also Sect. 2.2.4) there has been a quite extensive discussion on the relevance of translocation of nutrients in marine sessile organisms, for example stony corals. In for example the accretive growth models presented in Sect. 4.6 one of the basic assumptions is that translocation can be neglected and that local growth velocities can be directly related to locally absorbed food particles or local light intensities. It not yet clear how important is the role and the magnitude of translocation of nutrients in sponges and stony corals. In Sect. 2.2.3 it was demonstrated that fluid transport in the gastrovascular system of colonial hydrozoans controls morphological plasticity. One approach for investigating the importance of this phenomenon is to develop simulations in which a certain translocation parameter is included and to study the impact on morphogenesis. To be able to verify the assumed translocation factors, actual measurements are required.

GENETIC REGULATION OF GROWTH AND FORM. In the Sect. 2.2.4 on genetic regulation in sponges and stony corals it is demonstrated that there is, albeit fragmentary, knowledge available on the biological regulation of the growth process. In order to capture the genetic regulation of growth and form in simulation models, much more detailed information is required on the developmental biology of marine sessile organisms. Especially when it is possible to translate the biological regulation into a diffusive patterning mechanism, as for example was done in the preliminary experiments with the isomone model in Sect. 4.6.6, it seems to be feasible to include the effect of biological regulation in models. Of course one of the weak points in this example is that the physical carrier of the growth regulation, the isomone, is still to be discovered.

IMPACT OF HYDRODYNAMIC FORCES ON GROWTH AND FORM. From Sect. 2.1.1 it becomes clear that the flexibility in many of the marine sessile organisms and the physical stress due to hydrodynamic forces have a major impact on the growth process. Inclusion of hydrodynamic forces and mechanical effects, for example deformation and physical damage, in growth models may provide new insights into the role of these biomechanical effects in the growth process. In Sect. 4.3.1 on computational methods for modeling and simulating the influence of hydrodynamics, it was mentioned that one of the advantages of the lattice Boltzmann method is that during the computation of the flow velocities, the values of the local hydrodynamic forces are already more or less available "for free". This indicates that a model of the impact of hydrodynamic forces on the growth process, using the lattice Boltzmann method, and a comparison to in vivo in vitro measurements, could be an interesting option.

EFFECTS OF LIVING TISSUE IN GROWTH MODELS OF STONY CORALS. In Sect. 2.2.3 it was mentioned that corals cannot grow tissue without skeleton and they cannot grow skeleton without tissue. Also in Sect. 6.2 on the coral records the role of living tissue was mentioned. Until now in all growth models the effects of living tissue were neglected. Particularly in models of *Porites*, which have an important application in investigating bioarchives, it is clear that living tissue should also be included in the models.

BASIC TYPES OF GROWTH PROCESSES. When looking in detail at the level of cells (meristems in seaweeds) or skeleton elements in sponges and corals, there seems to be a large variation in basic growth processes, and consequently architectures, possible. For example the organisms shown in Fig. 1.4 are formed in a radiate accretive growth process. This type of growth process was also used as a case study in Sect. 4.6 on accretive growth models. It is possible to identify many other different types of basic growth processes and architectures in marine sessile organisms (see for example for sponges Wiedenmayer 1977). An example was also given in Fig. 2.19, of a skeleton with a dense axial condensation, resulting in a very different overall architecture. In seaweeds it is possible to distinguish a wide variety of different types of growth processes. Simulation models in which more of these basic growth patterns are included might enable us to further explore the morphospace of possible morphologies.

UNDERWATER LIGHT MODELS. Until now only highly simplified models of underwater light distributions have been applied in the growth models. In reality factors such as reflection and scattering (for example due to suspended material) for various wave lengths, the position of the sun during a day, etc., may have an important role. To investigate these questions, more elaborate light models are required.

In many marine sessile organisms, for example in stony corals, a combination is used of photosynthesis and filter feeding. In general the literature on these organisms gives no quantitative indication about the ratio between the two energy sources. This type of information is required for the construction of models of organisms using this mixed energy source. An alternative approach could be to explore the photosynthesis/particle feeding space systematically through simulation.

GROWTH IN TURBULENT FLOW. In all the simulated growth processes, presented in Sects. 4.5 and 4.6, a laminar flow regime was assumed. Although this assumption is valid to a certain extent for a part of the marine environment, in reality many of the marine sessile organisms inhabit an environment exposed to higher flow velocities. Consequently in simulations it is required to simulate turbulent flow in three dimensions. The method discussed in Sect. 4.3 is basically suitable for simulating flows for higher Reynolds numbers, although quite a few computational problems can be expected here. For example it might be necessary to scale up the dimensions of the simulation box, increase the simulation time, and adapt the tracer step. Turbulence can be expected to have a strong mechanical influence as well as a strong impact on the distribution of food particles. In simulations the Re and Pe (see (2.3) and (2.4)) parameter space can be explored and it becomes possible to construct morphodiagrams, similar to the photosynthesis/particle feeding parameter space discussed in the previous paragraph. Morphodiagrams give an overview of all potential morphologies and are for example used in evolutionary studies (see McGhee 1998).

THREE-DIMENSIONAL MORPHOLOGICAL ANALYSIS. In Chap. 3 we presented several methods for two-dimensional morphological analysis and some preliminary results for three-dimensional morphological analysis of marine sessile organisms. At least to our knowledge, there seems to be very little work done on the morphological analysis of modular organisms; work on the three-dimensional morphological analysis of these organisms is virtually nonexistent. Most methods applied to shape analysis in growth and form are based on landmarks (see for example Chaplain et al. 1999), which seem to be more suitable in unitary organisms.

MODELING THE PHYSIOLOGY. In Kooijman (2000) an interesting model is proposed for the relation between carbonate deposition and metabolism in stony coral. Inclusion of a model of the absorption of calcium carbonate could potentially provide insight into the role of reef organisms in the absorption of carbonate from their environment. There are almost no cases where models of what is happening inside the organism (the physiology) are extended to the morphology of the organisms.

7.1 Conclusion

We believe that the combination of in vivo, in vitro, and in silico experiments opens up several new ways to study growth and form in marine sessile organisms. This book might contribute by initiating new collaborations among scientists from different disciplines. Hopefully it will also contribute to a better understanding of the emergence of growth and form in these beautiful organisms and provide new methods for analyzing bioarchives and bio-monitoring studies.

References

A. Abelson, B.S. Galil, and Y. Loya. Skeletal modification in stony corals caused by indwelling crabs: hydrodynamical advantages for crab feeding. *Symbiosis*, 10:233–248, 1991.

A. Abelson, T. Miloh, and Y. Loya. Flow patterns induced by substrata and body morphologies of benthic organisms, and their roles in determining availability of food particles. *Limnol. Oceanogr.*, 38:1116–1124, 1993.

E.R. Abraham. The fractal branching of an arborescent sponge. *Mar. Biol.*, 138:503–510, 2001.

J.F. Adkins, H. Cheng, E.A. Boyle, E.R.M. Druffel, and R.L. Edwards. Deep-sea coral evidence for rapid change in ventilation of the deep north Atlantic 15,400 years ago. *Science*, 280:725–728, 1998.

L. Agassiz. *Report of Florida reefs*. Cambridge printed for the museum, Cambridge, 1880.

K. Althoff, C. Schütt, R. Steffen, R. Batel, and W.E.G. Müller. Evidence for a symbiosis between bacteria of the genus *Rhodobacter* and the marine sponge *Halichondria panicea*: harbor also for putatively-toxic bacteria? *Mar. Biol.*, 130:529–536, 1998.

R.S. Anderson and K.L. Bunnas. The mechanics of aeolian ripple sorting and stratigraphy as visualized through a cellular automaton model. *Nature*, 365:740–743, 1993.

W.E. Arntz, T. Brey, and V.A. Gallardo. Antarctic zoobenthos. *Oceangr. Mar. Biol. Ann. Rev.*, 32:241–304, 1994.

M.J. Atkinson and R.W. Bilger. Effects of water velocity on phosphate uptake in coral reef-flat communities. *Limnol. Oceanogr.*, 37(2):273–279, 1992.

D.J. Ayre and B.L. Willis. Population structure in the coral *Pavona cactus*: clonal genotypes show little phenotypic plasticity. *Mar. Biol.*, 99:495–505, 1988.

P. Bachman, P. Kornman, and K. Zetsche. Regulation der Entwicklung und des Stoffwechsels der Grünalge Urosporadurch die Temperatur. *Planta*, 128:241–245, 1976.

M.E. Baird and M.J. Atkinson. Measurement and prediction of mass transfer to experimental coral communities. *Limnol. Oceanogr.*, 42:1685–1693, 1997.

D.J. Barnes. Growth in colonial scleractinians. *Bull. Mar. Sci.*, 23:280–298, 1973.

D.J. Barnes and B.E. Chalker. Calcification and photosynthesis in reef building corals and algae. In: Z. Dubinsky, editor, *Ecosystems of the world, Vol. 25 Coral Reefs*, pp. 109–131, Elsevier, Amsterdam, 1990.

D.J. Barnes and M.J. Devereux. Variations in skeletal architecture associated with density banding in the hard coral *Porites*. *J. Exp. Mar. Biol. Ecol.*, 121:37–54, 1988.

D.J. Barnes and J.M. Lough. The nature of skeletal density banding in scleractinian corals: fine banding and seasonal patterns. *J. Exp. Mar. Biol. Ecol.*, 126:119–134, 1989.

D.J. Barnes and J.M. Lough. Systematic variations in the depth of skeleton occupied by coral tissue in massive colonies of *Porites* from the Great Barrier Reef. *J. Exp. Mar. Biol. Ecol.*, 159:113–128, 1992a.

D.J. Barnes and J.M. Lough. On the nature and causes of density banding in massive coral skeletons. *J. Exp. Mar. Biol. Ecol.*, 167:91–108, 1992b.

D.J. Barnes, R.B. Taylor, and J.M. Lough. On the inclusion of trace materials into coral skeletons. Part II. Distortions in skeletal records of annual climate cycles due to growth processes. *J. Exp. Mar. Biol. Ecol.*, 194:251–275, 1995.

R.D. Barnes. *Invertebrate Zoology*. W.B. Saunders Company, Philadelphia, 1974.

E. Ben-Jacob. From snowflake formation to growth of bacteria colonies I: diffusive patterning in azoic systems. *Contemporary Physics*, 34:247–273, 1993.

E. Ben-Jacob. From snowflake formation to growth of bacteria colonies II: cooperative formation of complex colonial patterns. *Contemporary Physics*, 38:205–241, 1997.

G.P. Bidder. The relation of the form of a sponge to its currents. *Quart. J. Microsc. Sci.*, 67:293–323, 1923.

R.W. Bilger and M.J. Atkinson. Anomalous mass transfer of phosphate on coral reef flats. *Limnol. Oceanogr.*, 37:261–272, 1992.

R. Black. The effects of grazing by the limpet, *Acmaea insessa*, on the kelp *Egregia laevigata*, in the intertidal zone. *Ecology*, 57:65–77, 1976.

N.W. Blackstone. Gastrovascular flow and colony development in two colonial hydroids. *Biol. Bull.*, 190:56–68, 1996.

N.W. Blackstone. A dose-response relationship for experimental heterochrony in a colonial hydroid. *Biol. Bull.*, 193:47–61, 1997.

N.W. Blackstone. Physiological and metabolic aspects of experimental heterochrony in colonial hydroids. *J. Evol. Biol.*, 11:421–438, 1998.

N.W. Blackstone. Redox control in development and evolution: evidence from colonial hydroids. *J. Exp. Biol.*, 202:3541–3553, 1999.

N.W. Blackstone and L.W. Buss. Shape variation in hydractiniid hydroids. *Biol. Bull.*, 180:394–405, 1991.

N.W. Blackstone and L.W. Buss. Treatment with 2,4-dinitrophenol mimics ontogenetic and phylogenetic changes in a hydractiniid hydroid. *Proc. Natl. Acad. Sci.*, 89:4057–4061, 1992.

N.W. Blackstone and L.W. Buss. Experimental heterochrony in hydractiniid hydroids: why mechanisms matter. *J. Evol. Biol.*, 6:307–327, 1993.

B.A. Bluhm, D. Piepenburg, and K. Juterzenka. Distribution, standing stock, growth, mortality and production of *Strongylocentrotus pallidus* (Echinodermata: Echinoidea) in the northern Barents Sea. *Polar Biol.*, 20:325–334, 1998.

D.W.J. Bosence. Ecological studies on two unattached coralline algae from western Ireland. *Palaeontology*, 19(2):365–395, 1976.

J.S. Bowerbank. *A Monograph of the British Spongiadae VIII*. London Royal Society, London, 1876.

J.C. Braekman and R.W.M van Soest. Results of the sponge natural products workshop. Technical report, University of Amsterdam, 1993. EC-MAST II Strategic Paper, Amsterdam.

M. Braverman. Studies on hydroid differentiation II. Colony growth and the initiation of sexuality. *J. Embryol. Exp. Morph.*, 11:239–253, 1963.

D.A. Brazeau and H.R. Lasker. Inter- and intraspecific variation in gorgonian morphology: quantifying branching patterns in arborescent animals. *Coral Reefs*, 7:139–143, 1988.

E. Brener, K. Kassner, and H. Müller-Krumbhaar. Pattern formation in first-order phase transitions. *International Journal of Modern Physics C*, 3(5):825–851, 1992.

P. Brien, C. Lévi, M. Sara, O. Tuzet, and J. Vacelet. *Traité de zoologie: anatomie, systématique, biologie, Tome III Spongiares, fascicule 1*. Masson et Cie Editeurs, Paris, 1973.

J.F. Bruno and P.J. Edmunds. Clonal variation for phenotypic plasticity in the coral *Madracis mirabilis*. *Ecology*, 78(7):2177–2190, 1997.

L.W. Buss. Growth by intussusception in hydractiniid hydroids. In: J.B.C. Jackson, S. Lidgad, and F.K. MacKinney, editors, *Pattern from Process in the Fossil Record*, University of Chicago Press, Chicago. In press.

L.W. Buss and J.B.C. Jackson. Planktonic food availability and suspension-feeder abundance: evidence of in situ depletion. *J. Exp. Mar. Biol. Ecol.*, 49:151–161, 1981.

H.S. Callahan, M. Pigliucci, and C.D. Schlichting. Developmental phenotypic plasticity: where ecology and evolution meet molecular biology. *BioEssays*, 19:519–525, 1997.

R.C. Carpenter, J.M. Hackney, and W.H. Adey. Measurements of primary productivity and nitrogenase activity of coral reef algae in a chamber incorporating oscillatory flow. *Limnol. Oceanogr.*, 36:40–49, 1991.

E. Carrington. Drag and dislodgement of an intertidal macroalga: Consequences of morphological variation in *Mastocarpus papillatus* Kutzing. *J. Exp. Mar. Biol. Ecol.*, 139:185–200, 1990.

P. Cartwright, J. Bowsher, and L. W. Buss. Expression of a Hox gene, Cnox-2, and the division of labor in a colonial hydroid. *Proc. Natl. Acad. Sci.*, 96:2183–2186, 1999.

P. Cartwright and L. W. Buss. Colony integration and the expression of the Hox gene, Cnox-2, in *Hydractinia symbiolongicarpus* (Cnidaria:Hydrozoa). *Journal of Experimental Zoology (Mol. Dev. Evol.)*, 285:57–62, 1999.

J.A. Chamberlain and R.R. Graus. Water flow and hydromechanical adaptations of branched reef corals. *Bull. Mar. Sci.*, 25:112–125, 1977.

M.A.J. Chaplain, G.D. Singh, and J.C. McLachlan. *On Growth and Form, Spatio-temporal Pattern Formation in Biology*. Wiley, New York, 1999.

S. Chen, Z. Wang, X. Shan, and G. Doolen. Lattice Boltzmann computational fluid dynamics in three dimensions. *Journal of Statistical Physics*, 68(3/4):379–400, 1992.

B. Chopard and M. Droz. *Cellular Automata Modeling of Physical Systems*. Cambridge Univerity Press, Cambridge, 1998.

T. Chopin, C.J. Bird, C.A. Murphy, J.A. Osborne, M.U. Patwary, and J.-Y. Floch. A molecular investigation of polymorphism in the North Atlantic red alga, *Chondrus crispus* (Gigartinales). *Phycological Research*, 44:69–80, 1996.

S.S. Cohen. Sponges, cancer chemotherapy, and cellular aging. *Perspect. Biol. Med.*, 6:215–227, 1963.

L. Collado-Vides, G. Gomez, V. Gomez, and G. Lechuga. Simulation of the clonal growth of *Bostrychia radicans* (Ceramiales Rhodophyta) using L-Systems. *BioSystems*, 42:19–27, 1997.

E.R. Cook and L.A. Kairiukstis. *Methods of Dendrochronology*. Kluwer, Dordrecht, 1990.

V. Cornish. *Waves of Sand and Snow*. T. Fisher Unwin, London, 1914.

G.R. Cui, B. Williams, and G. Kuczera. A stochastic Tokunaga model for stream networks. *Water Resour. Res.*, 35(10):3139–3147, 1999.

B.W. Dade. Near-bed turbulence and hydrodynamic control of diffusional mass transfer at the sea floor. *Limnol. Oceanogr.*, 38:52–69, 1993.

J.M. Dauget. Application of tree architectural models to reef-coral growth forms. *Mar. Biol.*, 111:157–165, 1991.

L.B. Dayton. Biological accommodation in the benthic community at McMurdo Sound, Antarctica. *Ecol. Monogr.*, 44:105–128, 1974.

P.K. Dayton. Observations of growth, dispersal and population dynamics of some sponges in McMurdo Sound, Antarctica. In: C. Levi and N. Boury-Esnault, editors, *Biologie des spongiaires*, pp. 271–302, Éditions du Centre National de la Recherche Scientifique, Paris, 1978.

R.J Delmas. Ice records of the past environment. *Science*, 143:17–30, 1994.

M.W. Denny. *Biology and the Mechanics of the Wave-Swept Environment*. Princeton University Press, Princeton, NJ, 1988.

M.W. Denny. Are there mechanical limits to size in wave-swept organisms? *J. Exp. Biol.*, 202:3463–3467, 1999.

M.W. Denny, T. Daniel, and M.A.R. Koehl. Mechanical limits to the size of wave-swept organisms. *Ecol. Monogr.*, 55:69–102, 1985.

M.W. Denny, B. Gaylord, B. Helmuth, and T. Daniel. The menace of momentum: Dynamic forces on flexible organisms. *Limnol. Oceanogr.*, 43:955–968, 1998.

P.S. Dodds and D.H. Rothman. Unified view of scaling laws for river networks. *Phys. Rev. E*, 59(5):4865–4877, 1999.

P.S. Dodds and D.H. Rothman. Statistical generalization of Horton's laws for river networks. In preparation.

M.J. Dring and K. Lüning. Induction of two-dimensional growth and hair formation by blue light in the brown alga, *Scytosiphon lomentaria*. *Z. Pflanzenphysiol.*, 75:107–117, 1975.

S. Dudgeon, A. Wagner, J.R. Vaisnys, and L. Buss. Dynamics of gastrovascular circulation in the hydroid *Podocoryne carnea*: the 1-polyp case. *Biol. Bull.*, 196:1–17, 1999.

S.R. Dudgeon and L. W. Buss. Growing with the flow: on the maintenance and malleability of colony form in the hydroid *Hydractinia*. *Am. Nat.*, 147:667–691, 1996.

J.A. Eddy and H. Oeschger (eds.). *Global Changes in the Perspective of the Past*. J. Wiley, London, 1993.

D. Erwin, J. Valentine, and D. Jablonski. The origin of animal body plans. *American Scientist*, 85:126–137, 1997.

K.E. Fabricius, A. Genin, and Y. Benayahu. Flow-dependent herbivory and growth in zooxanthellae-free soft corals. *Limnol. Oceanogr.*, 40:1290–1301, 1995.

P.G. Falkowski and Z. Dubinsky. Light-shade adaptation of *Stylophora pistillata*, a hermatypic coral from the Gulf of Eilat. *Nature*, 289:172–174, 1981.

J. Feder. *Fractals*. Plenum Press, New York, London, 1988.

J.D. Foley, A. van Dam, S.K. Feiner, and J.F. Hughes. *Computer Graphics: Principles and Practice*. Addison-Wesley, New York, 1990.

A.B. Foster. Environmental variation in skeletal morphology within the Caribbean reef corals *Montastea annularis* and *Siderastrea siderea*. *Bull. Mar. Sci.*, 30(3):678–709, 1980.

U. Frank, U. Oren, Y. Loya, and B. Rinkevich. Alloimmune maturation in the coral *Stylophora pistillata* is achieved through three distinctive stages, four months post metamorphosis. *Proc. R. Soc. Lond.*, B 264:99–104, 1997.

M. Frechette, C.A. Butman, and W.R. Geyer. The importance of boundary-layer flows in supplying phytoplankton to the benthic suspension feeder *Mytilus edulis* L. *Limnol. Oceanogr.*, 34(1):19–36, 1989.

U. Frisch, D. d'Humières, B. Hasslacher, P. Lallemand, Y. Pomeau, and J.P. Rivet. Lattice gas hydrodynamics in two and three dimensions. In: D. Doolen, editor, *Lattice gas methods for partial differential equations*, pp. 75–137, Addison-Wesley, New York, 1987.

H.C. Fritts. *Tree Rings and Climate*. Academic Press, London, 1976.

Y.C.B. Fung. *Biomechanics: Motion, Flow, Stress, and Growth*. Springer-Verlag, New York, 1990.

K.N. Ganeshaiah and T. Veena. Topology of the foraging trails of *Leptogenys processionalis* - why are they branched? *Behav. Ecol. Sociobiol.*, 29:263–270, 1991.

D.J. Garbary and J.D. Corbit. Lindenmayer-systems as models of red algal morphology and development. *Prog. Phycol. Res.*, 8:143–177, 1992.

D. Gateo, A. Israel, Y. Barki, and B. Rinkevich. Gastrovascular circulation in an octocoral: evidence of significant transport of coral and symbiont cells. *Biol. Bull.*, 194:178–186, 1974.

S. Gatti and T. Brey. Reproduction and growth rates of the Antarctic demosponge *Stylocordyla borealis*. In preparation.

B. Gaylord, C.A. Blanchette, and M.W. Denny. Mechanical consequences of size in wave-swept algae. *Ecol. Monogr.*, 64:287–313, 1994.

S. Ghorai and N.A. Hill. Wavelengths of gyrotactic plumes in bioconvection. *Bulletin of Mathematical Biology*, 62:429–450, 2000.

T.F. Goreau, N.I. Goreau, and C.M. Yonge. Reef corals: autotrophs or heterotrophs? *Biol. Bull.*, 141:247–266, 1971.

T.L. Goulet and M.A. Coffroth. A within colony comparison of zooxanthellae genotypes in the Caribbean gorgonian *Plexaura kuna*. In: H.A. Lessios and I.G. MacIntyre, editors, *Proceedings of the 8th International Coral Reef Symposium*, pp. 1331–1336, Smithsonian Tropical Research Institute, Panama, 1997.

L.E. Graham and L.W. Wilcox. *Algae*. Prentice-Hall, London, 2000.

R.R. Graus and I.G. Macintyre. Light control of growth form in colonial reef corals: computer simulation. *Science*, 193:895–897, 1976.

R.R. Graus and I.G. Macintyre. Variation in growth forms of the reef coral *Montastrea annularis* (Ellis and Solander): a quantitative evaluation of growth response to light distribution using computer simulation. *Smithson. Contr. Mar. Sci.*, 12:441–464, 1982.

R.W. Grigg. Coral reef development at high latitudes in Hawaii. *Proc. 4th Int. Coral Reef Symp., Manila*, 1:687–693, 1981.

R.W. Grigg. Paleoceanography of coral reefs in the Hawaiian-Emperor Chain – revisited. *Coral Reefs*, 16:S33–S38, 1997.

J.P. Grotzinger and D.H. Rothman. An abiotic model stromatolite morphogenesis. *Nature*, 383:423–425, 1996.

J. Gutt, A. Starmans, and G. Dieckmann. Impact of iceberg scouring on polar benthic habitats. *Mar. Ecol. Prog. Ser.*, 137:311–316, 1996.

E. Haeckel. *Atlas der Kalkschwämme*. Reimer, Berlin, 1872.

M.D. Hanisak and M.A. Samuel. Growth rates in culture of several species of *Sargassum* from Florida. *Hydrobiologia*, 151/152:399–404, 1987.

W.H. Harvey. *A Manual of the British Marine Algae*. John van Voorst, Paternoster Row, London, 1869.

K.B. Heidelberg, K.P. Sebens, and J E. Purcell. Effects of escape behavior and water flow on prey capture by the scleractinian coral, *Meandrina meandrites*. *Proc. 8th Int. Coral Reef Symp., Panama*, 2:1081–1086, 1997.

A.W.J. Heijs and C.P. Lowe. Numerical evaluation of the permeability and the Kozeny constant for two types of porous media. *Phys. Rev. E*, 51(5):4346–4351, 1995.

B. Helmuth and K. Sebens. The influence of colony morphology and orientation to flow on particle capture by the scleractinian coral *Agaricia agaricites* (Linnaeus). *J. Exp. Mar. Biol. Ecol.*, 165:251–278, 1993.

B.S.T. Helmuth, K.P. Sebens, and T.L. Daniel. Morphological variation in coral aggregations: branch spacing and mass flux to coral tissues. *J. Exp. Mar. Biol. Ecol.*, 209:233–259, 1997.

R.C. Highsmith. Reproduction by fragmentation in corals. *Mar. Ecol. Prog. Ser.*, 7:207–226, 1982.

T. Hincks. *British Hydroid Zoophytes*. John van Voorst, Paternoster Row, London, 1868.

P. Hoffman. Environmental diversity of middle precambrian stromatolites. In: M.R. Walter, editor, *Stromatolites*, pp. 599–611, Elsevier, Amsterdam, 1976.

R.J. Horodyski. Stromatolites from the middle proterozoic Altyn limestone, Belt Supergroup, Glacier National Park, Montana. In: M.R. Walter, editor, *Stromatolites*, pp. 585–597, Elsevier, Amsterdam, 1976.

R.E. Horton. Erosional development of streams and their drainage basins: Hydrophysical approach to quantitative morphology. *Bull. Geol. Soc. Am*, 56(3):275–370, 1945.

T. Hunter. Suspension feeding in oscillating flow: the effect of colony morphology and flow regime on plankton capture by the hydroid *Obelia longissima*. *Biol. Bull.*, 176:41–49, 1989.

C.L. Hurd, P.J. Harrison, and L.D. Druehl. Effect of seawater velocity on inorganic nitrogen uptake by morphologically distinct forms of *Macrocystis integrifolia* from wave-sheltered and exposed sites. *Mar. Biol.*, 126:205–214, 1996.

C.L. Hurd and C.L. Stevens. Flow visualization around single- and multiple-bladed seaweeds with various morphologies. *J. Phycol.*, 33:360–367, 1997.

G. Imsiecke, J. Münkner, B. Lorenz, W.E.G. Müller, and H.C. Schröder. Inorganic polyphosphates in the developing freshwater sponge *Ephydatia muelleri*: effect of stress by polluted waters. *Environ. Toxicol. Chem.*, 15:1329–1334, 1996.

J.B.C. Jackson. Morphological strategies of sessile animals. In: C. Larwood and B.R. Rosen, editors, *Biology and Systematics of Colonial Organisms Volume II*, pp. 499–555, Academic Press, London New York, 1979.

A.S. Johnson and M.A.R. Koehl. Maintainence of dynamic strain similarity and environmental stress factor in different flow habitats: Thallus allometry and material properties of a giant kelp. *J. Exp. Biol.*, 195:381–410, 1994.

A.S. Johnson and K.P. Sebens. Consequences of a flattened morphology: effect of flow on feeding rates of the scleractinian coral *Meandrina meandrites*. *Mar. Ecol. Prog. Ser.*, 99:99–114, 1993.

P.P. Jonker and A.M. Vossepoel. Mathematical morphology in 3D images: comparing 2D & 3D skeletonization algorithms. In: K.Wojciechowski, editor, *BENEFIT Summer School on Morphological Image and Signal Processing, Zakopane*, pp. 83–108, Silesian Technical University, ACECS, Gliwice, Poland, 1995.

E. Jordan and R.S. Nugent. Evaluacion poblacional de *Plexauralla homomalla* (Esper) en la costa Noroeste de la peninsula de Yucatan. *An. Centro Cienc. Mar. Limnol. Univ. Nal. Auton. Mexico*, 5:189–200, 1978.

P.A. Jumars and A.R.M. Nowell. Fluid and sediment dynamic effects on marine benthic community structure. *Am. Zool.*, 24:45–55, 1984.

J.A. Kaandorp. Modelling growth forms of the sponge *Haliclona oculata* (Porifera; Demospongiae) using fractal techniques. *Mar. Biol.*, 110:203–215, 1991.

J.A. Kaandorp. A formal description of radiate accretive growth. *J. Theor. Biol.*, 166:149–161, 1994a.

J.A. Kaandorp. *Fractal modelling: Growth and Form in Biology*. Springer-Verlag, Berlin, New York, 1994b.

J.A. Kaandorp. Analysis and synthesis of radiate accretive growth in three dimensions. *J. Theor. Biol.*, 175:39–55, 1995.

J.A. Kaandorp. Morphological analysis of growth forms of branching marine sessile organisms along environmental gradients. *Mar. Biol.*, 134:295–306, 1999.

J.A. Kaandorp and M.J. de Kluijver. Verification of fractal growth models of the sponge *Haliclona oculata* (Porifera; class Demospongiae) with transplantation experiments. *Mar. Biol.*, 113:133–143, 1992.

J.A. Kaandorp, C. Lowe, D. Frenkel, and P.M.A. Sloot. The effect of nutrient diffusion and flow on coral morphology. *Phys. Rev. Lett.*, 77(11):2328–2331, 1996.

J.A. Kaandorp and P.M.A. Sloot. Parallel simulation of accretive growth in 3-dimensions. *BioSystems*, 44(3):181–192, 1997.

J.A. Kaandorp and P.M.A. Sloot. Morphological models of radiate accretive growth and the influence of hydrodynamics. *J. Theor. Biol.*, 209:257–274, 2001.

W.M. Kays and M.E. Crawford. *Convective Heat and Mass Transfer, 3rd edition*. McGraw-Hill, New York, 1993.

G.H. Kim, I.K. Lee, and L. Fritz. The wound-healing responses of *Antithamnion japonicum* and *Griffithsia pacificum* (Ceramiales, Rhodophyta) monitored by lectins. *Phycological Research*, 43:161–166, 1995.

K. Kim and H.R. Lasker. Flow-mediated resource competition in the suspension feeding gorgonian *Plexaura homomalla* (Esper). *J. Exp. Mar. Biol. Ecol.*, 215:49–64, 1997.

R. King and C.F. Puttock. Morphology and taxonomy of *Bostrychia* and *Stictosiphonia* (Rhodomelaceae/Rhodophyta). *Austr. Syst. Bot.*, 21:1–73, 1989.

E.P van Klaveren and S. Verhoeven. Interview met Alexander Polyakov, Bitterzoete wetenschap. *Stroom*, Januari:5–8, 1995.

M.J. de Kluijver. Sublittoral hard substrate communities of the southern Delta area, SW Netherlands. *Bijdr. Dierk.*, 59(3):141–158, 1989.

R.A. Knutson, R.W. Buddemeier, and S.V. Smith. Coral chronometers: seasonal growth bands in reef corals. *Science*, 177:270–272, 1972.

M.A.R. Koehl. Physiological, ecological, and evolutionary consequences of the hydrodynamics of individual organisms. OEUVRE (published electronically),
http://www.joss.ucar.edu/joss_psg/project/oce_workshop/oeuvre/, 1998.

M.A.R. Koehl. Effects of sea anemones on the flow forces they encounter. *J. Exp. Biol.*, 69:87–105, 1977a.

M.A.R. Koehl. Mechanical organization of cantilever-like sessile organisms: Sea anemones. *J. Exp. Biol.*, 69:127–142, 1977b.

M.A.R. Koehl. The interaction of moving water and sessile organisms. *Scientific American*, 247(6):110–120, 1982.

M.A.R. Koehl. Seaweeds in moving water: Form and mechanical function. In: T. J. Givnish, editor, *On the Economy of Plant Form and Function*, pp. 603–634, Cambridge University Press, Cambridge, 1986.

M.A.R. Koehl. Ecological biomechanics: Life history, mechanical design, and temporal patterns of mechanical stress. *J. Exp. Biol.*, 202:3469–3476, 1999.

M.A.R. Koehl. Consequences of size change during ontogeny and evolution. In: J.H. Brown and G.B. West, editors, *Scaling in Biology*, pp. 67–86, Oxford University Press, New York, 2000.

M.A.R. Koehl and R.S. Alberte. Flow, flapping, and photosynthesis of *Nereocystis luetkeana*: a functional comparison of undulate and flat blade morphologies. *Mar. Biol.*, 99:435–444, 1988.

M.A.R. Koehl and S.A. Wainwright. Mechanical design of a giant kelp. *Limnol. Oceanogr.*, 22:1067–1071, 1977.

V.M. Koltun. Spicules of sponges as an element of bottom sediments of the Antarctic. In: *SCAR symposium on Antarctic oceanography*, pp. pp 121–123, Scott Polar Research Institute, Cambridge, 1968.

S.A.L.M. Kooijman. *Dynamic Energy and Mass Budgets in Biological Systems*. Cambridge University Press, Cambridge, 2000.

G.P. Korotkova. Regeneration and somatic embryogenesis in sponges. In: W.G. Fry, editor, *The Biology of Porifera, Symp. Zool. Soc. Lond. Vol 25*, pp. 423–436, Academic Press, London, 1970.

A. Krasko, B. Lorenz, R. Batel, H.C. Schröder, I.M. Müller, and W.E.G. Müller. Expression of silicatein and collagen genes in the marine sponge *Suberites domuncula* is controlled by silicate and myotrophin. *Europ. J. Biochem.*, 267:4878–4887, 2000.

J.E. Kübler and S.R. Dudgeon. Temperature dependent change in the complexity of form of *Chondrus crispus* fronds. *J. Exp. Mar. Biol. Ecol.*, 207:15–24, 1996.

M. LaBarbera. Feeding currents and particle capture mechanisms in suspension feeding animals. *Amer. Zool.*, 24:71–84, 1984.

A.J.C. Ladd. Numerical simulations of particulate suspensions via a discretized Boltzmann equation Part I. Theoretical foundation. *J. Fluid Mech.*, 271:285–309, 1994.

H.R. Lasker. A comparison of the particulate feeding abilities of three species of gorgonian soft coral. *Mar. Ecol. Prog. Ser.*, 5:61–67, 1981.

H.R. Lasker. Asexual reproduction, fragmentation and skeletal morphology of a plexaurid gorgonian. *Mar. Ecol. Prog. Ser.*, 19:261–268, 1984.

J.J. Leichter and J.D. Witman. Water flow over subtidal rock walls: relation to distributions and growth rates of sessile suspension feeders in the Gulf of Maine. *J. Exp. Mar. Biol. Ecol.*, 209:293–307, 1997.

M.P. Lesser. Elevated temperatures and ultraviolet radiation cause oxidative stress and inhibit photosynthesis on symbiotic dinoflagellates. *Limnol. Oceanogr.*, 41:271–283, 1996.

M.P. Lesser, V.M. Weis, M.R. Patterson, and P.L. Jokiel. Effects of morphology and water motion on carbon delivery and productivity in the reef coral, *Pocillopora damicornis* (Linnaeus): diffusion barriers, inorganic carbon limitation, and biochemical plasticity. *J. Exp. Mar. Biol. Ecol.*, 178:153–179, 1994.

J.B. Lewis. Experimental tests of suspension feeding in Atlantic reef corals. *Mar. Biol.*, 36:147–150, 1976.

J.C. Lewis, T.F. Barnowski, and G.J. Telenski. Characteristics of carbonates of gorgonian axes (Coelenterata, Octocorallia). *Biol. Bull.*, 183:278–296, 1992.

A. Lindenmayer. Mathematical models for cellular interactions in development, Part I and II. *J. Theor. Biol.*, 18:280–315, 1968.

A. Lindenmayer. Developmental systems without cellular interaction, their languages and grammars. *Journal of Theoretical Biology*, 30:455–484, 1971.

A. Lindenmayer. Algorithms for plant morphogenesis. In: R. Sattler, editor, *Theoretical Plant Morphology*, pp. 37–81, Leiden Univ. Press, Leiden, 1978. Suppl. to Acta Biotheoretica, Vol. 27.

A. Lindenmayer. Models for multicellular development: Characterization, inference and complexity of L-systems. In: A. Kelemenová and J. Kelemen, editors, *Trends, Techniques and Problems in Theoretical Computer Science*, Lecture Notes in Computer Science 281, pp. 138–168. Springer-Verlag, Berlin, 1987.

A. Lindenmayer and H. Jürgensen. Grammars of development: Discrete-state models for growth, differentiation and gene expression in modular organisms. In: G. Rozenberg and A. Salomaa, editors, *Lindenmayer Systems: Impacts on Theoretical Computer Science, Computer Graphics, and Developmental Biology*, pp. 3–21. Springer-Verlag, Berlin, 1992.

M.M. Littler and D.S. Littler. The evolution of thallus form and survival strategies in benthic marine macroalgae: field and laboratory tests of a functional form model. *Am. Nat.*, 116:25–44, 1980.

W.E. Lorensen and H.E. Cline. Marching cubes: a high resolution 3D surface construction algorithm. *ACM Computer Graphics*, 21(4):163–169, 1987.

B. Lorenz, R. Batel, N. Bachinski, W.E.G. Müller, and H.C. Schröder. Purification of two exopolyphosphatases from the marine sponge *Tethya lyncurium*. *Biochim. Biophys. Acta*, 1245:17–28, 1995.

J.M. Lough and D.J. Barnes. Comparison of skeletal density variations in *Porites* from the central Great Barrier Reef. *J. Exp. Mar. Biol. Ecol.*, 155:1–25, 1992.

J.M. Lough and D.J. Barnes. Several centuries of variation in skeletal extension, density and calcification in massive *Porites* colonies from the Great Barrier Reef: a proxy for seawater temperature and a background of variability against which to identify unnatural change. *J. Exp. Mar. Biol. Ecol.*, 211:29–67, 1997.

J.M. Lough and D.J. Barnes. Comparison of skeletal density variations in *Porites* from the central Great Barrier Reef. *J. Exp. Mar. Biol. Ecol.*, 245:225–243, 2000.

J. Love, C. Brownlee, and A.J. Trewavas. Ca2+ and calmodulin dynamics during photopolarization in *Fucus serratus* zygotes. *Plant Physiol.*, 115:249–261, 1997.

L. Lovett Doust. Population dynamics and local specialization in a clonal perennial (*Ranunculus repens*). *Journal of Ecology*, 69:743–755, 1981.

Y. Loya. Skeletal regeneration in a Red Sea scleractinean coral population. *Nature*, 261:490–491, 1976.

H. B. Lück and J. Lück. Cell number and cell size in filamentous organisms in relation to ancestrally and positionally dependent generation times. In: A. Lindenmayer and G. Rozenberg, editors, *Automata, Languages, Development*, pp. 109–124. North-Holland, Amsterdam, 1976.

K. Lüning. *Seaweeds: Their Environment, Biogeography and Ecophysiology.* John Wiley & Sons, New York, 1990.

J. Machta. The computational complexity of pattern formation. *Journal of Statistical Physics*, 70(3/4):949–967, 1993.

B.B. Mandelbrot. *The Fractal Geometry of Nature.* Freeman, San Francisco, 1983.

A. Marani, R. Rigon, and A. Rinaldo. A note on fractal channel networks. *Water Resources Research*, 27:3041–3049, 1991.

L. Margulis. *Symbiosis in Cell Evolution.* W.H. Freeman and Company, New York, 1993.

R. Mariscal and C. Bigger. A comparison of putative sensory receptors associated with nematocysts in an anthozoan and scyphozoan. In: G.O. Mackie, editor, *Coelenterate Ecology and Behaviour*, pp. 559–568, Plenum Press, New York, 1975.

A. Masselot. *A New Numerical Approach to Snow Transport by Wind: a Parallel Lattice Gas Model.* PhD thesis, UNIGE, 2000.

A. Masselot and B. Chopard. A lattice Boltzmann model for particle transport and deposition. *Europhys. Lett.*, 42:259–264, 1998.

T. Matsuyama and M. Matsushita. Fractal morphogenesis by a bacterial cell population. *Critical Reviews in Microbiology*, 19(2):117–135, 1993.

C.S. McFadden. Colony fission increases particle capture rates of a soft coral: Advantages of being a small colony. *J. Exp. Mar. Biol. Ecol.*, 103:1–20, 1986.

G.R. McGhee, Jr. *Theoretical Morphology: The Concept and its Applications.* Columbia University Press, New York, 1998.

F.K. McKinney and J.B.C. Jackson. *Bryozoan Evolution.* The University of Chicago Press, Chicago, 1991.

H. Meinhardt. *The Algorithmic Beauty of Sea Shells.* Springer-Verlag, Berlin, New York, 1998. Second Enlarged Edition.

N. Mitchell, M. Dardeau, and W. W. Schroeder. Colony morphology, age structure, and relative growth of two gorgonian corals, *Leptogorgia hebes* (Verill) and *Leptogorgia virgulata* (Lamarck), from the northern Gulf of Mexico. *Coral Reefs*, 12:65–70, 1993.

G.J. Mitchison and M. Wilcox. Rule governing cell division in *Anabaena*. *Nature*, 239:110–11, 1972.

C. Moore and M.G. Nordhal. Lattice gas prediction is P-Complete. Technical report, Santa Fe Institute for Complex Studies, 1997. SFI 97-04-043.

R.A. Morelli, R.E. Walde, E. Akstin, and C.W. Schneider. L-system representation of speciation in the red algal genus *Dipterosiphonia* (Ceramiales, Rhodomelaceae). *J. Theor. Biol.*, 149:453–465, 1991.

S. Muko, K. Kawasaki, K. Sakai, F. Takasu, and N. Shigesada. Morphological plasticity in the coral *Porites sillimaniani* and its adaptive significance. *Bull. Mar. Sci*, 66:225–239, 2000.

W.E.G. Müller. Molecular phylogeny of metazoa (animals): monophyletic origin. *Naturwissenschaften*, 82:321–329, 1995.

W.E.G. Müller. Origin of metazoan adhesion molecules and adhesion receptors as deduced from their cDNA analyses from the marine sponge *Geodia cydonium*. *Cell and Tissue Res.*, 289:383–395, 1997.

W.E.G. Müller. Origin of Metazoa: sponges as living fossils. *Naturwissenschaften*, 85:11–25, 1998.

W.E.G. Müller, B. Blumbach, and I.M. Müller. Evolution of the innate and adaptive immune systems: Relationships between potential immune molecules in the lowest metazoan phylum [Porifera] and those in vertebrates. *Transplantation*, 68:1215–1227, 1999a.

W.E.G. Müller, M. Wiens, R. Batel, R. Steffen, R. Borojevic, and M.R. Custodio. Establishment of a primary cell culture from a sponge: primmorphs from *Suberites domuncula*. *Mar. Ecol. Progr. Ser.*, 178:205–219, 1999b.

K. Nakanishi, M. Nishijima, M. Nishimura, K. Kuwano, and N. Saga. Bacteria that induce morphogenesis in *Ulva pertusa* (Chlorophyta) grown under axenic conditions. *Journal of Phycology*, 32:479–482, 1996.

L. Newton. *A Handbook of the British Seaweeds*. British Museum (Natural History), London, 1931.

C. Nüsslein-Volhard. Gradients that organize embryo development. *Scientific American*, August:38–43, 1996.

J.K. Oliver, B.E. Chalker, and W.C. Dunlap. Bathymetric adaptations of reef-building corals at Davies Reef, Great Barrier Reef, Australia. I. Long-term growth responses of *Acropora formosa*. *J. Exp. Mar. Biol. Ecol.*, 73:11–35, 1983.

U. Oren, B. Rinkevich, and Y. Loya. Oriented intra-colonial transport of C-14 labeled materials during coral regeneration. *Mar. Ecol. Prog. Ser.*, 161:117–122, 1997.

A. Ortmann. Studien über Systematik und geographische Verbreitung der Steinkorallen. *Zool. Jahrb. Abt. Syst.*, 3:143–788, 1888.

R. Osinga, P.B. de Beukelaer, E.M. Meijer, J. Tramper, and R.H. Wijffels. Growth of the sponge *Pseudosuberites (aff.) andrewsi* in a closed system. *Journal of Biotechnology*, 70:155–161, 1999.

S. Pain. Hostages of the deep. *New Scientist*, 1701:38–42, 1996.

H.S. Parker. Influence of relative water motion on the growth, ammonia uptake and carbon and nitrogen composition of *Ulva lactuca* (Chlorophyta). *Mar. Biol.*, 63:309–318, 1981.

M.R. Patterson. Patterns of whole colony prey capture in the octocoral, *Alcyonium siderium*. *Biol. Bull.*, 167:613–629, 1984.

M.R. Patterson and K.P. Sebens. Forced convection modulates gas exchange in cnidarians. *Proc. Natl. Acad. Sci*, 86:8833–8836, 1989.

M.R. Patterson, K.P. Sebens, and R.R. Olson. In situ measurements of flow effects on primary production and dark respiration in reef corals. *Limnol. Oceanogr.*, 36(5):936–948, 1991.

S.D. Peckham. New results for self-similar trees with applications to river networks. *Water Resour. Res.*, 31(4):1023–1029, 1995.

S.D. Peckham and V.K. Gupta. A reformulation of Horton's laws for large river networks in terms of statistical self-similarity. *Water Resour. Res.*, 35(9):2763–2777, 1999.

J.B. Pettigrew. *Design in Nature*. Longmans, Green, and Co., London, 1908.

M. Pigliucci. How organisms respond to environmental changes: from phenotypes to molecules (and vice versa). *TREE*, 11:168–173, 1996.

A.J. Pile, M. Savarese, and V.I. Chernykh. Trophic effects of sponge feeding within Lake Baikal's littoral zone. 2. Sponge abundance, diet, feeding efficiency, and carbon flux. *Limnol. Oceanogr.*, 42(1):178–184, 1997.

J.W. Porter. Autotrophy, heterotrophy, an resource partitioning in Caribbean reef-building corals. *The American Naturalist*, 110(975):731–742, 1976.

W.H. Press, B.P. Flannery, S.A. Teukolsky, and W.T. Vetterling. *Numerical recipes in C*. Cambridge University Press, Cambridge, 1988.

L. Provasoli and I.J. Pintner. Bacteria induced polymorphism in an axenic laboratory strain of *Ulva lactuca* (Chlorophyceae). *Journal of Phycology*, 16:196–201, 1980.

P. Prusinkiewicz and J. Hanan. Visualization of botanical structures and processes using parametric L-systems. In: D. Thalmann, editor, *Scientific Visualization and Graphics Simulation*, pp. 183–201. J. Wiley & Sons, Chichester, 1990.

P. Prusinkiewicz and L. Kari. Subapical bracketed L-systems. In: J. Cuny, H. Ehrig, G. Engels, and G. Rozenberg, editors, *Graph Grammars and their Application to Computer Science; Fifth International Workshop*, Lecture Notes in Computer Science 1073, pp. 550–564. Springer-Verlag, Berlin, 1996.

P. Prusinkiewicz and A. Lindenmayer. *The Algorithmic Beauty of Plants*. Springer-Verlag, New York, Berlin, 1990.

R.A. Raff. *The Shape of Life: Genes, Development, and the Evolution of Animal Form*. The University of Chicago Press, Chicago, 1996.

H.M. Reiswig. Water transport, respiration, and energetics of three tropical marine sponges. *J. Exp. Mar. Biol. Ecol.*, 14:231–249, 1974.

M. Ribes, R. Coma, and J. Gili. Heterogeneous feeding in benthic suspension feeders: the natural diet and grazing rate of the temperate gorgonian *Paramuricea clavata* (Cnidaria: Octocorallia) over a year cycle. *Mar. Ecol. Prog. Ser.*, 183:125–137, 1999.

A. Rinaldo, I. Rodriguez-Iturbe, R. Rigon, R. Bras, and E. Ijasz-Vasquez. Minimum energy and fractal structures of drainage networks. *Water Resources Research*, 28:2183–2195, 1992.

B. Rinkevich. The contribution of photosynthetic products to coral reproduction. *Mar. Biol.*, 101:259–263, 1989.

B. Rinkevich. A long-term compartmental partitioning of photosynthetically fixed carbon in a symbiotic reef coral. *Symbiosis*, 10:175–194, 1991.

B. Rinkevich. Steps towards the evaluation of coral reef restoration by using small branch fragments. *Mar. Biol.*, 136:807–812, 2000.

B. Rinkevich and Y. Loya. The reproduction of the Red Sea coral *Stylophora pistillata* II. Synchronization in breeding and seasonality of planulae shedding. *Mar. Ecol. Prog. Ser.*, 1:145–152, 1979.

B. Rinkevich and Y. Loya. Short term fate of photosynthetic products in a hermatypic coral. *J. Exp. Mar. Biol. Ecol.*, 73:175–184, 1983a.

B. Rinkevich and Y. Loya. Oriented translocation of energy in grafted reef corals. *Coral Reefs*, 1:243–247, 1983b.

B. Rinkevich and Y. Loya. Coral illumination through an optic glass fiber. Incorporation of 14 C material. *Mar. Biol.*, 80:7–15, 1984.

B. Rinkevich and Y. Loya. Coral isomone: a proposed chemical signal controlling interclonal growth patterns in a branching coral. *Bull. Mar. Sci.*, 36:319–324, 1985a.

B. Rinkevich and Y. Loya. Intraspecific competition in a reef coral: Effects on growth and reproduction. *Oecologia*, 66:100–105, 1985b.

B. Rinkevich and Y. Loya. Senescence and dying signals in a reef building coral. *Experientia*, 43:320–322, 1986.

B. Rinkevich and Y. Loya. Variability in the patterns of sexual reproduction of the coral *Stylophora pistillata* at Eilat, Red Sea: A long-term study. *Biol. Bull.*, 173:335–344, 1987.

B. Rinkevich and Y. Loya. Reproduction in regenerating colonies of the coral *Stylophora pistillata*. In: E. Spanier, Y. Steinberger, and M. Luria, editors, *Environmental Quality and Ecosystem Stability. Vol. IVB, Environmental quality*, pp. 259–269, John Wiley, ISEEQS Publ., 1989.

B. Rinkevich and I.L. Weissman. Chimeras in colonial invertebrates: a synergistic symbiosis or somatic and germ cell parasitisms? *Symbiosis*, 4:117–134, 1987.

B. Rinkevich, Z. Wolodarsky, and Y. Loya. Coral crab association: a compact domain of multilevel trophic system. *Hydrobiologia*, 216:279–284, 1991.

P.J. Roache. *Computational Fluid Dynamics*. Hermosa Publishers, Alberque, 1976.

R.J. Roark and W.C. Young. *Formulas for Stress and Strain*. McGraw-Hill, New York, 1975.

P.J. Roos. *Growth and Occurrence of the Reef Coral Porites astreoides Lamarck in Relation to Submarine Radiance Distribution*. PhD thesis, University of Amsterdam, 1967.

A. Rosenfeld and A.C. Kak. *Digital Picture Processing*. Academic Press, New York, 1982.

L.K. Rosenvinge. *The Marine Algae of Denmark*. Bianos Lunos Bogtrykkeri, Copenhagen, 1923.

D.H. Rothman and S. Zaleski. *Lattice-Gas Cellular Automata*, volume 5 of *Collection Aléa-Saclay*. Cambridge University Press, Cambridge, UK, 1997.

D.I. Rubenstein and M.A.R. Koehl. The mechanisms of filter feeding: some theotetical considerations. *The American Naturalist*, 111:981–994, 1977.

C.W. Schneider and R.E. Walde. L-system computer simulations of branching divergence in some dorsiventral members of the tribe Polysiphonieae (Rhodomelaceae, Rhodophyta). *Phycologia*, 31(6):581–590, 1992.

C.W. Schneider, R.E. Walde, and R.A. Morelli. L-system computer models generating distichous spiral organisation in the Dasyaceae (Ceramiales, Rhodophyta). *Eur. J. Phycol.*, 29:165–170, 1994.

H.C. Schröder, A. Krasko, R. Batel, A. Skorokhod, S. Pahler, M. Kruse, I.M. Müller, and W.E.G. Müller. Stimulation of protein (collagen) synthesis in sponge cells by a cardiac myotrophin-related molecule from *Suberites domuncula*. *FASEB J.*, 2000a. In press.

H.C. Schröder, M. Kruse, R. Batel, I.M. Müller, and W.E.G. Müller. Cloning and expression of the sponge longevity gene SDLAG. *Mechanisms of Development*, 95:219–220, 2000b.

W. Schroeder, K. Martin, and B. Lorensen. *The Visualization Toolkit: An Object-Oriented Approach To 3D Graphics*, 2nd edition. Prentice Hall, New Jersey, 1997.

F.E. Schulze. Report on the Hexactinellida. *Challenger's Reports Zoology*, 21:1–513, 1887.

H. Schumacher. *Korallenriff ihre Verbreitung, Tierwelt und Ökologie*. BLV Verlagsgesellschaft, München, 1976.

S.A. Schumm. Evolution of drainage systems and slopes in badlands at Perth Amboy, New Jersey. *Bull. Geol. Soc. Am*, 67:597–646, 1956.

F.H. Schweingruber. *Tree Rings: Basics and Applications of Dendrochronology*. Reidel, Dordrecht, 1988.

T.P. Scoffin, A.W. Tudhope, B.E. Brown, H. Chansang, and R.F. Cheeney. Patterns and possible environmental controls of skeletogenesis of *Porites lutea*, South Thailand. *Coral Reefs*, 11:127–130, 1992.

K.P. Sebens and A.S. Johnson. Effects of water movement on prey capture and distribution of reef corals. *Hydrobiologia*, 226:91–101, 1991.

K.P. Sebens, J. Witting, and B. Helmuth. Effects of water flow and branch spacing on particle capture by the reef coral *Madracis mirabilis* (Duchassaing and Michelotti). *J. Exp. Mar. Biol. Ecol.*, 211:1–28, 1997.

M. Seimiya, M. Naito, Y. Watanabe, and Y. Kurosawa. Homeobox genes in the freshwater sponge *Ephydatia fluviatilis*. *Progr. Molec. Subcell. Biol.*, 19:132–155, 1998.

J. Shimeta and P.A. Jumars. Physical mechanisms and rates of particle capture by suspension feeders. *Oceanogr. Mar. Biol. Annu. Rev.*, 29:191–257, 1991.

K. Shimzu, J. Cha, G.D. Stucky, and D.E Morse. Silicatein a: cathepsin L-like protein in sponge biosilica. *Proc. Natl. Acad. Sci. USA*, 95:6234–6238, 1998.

R.W.M. van Soest and J.C. Braekman. Chemosystematics of Porifera: a review. *Memoirs of the Queensland Museum*, 44:569–589, 1999.

D. St Johnston and C. Nüsslein-Volhard. The origin of pattern and polarity in the *Drosophila* embryo. *Cell*, 68:201–209, 1992.

C.W. Stearn and R. Riding. Forms of the hydrozoan *Millepora* on a recent coral reef. *Lethaia*, 6:187–200, 1973.

T.A. Stephenson. Development and the formation of colonies in *Pocillopora* and *Porites*. Part I. *Sci. Rep. Gt. Barrier Reef Exped.*, 3:113–134, 1931.

H.L. Stewart. *The Effects of Water Flow on the Relationship between Photosynthesis and Morphology of Marine Macroalgae*. MS Thesis. California State University, Northridge, 1999.

A.N. Strahler. Quantitative analysis of watershed geomorphology. *EOS Trans. AGU*, 38(6):913–920, 1957.

J. Stratman, G. Paputsogle, and W. Oretel. Differentiation of *Ulva mutabilis* (Chlorophyta) gametangia and gamete release are controlled by extracellular inhibitors. *Journal of Phycology*, 32:1009–1021, 1996.

D.L. Taylor. Intra-colonial transport of organic compounds and calcium in some Atlantic reef corals. *Proceedings of the 3th International Coral Reef Symposium*, 2:432–436, 1977.

R.B. Taylor, D.J. Barnes, and J.M. Lough. Simple models of density band formation in massive corals. *J. Exp. Mar. Biol. Ecol.*, 167:109–125, 1993.

R.B. Taylor, D.J. Barnes, and J.M. Lough. On the inclusion of trace materials into coral skeletons. Part I. Materials occuring in the environment in short pulses. *J. Exp. Mar. Biol. Ecol.*, 183:255–278, 1995.

D'A.W. Thompson. *On Growth and Form*. Cambridge University Press, Cambridge, 1917. Abridged edition by J.T. Bonner, 1961.

L.G. Thompson, E. Mosley-Thompson, M.E. Davies, P.N. Lin, K.A. Henderson, J. Cole-Dai, J.F. Bolzan, and K.B. Liu. Late glacial stage and Holocene tropical ice core records from the Huascaran, Peru. *Science*, 269:46–50, 1995.

M.D'A.A. Le Tissier, B. Clayton, B.E. Brown, and P. Spencer Davies. Skeletal correlates of coral density banding and an evaluation of radiography as used in scelerochronology. *Mar. Ecol. Prog. Ser.*, 110:29–44, 1994.

E. Tokunaga. The composition of drainage network in Toyohira River Basin and the valuation of Horton's first law. *Geophys. Bull. Hokkaido Univ.*, 15:1–19, 1966. (In Japanese with English summary.).

E. Tokunaga. Consideration on the composition of drainage networks and their evolution. *Geogr. Rep., Tokyo Metrop. Univ.*, 13:1–27, 1978.

E. Tokunaga. Ordering of divide segments and law of divide segment numbers. *Trans. Jpn. Geomorphol. Union*, 5(2):71–77, 1984.

Y.F. Tsao and K.S. Fu. A parallel thinning algorithm for 3-D pictures. *Computer Graphics and Image Processing*, 17:315–331, 1981.

D.L. Turcotte, J.D. Pelletier, and W.I. Newman. Networks with side branching in biology. *J. Theor. Biol.*, 193:577–592, 1998.

J. Uchmanski, V. Grimm, and T. Wyszomirski. Individual-based approach in ecology. In: A. farina, editor, *Perspectives in Ecology*, pp. 187–195, Backhuys Publishers, Leiden, 1999.

C. Upson. Volumetric visualization techniques. In: D.F. Rogers and R.A. Earnshaw, editors, *State of the Art in Computer Graphics*, pp. 313–350, Springer-Verlag, Berlin, 1991.

M.J.A. Vermeij, J.A. Kaandorp, R.P.M. Bak, L.E.H. Lampmann, and P.M.A. Sloot. Three-dimensional analysis of growth forms of *Madracis* species. In preparation.

J.E.N. Veron. *Corals in Space and Time: The Biogeography and Evolution of Scleractinia*. Cornell University Press, Cornell, 1995.

J.E.N. Veron and M. Pichon. *Scleractinia of Eastern Australia Part I: Families Thamnasteriidae, Astrocoeniidae, Pocilloporidae*. Australian Government Publishing Service, Canberra, 1976.

J.E.N. Veron and C.C. Wallace. *Scleractinia of Eastern Australia. Part V. Family Acroporidae*. Australian Institute of Marine Science and Australian National University Press, Canberra, 1984.

S. Vogel. Current-induced flow through the sponge, *Halichondria*. *Biol. Bull.*, 147:443–456, 1974.

S. Vogel. Drag and flexibility in sessile organisms. *Am. Zool.*, 24:28–34, 1984.

S. Vogel. *Life's devices: The Physical World of Animals and Plants*. Princeton University Press, Princeton, 1988.

S. Vogel. *Life in Moving Fluids: The Physical Biology of Flow, Second Edition*. Princeton University Press, Princeton, 1994.

C.J. Vosmaer. On the distinction between the genera *Axinella*, *Phakelia*, *Acanthella* a.o. In: J.W. Spengel, editor, *Abdruck aus den Zoologischen Jahbüchern*, pp. 307–322, verlag von Gustav Fischer, Jena, 1912.

J. Voß. Zoogeography and community analysis of macrozoobenthos of the Weddell Sea (Antarctica). *Ber. Polarforsch*, 45:1–144, 1988.

S.D. Waaland and R.E. Cleland. Cell repair through cell fusion in the red alga, *Griffithsia pacifica*. *Protoplasma*, 79:185–196, 1974.

C.H. Waddington. Canalization of development and the inheritance of acquired characteristics. *Nature*, 150:563–565, 1942.

A. Wagner, S.R. Dudgeon, J.R. Vaisnys, and L.W. Buss. Non-linear oscillations in polyps of the colonial hydroid *Podocoryne carnea*. *Naturwissenschaften*, 85:1–5, 1998.

S.A. Wainwright, W.D. Biggs, J.D. Currey, and J.W. Gosline. *Mechanical Design in Organisms*. Princeton University Press, Princeton, 1976.

S.A. Wainwright and J.R. Dillon. On the orientation of sea fans (genus *Gorgonia*). *Biol. Bull.*, 136:130–139, 1969.

M.R. Walter. *Stromatolites*. Elsevier, Amsterdam, 1976.

W.H de Weerdt. Transplantation experiments with Caribbean *Millepora* species (Hydrozoa, Coelenterata), including some ecological observations on growth forms. *Bijdr. Dierk.*, 51(1):1–19, 1981.

W.H. de Weerdt. A systematic revision of the north-eastern Atlantic shallow-water Haplosclerida (Porifera, Demospongiae), part II: Chalinidae. *Beaufortia*, 36(6):81–165, 1986.

G.B. West, J.H. Brown, and B.J. Enquist. A general model for the origin of allometric scaling laws in biology. *Science*, 276:122–126, 1997.

F.M. White. *Heat and Mass Transfer*. Addison-Wesley, New York, 1988.

F.M. White. *Fluid Mechanics*. McGraw-Hill, New York, 1994.

F. Wiedenmayer. *Shallow-Water Sponges of the Western Bahamas*. Birkhäuser, Basel, 1977.

M. Wiens, A. Krasko, I.M. Müller, and W.E.G. Müller. Increased expression of the potential proapoptotic molecule DD2 and increased synthesis of leukotriene B4 during allograft rejection in a marine sponge. *Cell Death and Differentiation*, 7:461–469, 2000a.

M. Wiens, A. Krasko, I.M. Müller, and W.E.G. Müller. Molecular evolution of apoptotic pathways: cloning of key domains from sponges (Bcl-2 homology domains and death domains) and their phylogenetic relationships. *J. Molec. Evolution*, 20:520–531, 2000b.

D. Wildish and D. Kristmanson. *Benthic Suspension Feeders and Flow*. Cambridge University Press, Cambridge, 1997.

B.L. Willis. Phenotypic plasticity versus phenotypic stability in the reef corals *Turbinaria mesenteriana* and *Pavona cactus*. *Proc. 5th Int. Coral Reef Congr.*, 4:107–112, 1985.

B.L. Willis and D.J. Ayre. Asexual reproduction and genetic determination of growth form in the coral *Pavona cactus*: biochemical genetic and immunogenic evidence. *Oecologia*, 65:516–525, 1985.

T.A. Witten and L.M. Sander. Diffusion-limited aggregation, a kinetic critical phenomenon. *Phys. Rev. Lett.*, 47(19):1400–1403, 1981.

L. Wolpert, R. Beddinton, J. Brooks, and T. Jessel. *Principles of Development*. Current Biology Ltd, Oxford University Press, Oxford, 1998.

M. Zamir. On fractal properties of arterial trees. *J. Theor. Biol.*, 197:517–526, 1999.

T.Y. Zhang and C.Y. Suen. A fast parallel algorithm for thinning digital patterns. *Communications of the ACM*, 27(3):236–239, 1984.

C. Zilberberg and P.J. Edmunds. Patterns of skeletal structure variability in clones of the reef coral *Montastraea franksi. Bull. Mar. Sci.*, 64:373–381, 1999.

Subject Index

absorption patterns 142
acceleration reaction force
- equation 20
accretive growth 42, 125
accretive growth model
- approximated growth velocities 128
- basic size skeleton element 128
- comparison transplantation experiments 153
- geometrical model 127
- growth axes 130
- growth function 128
- initial parameters 130
- isomone 139
- light-driven 137, 141
- nutrient-driven 131, 141 algorithm 133
- triangulation 130
Acropora formosa 148
Acropora palmata 141
aggregates
- absorption 124
- asymmetry 123
- compactness 124
- radial symmetry 123
aggregation model
- algorithm 120
- equation growth 117
alcyonaceans 48
allografts 57
allometric constraints
- octocorals 48
amount of contact with the environment 128
Anabaena catenula 92
anastomosis 72, 78
- stony corals 65
anisotropy 139
apoptosis 60
aquacultures 11, 159
aquiferous system 58

- sponges 42
aragonite 55
architectures
- growth processes 171
autografts 57
autotrophy 54
Axinella polypoides 141

bacterial colonies 5
b_angle branching angle 78
bidirectional flow 119
Bilateria 9
bioarchives 11, 12, 160, 170
biomonitoring 11, 169
body axes 8
body plan 8
- cnidarians 43
- sponges 57
Boltzmann molecular chaos assumption 102
Bostrychia radicans 95
boundary layer 17, 22
branch spacing 76
- *br_spacing* 79
- *Madracis mirabilis* 29
branching networks 67
branching patterns
- gorgonians 74
breakability
- marine sessile organisms 22
brown seaweeds 5, 30

Calcarea 59
calcification rates
- stony corals 66
Callithamnion roseum 2
Cambrian explosion 56
cancer 11, 60
- cytotoxic compounds 159
catabolic enzymes 60
cellular automata 100

center of gravity 136
- aggregates 119
Chaetomorpha linum 94
chimera 64
Chlamydomonas nivalis 143
Chondrus crispus 97
cnidarians 5
coenenchyme 48
collision operator
- FHP model 104
Computed Tomography scanning 87
continuity equation 15
- lattice gas 103
coral bleaching 10
coralline algae 3, 5, 11
corallite 55
Corallium rubrum 6

da maximal thickness of a branch 78
damage during growth
- sponges 37
db minimal thickness of a branch 78
demosponges 5, 59
density
- lattice gas 101
density bands 55, 163
density variations
- stony corals 88
deposition processes
- marine sessile organisms 7
- physical 5
Desmophyllum cristagalli 11
determinate growth 1
dichotomous branching 112
diffuse reflection 138
diffusion coefficient D 143
diffusion equation 110
diffusion limited aggregation 12, 115, 116
- growth equation 116
diffusive boundary layer 23

dissepiments 55, 164
drag
– equation 18
Drosophila 7

electric discharge patterns 6, 110
erosion
– lattice Boltzmann model 107
Euler equation 103
evolution
– lattice gas 102

fan-like growth form 112
FHP model 103
Fick's 1st Law of Diffusion 22
flattened growth forms 139
flexible organisms 10, 18
flow fields
– visualization 25
flow and diffusion
– porous media 99
flow tank
– experiments 29
food particles
– simulation 127
fractal analysis
– box dimension 80
– *Raspailia inaequalis* 75
fractal branching 74
fractal dimension 136
– aggregates 118
– box-counting method 118
– uniformity nutrient 119, 136
fractals
– iterative 98
fusing of branches 65

g_angle geotropy angle 78
gastrovascular cavity 43
gastrovascular flow patterns 46
gemmules 60
genetic regulation
– seaweeds 56
– sponges 56
– stony corals 62
genetic regulation morphogenesis 170
genetic variation
– sponges 37
geographic variation
– seaweeds 33
global warming 162
gorgonians 48
green seaweeds 5, 30
growth axes 141, 143

growth function
– anisotropic 128
growth processes
– abiotic and biotic 6
– complex systems 13
– intractability 12
– sponges 35
growth rates
– sponges 35

Haliclona oculata 7, 38, 76, 77, 114, 140, 151
Haliclona simulans 42
hermatypic 54
heterotrophy 54
hexacorals 43
hexactinellids 5, 59
homeobox genes
– sponges 57
Horton ratios 69
Horton statistics 69
Horton–Strahler ordering 68
– *Raspailia inaequalis* 73
Hox genes 7
HPP model 103
hybridization
– stony corals 54
Hydra 56
Hydractinia 45
Hydractinia symbiolongicarpus 156
hydractiniid polyps 144, 155
– dynamic model 144
hydrodynamic forces 20, 21, 170
– lattice Boltzmann model 107
hydrodynamics 99
– equations 15
hydrozoans 5, 43, 144

immortal cells 60
immune molecules 57
indeterminate growth 1
integrins 57
isomone 65, 86, 126, 139, 155
– decay 139
– gradient 139
– model 144
– simulated 127
isotropy
– lattice gas 103

L-system 91, 95
– parametric 92
Lambert–Beer Law 16
laminar flow 23

Laplace equation 110
– numerical model 110
Laplacian growth 109
large-scale computing 9
lattice Boltzmann equation 105
lattice Boltzmann method 9, 99, 115, 127
– coupling accretive growth model 132
– tracer step 115
lattice gas 9
– collision operator 104
– Galilean invariance 104
– noisy dynamics 104
Lattice Gas Automata 100
Lattice-BGK model 105
Leucosolenia primordialis 58
life cycles
– seaweeds 33
lift
– equation 19
light intensity
– simulated 127
light model 137
– equation 16
Lithothamnion coralliodes 7
lobed growth form 141
Lophelia pertusa 54

macroscopic behavior
– lattice gas 103
Madracis decactis 29
Madracis mirabilis 29, 114, 124, 143
marching cubes method 88, 117
marking experiments growth process
– sponges 42
mass conservation
– lattice gas 101
mass transfer 22
– metabolic rate 24
– models 23
– phosphorus uptake 26
– rough surfaces 24
– size 26
Mastocarpus stellatus 33
mesoglea 43, 55
Millepora alcicornis 76, 77, 114
modular organisms 1
module 1
momentum, lattice gas 101
momentum boundary layer 23
momentum conservation
– lattice gas 101
momentum conservation equation
– lattice gas 103

monodirectional flow   123
Montastrea annularis   51, 125, 141, 148
morphogens   7
– sponges   57
morphological analysis
– three-dimensional   171
morphological plasticity
– cnidarians   43
– genetic control   63
– hydrozoans   43, 48
– physical environment   76
– seaweeds   33
– sponges   34, 38
– stony corals   51
– temperature   98
morphological skeleton   72, 78
– three-dimensional   89
Muriceopsis   51
myotrophin   58

Navier–Stokes equation   15
Nereocystis luetkeana   21, 24
nutrient gradient   133

octocorals   5, 43, 48
orientation octocorals unidirectional flow   49
oscillatory flow   23, 30

P-completeness   100
Paenibacillus dendritiformis   5
particle capturing
– corals   30
– flow speed   29
Péclet number   16, 99, 115, 143
perturbation experiments   156
phalanx strategy   97
physical oceanography in the past   11
plexaurid gorgonians   49
Pocillopora damicornis   1, 76, 77, 114, 124
Podocoryne   45, 145
Podocoryne carnea   155
polymorphic polyps   44
polyp
– cnidarians   43
– oscillations   45
– stony coral   55
polyphosphate   60
population dynamics
– individual-based   169
Porites   165
Porites porites   7
Porites sillimaniani   51, 125, 149

primmorphs   60
Pseudopterogorgia   51
Pythagoras tree   98

radial symmetry
– accretive growth model   137
radiate accretive architecture   40, 125, 141
radius of curvature   127, 129
Raspailia inaequalis   34, 72, 110
rb branching rate   78
recurrent basic shapes   2, 3
red seaweeds   5, 30
– L-system models   94
regeneration, stony corals   66
Reynolds number   16, 23, 99
rhodomorphin   56
rigid organisms   10
ripple formation   109
river networks   68
river systems   74
rotation experiment   154
runner genotypes   45

scaling
– marine sessile organisms   21
Schmidt number   23
scleractinians   5, 43
seawater temperature   168
sedimentation
– lattice Boltzmann model   106
self-shading   138
self-similarity branching networks   70
sheet phenotypes   45
Sherwood number   23
shrubby growth form   112
skeletal growth
– sponges   60
– stony corals   55
skeleton architecture, sponges   40, 42
sphere, triangulated   130
spicules   40, 59
sponges   5
– Antarctica   160
– anticancer compounds   159
– growth rates   152, 162
– suspension feeding   125
Stanton number   23
stolon   44
stony corals   43
– calcification rate   165
– extension rate   165
– living tissue   170
– model physiology   172

– photosynthesis   125
– radial symmetry   126
– tissue layer   164
– translocation nutrients   125
– transplantation   147
stromatolites   3, 5
– section   7
Stylocordyla borealis   161
Stylophora pistillata   63, 126, 155
Suberites domuncula   57
surface area volume ratio   25
surface normal deposition   125
surface rendering   88
suspension-feeding   10, 26, 114
– corals   27
– octocorals   49
– sponges   27
Sycon raphanus   58
Symbiodinium microadriaticum   54
symbiosis   3
– octocorals   49
– sponges   60
– stony corals   54

telomerase activity   60
thinning algorithm   78
– three-dimensional   89
Tokunaga ratios
– *Raspailia inaequalis*   73
Tokunaga statistics   70
Tokunaga's law   70
translocation of nutrients   170
translocation of photosynthates   65
transplantation experiments   147
transport suspensions
– lattice Boltzmann model   106
tree rings   160, 168
turbulent flow   23, 171

underwater light models   171
underwater light intensities
– equations   16
unidirectional flow   17, 18, 30
unitary organisms   1

vascular networks   68
viscous-fingering   6
volume rendering   88

wave tank
– experiments   21
waves   19

zooxanthellae   54